Network Routing

Network Routing

Fundamentals, Applications, and Emerging Technologies

Sudip Misra
Indian Institute of Technology, Kharagpur, India

Sumit Goswami
Defence Research and Development Organization
New Delhi, India

Registered Office
John Wiley & Sons Ltd, The Atrium, Southern Gate, Chichester, West Sussex, PO19 8SQ, United Kingdom

For details of our global editorial offices, for customer services and for information about how to apply for permission to reuse the copyright material in this book please see our website at www.wiley.com.

Library of Congress Cataloging-in-Publication Data

Names: Misra, Sudip, author. | Goswami, Sumit, author.
Title: Network routing : fundamentals, applications and emerging technologies /
    Sudip Misra, Sumit Goswami.
Description: Chichester, West Sussex, United Kingdom : John Wiley & Sons,
    Inc., [2017] | Includes bibliographical references and index.
Identifiers: LCCN 2016028174 | ISBN 9780470750063 (cloth) | ISBN 9781119029380
    (epub) | ISBN 9781119029397 (ePdf)
Subjects: LCSH: Routing (Computer network management)
Classification: LCC TK5105.5487 .M57 2017 | DDC 004.6–dc23 LC record available
    at https://lccn.loc.gov/2016028174

A catalogue record for this book is available from the British Library.

Cover Design: Wiley
Cover Images: (Earth) Harvepino/Gettyimages; (Background) amgun/Gettyimages

Set in 10/12pt Warnock by SPi Global, Pondicherry, India
Printed and bound in Malaysia by Vivar Printing Sdn Bhd

10 9 8 7 6 5 4 3 2 1

Sudip dedicates this book to his family.
Sumit dedicates this book to the organization in which his father served and in which he grew up – the Border Security Force, India.

Contents

About the Authors

Dr Sudip Misra is an Associate Professor in the Department of Computer Science and Engineering at the Indian Institute of Technology, Kharagpur, India. Prior to this he was associated with Cornell University (USA), Yale University (USA), Nortel Networks (Canada), and the Government of Ontario (Canada). He received his PhD degree in Computer Science from Carleton University, Ottawa, Canada, and Master's and Bachelor's degrees, respectively, from the University of New Brunswick, Fredericton, Canada, and the Indian Institute of Technology, Kharagpur, India. Dr Misra has several years of experience working in academia, government, and the private sector in research, teaching, consulting, project management, software design, and product engineering roles.

His current research interests include mobile ad hoc and sensor networks, the Internet of Things (IoT), computer networks, and learning systems. Dr Misra is the author of over 260 scholarly research papers, of which over 150 have been published in distinguished journals. He has won nine research paper awards in different international conferences. He was awarded 3rd Prize in the Samsung Innovation Award (2014) at IIT, Kharagpur, and also the IEEE ComSoc Asia Pacific Outstanding Young Researcher Award at IEEE GLOBECOM 2012, Anaheim, California, USA. He is also the recipient of several academic awards and fellowships, such as the Young Scientist Award (National Academy of Sciences, India), the Young Systems Scientist Award (Systems Society of India), the Young Engineers Award (Institution of Engineers, India), the (Canadian) Governor General's Academic Gold Medal at Carleton University, the University Outstanding Graduate Student Award at Doctoral Level at Carleton University, and the National Academy of Sciences, India – Swarna Jayanti Puraskar (Golden Jubilee Award). He was also awarded the Canadian Government's prestigious NSERC Postdoctoral Fellowship and the Humboldt Research Fellowship in Germany.

Dr Misra is the Editor-in-Chief of the *International Journal of Communication Networks and Distributed Systems* (IJCNDS), Inderscience, UK. He has also served (is serving) as the Associate Editor of *IEEE Transactions on Mobile Computing, Telecommunication Systems Journal* (Springer), *Security and Communication Networks Journal* (Wiley), *International Journal of Communication Systems* (Wiley), and *EURASIP Journal of Wireless Communications and Networking*. He is also an Editor/Editorial

Board Member/Editorial Review Board Member of the *IET Communications Journal,* *IET Wireless Sensor Systems,* and *Computers and Electrical Engineering Journal* (Elsevier).

Dr Misra has published ten books in the areas of opportunistic networks, wireless ad hoc networks, wireless sensor networks, wireless mesh networks, communication networks and distributed systems, network reliability and fault tolerance, and information and coding theory, published by reputed publishers such as Wiley, Springer, Cambridge University Press, and World Scientific.

He has been invited to chair several international conference/workshop programs and sessions. He has served on the program committees of several international conferences. Dr Misra has also been invited to deliver keynote/invited lectures in over 30 international conferences in the USA, Canada, Europe, Asia, and Africa.

Dr Sumit Goswami is a scientist with the Defence Research and Development Organization (DRDO), Ministry of Defence, Government of India. He has worked in the field of information security, wide area networks, website hosting, network management, and information extraction. He gained his PhD degree and Master's degree in Computer Science and Engineering from the Indian Institute of Technology, Kharagpur, India. He also holds a Postgraduate Diploma in Journalism and Mass Communication, a Bachelor's Degree in Library and Information Science, and a BTech Degree in Computer Science and Engineering.

He has made significant research contributions in the area of public key infrastructure, mobile ad hoc and sensor networks, stylometric analysis, the Internet of Things, and machine learning. He has published more than 70 papers/chapters in various journals, books, data competitions, conferences, and seminars. He also has experience in Engineering Design and Technology Development related to mailing services, website security, webcasting, and intranets. For excellence of work, he has twice been awarded the DRDO's Medal and Commendation Certificate. He also has expertise in techno-managerial analysis and monitoring of projects. He was a member of the Team award by the Defence Minister for Best Techno-Managerial Services – 2012.

Prior to joining DRDO, he worked with CMC Limited, New Delhi, as Engineer – System Integration. At CMC Limited he worked on a range of network products and technologies, some of which were legacy networks while others were contemporary. He has experience of designing and managing networks based on full duplex fast Ethernet and ATM on local area networks and satellite, microwave, ISDN, MPLS, and STM link-based wide area networks.

To promote computer science among the masses, he writes general computer science articles in magazines and has contributed to more than 15 such topics. He is also a prolific instructor and has been the guest lecturer in various engineering colleges, as well as a regular speaker in a number of courses within his organization. He has also served on the program committees of several international conferences. He is a life member of the Computer Society of India. He is presently appointed as Counsellor (Defence Technology) in the Embassy of India, Washington D.C., USA.

Foreword

Network routing has evolved over the past 65 years, and this book systematically makes its readers traverse this journey of evolution of network routing through its 500 plus pages. With 'nano' on the anvil, an in-depth understanding of present and past routing technologies is essential to explore future innovations and discoveries. The book will uniquely prepare its readers to face the toughest routing challenges in all varieties of networks owing to the tremendous effort put into presenting detailed illustrations throughout the book to aid learning.

The book delivers its contents in five parts: fundamental concepts, routing in QoS and traffic engineering, routing on the Internet, other routing contexts (ATM, cellular wireless networks, wireless ad hoc networks, wireless sensor networks, 6LoWPAN), and advanced concepts pertaining to security and reliability. It demonstrates the commitment of the authors to connect from history to the future through the contemporary. There are a total of 14 chapters, with many figures, questions, and exercises. The core of the book lies in its coverage of all routing mechanisms under a single source, with special mention of its unprecedented collection of routing techniques in the entire range of wireless networks. IoT is a game changer, and this book has a chapter on 6LoWPAN too. Demands also remain among the network administrators for network management and network security, which together form the backbone for supporting efficient and fault-free routing. This book will definitely build on the confidence of network researchers and administrators in this area too. In addition, it introduces a few topics of upcoming interest – attack surfaces, smart systems, mobile agents, networked battlefields, and cognitive security.

While the book has focused on present-day routing techniques, it has not lost sight of the basic routing protocols and algorithms, which have been explained in depth. Teaching materials in the form of presentations in sync with the chapters are also made available by the authors. Answers to all the questions have been provided. However, solution to only selective exercise questions have been uploaded onto the website of the book, so as to ensure sustained memory performance leading to different possible solutions.

The book will steward its readers on the journey from legacy networks to future-generation networks.

Dr Khaled B. Letaief
Fellow of IEEE,
Chair Professor, Department of Electronic and Computer Engineering,
Hong Kong University of Science and Technology
https://www.ece.ust.hk/~eekhaled/

Preface

Overview

Coverage of routing techniques in various wired and wireless networks is the unique proposition of this book. Routing protocols and algorithms are the brains of any network. The selection of topics in this book is clear: we attempt to explain routing in its entirety, starting from fundamental concepts, then moving through routing on the Internet, and finally cutting across the recent-day cellular, ad hoc, and wireless networks. At the same time, the book has significant coverage of related topics, i.e. network reliability, management, and security. The core concepts elaborated in the book provide a foundation for understanding the next-generation networks and pushing them to their safe boundary limits. These concepts are integrated with illustrations and flow diagrams that will enable the readers to experience a fly-through of the routing processes over the devices.

In this age of rapidly evolving networks, this book stands at the intersection of historical network routing techniques and evolving concepts that the world is working on. The book builds on the foundation to create a 'network analyst' and a 'routing strategist'.

Organization of the Book

The book is organized into five parts, starting from a basic introduction and ending with advanced concepts.

The first part of the book presents the fundamental concepts of networks and routing. Chapter 1 provides basic knowledge about networks, addressing schemes, architectures, and standards so as to act as a foundation for those readers who have not done a basic course in computer networks. Routing algorithms based on various strategies are described in Chapter 2. All the major categories of fundamental routing protocols are covered in Chapter 3.

The second part of the book presents routing with quality of service and traffic engineering. Chapter 4 has complete coverage of QoS measures, terminologies, algorithms, and protocols. Chapter 5 is dedicated to traffic engineering and describes multiprotocol label switching and TE routing algorithms.

The third part of the book presents routing on the Internet. This part starts with exhaustive coverage of two major kinds of interior gateway protocol in Chapter 6, and ends with the detailed evolution history of exterior gateway protocol along with its operational details in Chapter 7.

The fourth part of the book presents all other routing contexts, from legacy networks to future-generation networks. The heritage ATM network is covered in Chapter 8, including frame format, architecture, service categories, and routing. Chapters 9, 10, and 11 present the characteristics, followed by comprehensive coverage of most of the routing techniques in contemporary networks, i.e. cellular wireless networks, wireless ad hoc networks, and wireless sensor networks. Chapter 12 is dedicated to the fundamentals, applications, and routing in upcoming networks for the future generation, i.e. 6LoWPAN.

The fifth part of the book presents advanced concepts related to network routing. Chapter 13 is dedicated to one of the major concerns in the area of network routing, i.e. security. The chapter contains sections on various kinds of attack, metrics to calculate exposure to attacks, security in battlefield networks, mobile agents for network management, and the upcoming area of cognitive security. Chapter 14 presents the fundamentals of network reliability, fault tolerance, and delay-tolerant networks.

Organization of the Chapters

All the chapters in the book are organized in a similar fashion. A chapter begins with the introduction of the topic, which includes the history or background and an overview of the topic. This is followed by a general description of the common terms used in the topic, for better contextual understanding, and then an in-depth description of the theoretical aspects. Applications of the topic under study may be covered at the beginning or at the end, depending on the ease of understanding for the reader as felt by the authors. Contemporary research being done in the field is also given appropriate coverage where deemed necessary. Each chapter has its own list of references, followed by a list of the abbreviations employed, for ready reference. This is followed by a set of questions to be used by instructors and students to test the understanding of the chapter. The chapter ends with some exercise questions, which the students are encouraged to attempt.

How to use this Book in a Course

The book will definitely be of help to computer and electronics engineers, researchers, network designers, routing analysts, and security professionals, who will be able to pick and choose between chapters and sections as per their requirements – to gain knowledge of the historical background, theoretical base, configuration details, ongoing research, or application areas. For those who have a basic background knowledge of wired or wireless networks, all the chapters are self-contained, and hence any chapter or a section therein can be selected at random for study.

For academicians, the suggested strategy for handling the book is a sequential approach with minor exclusions. The faculty can use this book in three different types of course.

Firstly, it can be used as a textbook for a course in network routing. The course can be offered to graduate or senior undergraduate students. A prerequisite course on networks or wireless networks is desirable but not essential. This book can even be introduced as a textbook for first exposure to networks in various branches of engineering, such as computer science, information technology, electronics, instrumentation, electrical engineering, or reliability. This approach has been tested with our student interns, who had not previously undertaken any course on computer networks and read this book so as to gain background knowledge of networks to work on the projects.

For a complete semester course in network routing, given the fixed number of instruction hours, a few chapters/sections from the book may be skipped, and students may read these either out of interest or to fulfil a project/assignment for the course. Chapters 5, 8, 13, and 14 may be skipped in their entirety. Chapters 5 and 8 cover legacy networks and are targeted at network professionals still managing these networks. Chapters 13 and 14 are intended for security professionals, academic researchers, and routing analysts. Certain sections from a few chapters also may not be taught in class. The section on exterior gateway protocol in Chapter 7 is an exposure to historical routing protocol, the coverage of challenges in mobile computing in Chapter 9 is written for the research community, and the sections on interoperability, applications, and security in Chapter 12 have primarily been written for network designers and researchers from the industry, and thus these sections may not be covered in the one-semester course curriculum.

Secondly, it can be used as a reference book for any course in computer networks, data communication, wireless networks, and sensor networks.

Thirdly, based on certain sections on contemporary topics in the book, it can also be used as a reference book in certain courses other than on computer networks, such as reliability and fault tolerance, cognition, mobile agents, unified modeling language, and tactical networks.

Supplementary Resources

The following supplementary resources have been prepared along with the book:

- detailed presentation slides for all the chapters,
- answers to all the questions,
- solutions to selective exercises.

Faculty and other readers of the book may contact the publisher to receive a copy of the supplementary resources or may access it from the website www.wiley.com/go/misra2204. Regular updates of supplementary resources, with addendums and corrigendum, if any, will be uploaded, and hence please visit the website once the course is on.

The presentation can be used by the faculty for classroom teaching. This presentation can thereafter be used by students for quick revision of the contents of the chapters. Even though the contents of the book give an insight into what the book holds, these presentations also provide an opportunity for researchers and academicians to take a quick peep at the contents of the book for selective reading of the chapters of interests for any particular requirement.

The answers to all of the questions posed at the end of each of the chapters have been provided. However, solutions to only selective exercises have been added in the supplementary material. Among the others, there are certain exercises that do not have a single solution. The solution to such exercises will vary depending on the profile of the reader and the ecosystem where the course is being taught, and hence may be attempted accordingly. Evaluation of these exercises should be based on the stepwise solution approach adopted by the students and not for binary marking.

Acknowledgement

We would like to thank our families for their support, as the time spent writing this book was carved out of time that might otherwise have been shared.

We thank our colleagues at the Indian Institute of Technology (IIT) Kharagpur, India, and the Defence Research and Development Organization (DRDO), Ministry of Defence, New Delhi, India, for their encouragement, suggestions, and help.

Special thanks to the publishers who granted us copyright permission (including all languages, all editions) without charge to use verbatim the authors' work published by them – IGI Global, Allied Publishers, New Delhi, Defence Scientific Information and Documentation Center DESIDOC, New Delhi [for *Defence Science Journal*, *DESIDOC Journal of Library & Information Technology* (DJLIT), and *DRDO Newsletter*], CyberMedia (India), New Delhi (for PC Quest).

Friends, indeed, helped in time of need. Chaynika Taneja, Mukesh and Pooja Sonkarr were always there for all kinds of help. Rahul Sangore and Rashi Arora provided a few illustrations. Anjali Madan, Kessar Singh, Sahil Srivastava, Ishita Kathuria, Preeti Kumari, and Onkar Rai helped us to prepare the supplementary material. A few student interns at both Institutes also gave their support, mainly contributing to the literature survey.

Wiley Publishers put their faith in our work and took on the task of publishing it and continuing its circulation worldwide. We are grateful to the publication team at Wiley who worked so swiftly to convert our manuscript into this book. The team comprised Sandra Grayson (Associate Book Editor), Yamuna Jayaraman (Production Editor), and Paul Curtis (Copy Editor), with whom we worked directly, and all those who contributed to the production of the book at the publishing house. Apart from publication assistance, it was the book editors from Wiley who relentlessly drove us to complete this book by regularly interacting with us.

Our heartfelt thanks to all those professors, technocrats, and academicians who went through our manuscript and endorsed the book. We express in advance our gratitude to all those professors who will use this book in their course curriculum, the students who will study the book, and the researchers and professionals who will refer the book. We look forward to receiving your reviews and suggestions for forthcoming reprints and editions.

13 January 2017

Sudip Misra
Sumit Goswami

About the Companion Website

This book is accompanied by a website:

www.wiley.com/go/misra2204

The website includes:
- Detailed presentation slides for the Chapters
- Answers to the questions at the end of each chapter
- Solutions to the selective exercises at the end of each chapter

Part I

Fundamental Concepts

1

Introduction to Network Routing

1.1 Introduction to Networks

A computer network supports data communication between two or more devices over a transmission medium. The transmission medium can either be wired or wireless. The network is established and data is transmitted over it with the support of networking hardware and the software running on the hardware. Network hardware comprises equipment that generates the signal at the source, transmits the signal over the transmission medium, and receives and processes the signal at the destination. The software comprises protocols, standards, instructions, and algorithms that support transmission services over the network. The essentiality of networks has increased over time, along with advancement in network hardware, software, and support applications. There are huge variations in the size of a network in use; there can be small networks confined to an office or home, and at the same time there are networks spread across cities and countries. The spread of the network can be described in various terms, such as distance covered and the number of computers and other resources connected to the network. A local area network confined to a building may connect thousands of computers, such as in a software development center, a call center, or a stock exchange. Alternatively, a network spread across continents may connect only a handful of computers; for example, a network from a country to its base station in Antarctica may cover a few thousand miles but connect only a few computers.

The purpose of a network is to enable transmission of information between two or more networked nodes. The networked nodes can be computing devices, storages,

Network Routing: Fundamentals, Applications, and Emerging Technologies, First Edition.
Sudip Misra and Sumit Goswami.
© 2017 John Wiley & Sons Ltd. Published 2017 by John Wiley & Sons Ltd.
Companion website: www.wiley.com/go/misra2204

networking devices, or network-enabled peripherals. The computing devices can be desktop computers, laptops, or servers. Network-enabled peripherals can be printers, FAX, or scanners, and the networking devices are switches, routers, or gateways. Any other network-enabled device capable of sending or receiving data over the transmission medium can be a part of the network. A network system comprises a source, a destination, and the transmission system in-between. The source prepares data for transmission over the transmission medium. The preparation involves transformation of data, striping it into smaller parts, encapsulation, encoding, modulation, and multiplexing for converting bit streams into electrical signals or electromagnetic or radio waves. The transmission medium comprises the network connecting different nodes. The transmission medium can support unidirectional flow of data (simplex), bidirectional flow of data (duplex), or flow of data in either direction at one time (half-duplex). It can also be wired or wireless, providing point-to-point connectivity, or it can work in a one-to-many broadcast mode. The transmission medium may directly connect the source to destination, or it may be through intermediate network nodes. Thus, a transmission medium can be in various forms, utilizing different technologies and encompassing a variety of architectures. The destination receives data from the transmission medium, demultiplexes, demodulates, and retrieves the original data after decoding, rearranging, and merging. The transmission medium is a complex system as it can be shared between various network devices and has to run identification, channel utilization, security, congestion control, and bandwidth assurance services on it.

In addition to the source, destination, and transmission medium, a network system also comprises a few services such as exchange management, error detection and correction, flow control, addressing and routing, recovery, message formatting, and network management [1]. *Exchange management* deals with the mutually agreed conventions for data format and transmission rules between the sender and the receiver. The network system is prone to errors due to signal distortion, introduction of noise in the data signal, and bit flips during transmission, which may lead to receiving incorrect data, data loss, and data alteration. These are handled by *error detection and correction* techniques. *Recovery* is the process through which a network system is able to resume its activity even after a failure. The recovery may be from the point of failure or from a restore point prior to the failure. *Flow control* helps in synchronizing the rate of transmission from the sender, its flow through the network, and the rate at which the data is received. Flow control ensures that the data is transmitted at a mutually agreed rate to take care of the difference in the processing speed or variation in the network bandwidth of the sender and the receiver. *Addressing* is used uniquely to identify a network resource, and *routing* helps in deciding the optimum path for the data to flow from the source to the destination through the intermediate network. *Message formats* are the mutually agreed form of data. *Network management* is to monitor the network system, detect points of failure, and monitor the health of the system in terms of bandwidth utilization and load on network nodes. This helps to predict probable points of failure in future and enables enhancement or change in resources to avoid any network outage. The network management system also helps in version control of the software running on the nodes, its centralized upgradation, patch management, and inventory control of software and hardware.

The Internet is the largest network in terms of its geographical spread as well as the number of connected computers. The Internet has become the de facto network for people as well as organizations worldwide owing to its capability to act as a connectivity medium across geographical regions. The common applications running over the

Internet are electronic mail, electronic commerce, and Web access. In the 1960s, when the networks were being conceptualized and experimented, vendors designed and developed proprietary network equipment and protocols. This led to competitiveness among the vendors for faster development of network protocols and devices for having a competitive edge by providing an advanced and more scalable network. However, it restricted interconnectivity among networks from different vendors such as Microsoft, Novell, Banyan, Xerox, IBM, and DEC. A network based on equipment and software of one vendor could not connect or exchange data with a network based on the products of another vendor. Thus, if some computers in an organization were on the Novell network, they could not share data with other computers of the same organization that were on the IBM network.

Introduction of the seven-layered open system interconnection (OSI) as an implementation of the ISO standard led to the establishment of a framework for multivendor network compatibility, connectivity, and interoperability. Based on the ISO standard, the vendors provided interconnectivity options over their proprietary networking protocols. However, the Transmission Control Protocol/Internet Protocol (TCP/IP) became a de facto network connectivity standard preferred over interconnection suites of the proprietary networks, and slowly the vendors moved on to support TCP/IP [2]. TCP/IP was an outcome of research in ARPANET, a United States Department of Defense (DoD) project and hence sometimes known as the 'DoD model'. TCP/IP is also commonly known as the 'public networking model', as the Internet Engineering Task Force (IETF) maintains the protocol with the involvement of representatives from various networking companies for evolution of TCP/IP standards. The Internet is built on TCP/IP providing connectivity between heterogeneous physical networks and protocols.

The present-day network is used for transmission of data, voice, video, and share resources. With time, there has been a rapid increase in the bandwidth supported by wired as well as wireless networks. The bandwidth availability has increased to cater for data sharing between computers and servers with high memory and processing power as well as voice and video applications. Concurrently, the problem of traffic congestion is evident owing to an increasing demand for network bandwidth. Congestion is also caused by scaling of the network without considering the available network resources in place. There has been a reduction in the cost of networking devices as well as computing devices. So, a faster and scalable network can be established at a much lower cost. However, the reduction in computing cost and increase in the number and type of computer applications lead to increase in the rate at which the devices push data into the network. Convergence of voice, video, and data into a single application and its transmission through a common integrated channel also increases the bandwidth utilization. With a high degree of office automation and dependence on the network for real-time or near-real-time data transmission and updates, slow and congested networks are unacceptable.

In order to avoid congestion, a network is generally broken down into smaller segments using networking devices called bridges, switches, and routers. The contents of the data being transmitted over a network are not of any interest to these networking devices. These networking devices look only into the origin, destination, and control information related to the data in transit so as to enable its effective delivery to the destination [3]. The effectiveness of the delivery varies with the application and may be optimized in terms of transmission time, secured delivery, reliable delivery, acknowledgement, delivery only through a dedicated path, or assurance of a minimum bandwidth throughout the transmission link, ensuring quality of service (QoS).

A network is generally gauged by three major criteria – performance, reliability, and security [4]. The performance of a network is dependent on a number of factors, such as the number of nodes connected to the network, the bandwidth of the transmission medium, the protocol used, the software overlay, and the amount of memory and processing capability of the networking hardware and the nodes. The network performance is evaluated in terms of throughput and delay, which are inversely proportional to each other. Transit time and response time are the two parameters used to measure the performance of the network. Transit time is the amount of time a message spends in the network after its transmission from the source until it reaches the destination. Response time is the total time between sending a query through the network and receiving its response. Reliability relates to the duration for which a network remains operational without failure, which is different from availability. A network that goes down every hour just for a second will be highly available, but its reliability will be low. Thus, reliability can be measured with the help of mean time between failures (MTBF). Network security relates to implementation of access policies, restricting the data from unauthorized access, protection from change of data (integrity), preventing damage or loss of data, detection of security breaches, and procedures for data and network recovery in case of security attack.

1.2 Network Architecture and Standards

Network architecture is the logical and structural layout of the network that assists and guides the network designer in implementing an optimum network. The network architecture also supports the network administrator in managing the network and troubleshooting the point of failure in case of a breakdown. The network architecture is an essential component for working on the security of the network and implementing access policies. Network communication is a multilayer task wherein each activity is accomplished at a particular layer of the architecture. The layering makes the architecture simple to develop and implement. Each product and protocol is designed to work in a particular layer or across a few layers with standard interlayer interfaces supporting interoperability among the products and protocols. Although a network may be designed and implemented in various forms, the OSI layer divides it into three basic categories. The data transmission uses the physical layer of the OSI model, the network devices operate at the data link layer and network layer, and the applications use the session layer, presentation layer, and application layer of the OSI model.

The network architecture is closely associated with the topology of the network. The topology is the logical design of the network, showing the interconnection of the networked nodes. The topology planned for a network is based on cost, scalability, application, criticality, size, and type of network. The commonly used topologies are star, bus, ring, and mesh. Various combinations or minor modifications of these common topologies can be used to evolve other topologies such as tree, distributed bus, extended star, distributed star, partial mesh, or hybrid.

In a star topology, the network nodes have a point-to-point connection with the central hub. In a bus topology, also known as a line network, each network node is connected to a single cable. In a ring topology, the network nodes are set up in a circular fashion in which the data transmission takes place around a ring in one direction and each neighboring node, either to the left or to the right, works as a repeater in order to maintain the strength of the signal as it is transmitted in a loop. This topology is also known as a loop network. In a mesh network, each node is connected point-to-point with all other nodes in the network. When every node is connected to all the other nodes in the network, it is known as a complete mesh. When some of the links in a complete mesh network are removed to reduce redundancy, this leads to the creation of a partial mesh.

Visual representation of the common topologies is given in Figures 1.1 to 1.4.

Each topology has its advantages, disadvantages, and applications, which are set out in Table 1.1.

A network is also classified according to the geographical spread of the nodes. Based on size, the classical types of network are typically the local area network (LAN), the metropolitan area network (MAN), and the wide area network (WAN). A LAN connects the networking devices within a short span of area and is generally controlled, maintained, and administered by a single person or a company. A MAN is an

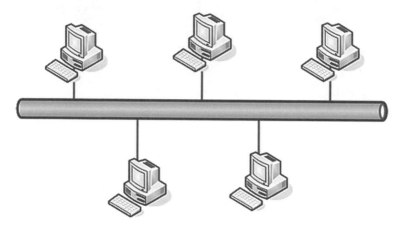

Figure 1.1 Bus topology.

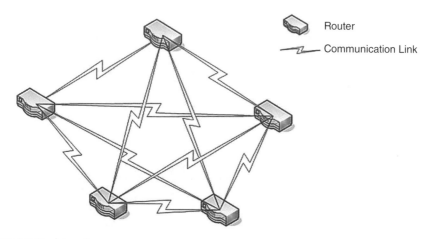

Figure 1.2 Mesh topology.

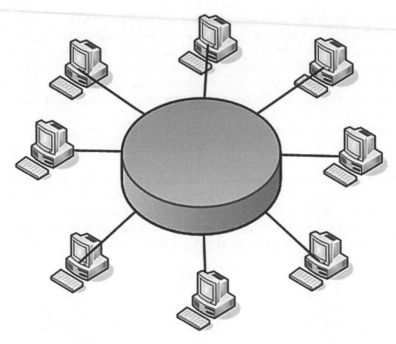

Figure 1.3 Ring topology.

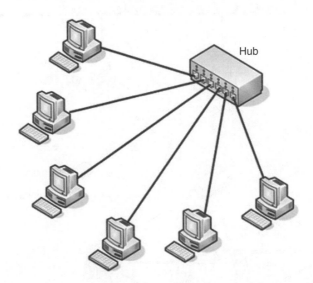

Figure 1.4 Star topology.

intermediate-sized network and in terms of outreach can be placed between a LAN and a WAN as it covers a large span of physical area such as a metropolitan city, which is larger in size than a LAN but smaller than a WAN. A WAN connects the networking devices and a collection of LANs that are distributed over a large geographical area, which may spread across cities or even continents. In addition to the classical types of

Table 1.1 Comparison of network topologies.

Topology	Architecture	Operation	Scalability	Point of failures	Advantage	Disadvantage
Bus	Each node is connected to a single backbone cable.	Information is transmitted from one node to another through the backbone cable.	High scalability at low cost as a node has to be directly connected to the backbone cable.	A break in the backbone cable disrupts the entire network.	Easy to install and cost effective.	Only one node can transmit at a time.
Star	Each node is connected to a central hub.	Information passes from one node to another through the central hub.	High scalability at optimum cost as a node has to be directly connected to the hub.	Failure of the hub disrupts the entire network.	Easy to install and cost effective; if a switch is used as a central node, multiple nodes can communicate with each other concurrently.	Single point of failure.
Ring	Each node is connected to a single backbone cable configured as a ring.	Information moves from the source node in a unidirectional manner along the ring until it reaches the destination.	Generally implemented using fiber cables. Hence relatively higher cost.	Even if there is a break in the cable, information can be transmitted through the rest of the ring.	Avoids a single point of failure as the data can move in another direction to reach the destination from any source in the case of a break in the ring.	Not in common use.
Mesh	Each node is connected to every other node.	Information is directly transmitted from source to destination without intermediate nodes and in a single hop.	Limited scalability as each new node has to be connected to every other existing node.	A cable break will disrupt the direct connectivity only between a pair of nodes. However, alternative routes will exist.	Highly redundant, and reliable and fast network.	Implementation is costly and complicated.

network, a few more network types have now emerged, which are defined according to the size and application domain of the network. Some of these are as follows:

Campus area network (CAN). These are the networks spread across the campus of large institutes, academic centers, research organizations, or industrial complexes.

System area network (SyAN). These networks connect the high-performance systems in a network. The network has low latency and high speed of the order of 1 GB/s. These are also known as cluster area networks as they provide a high-speed data and communication interconnection framework to workstations and PC clusters [5].

Storage area network (SAN). SAN is a network of storage devices and associated servers redundantly interconnected with switches using fiber connectivity providing high bandwidth and parallel links. The storage media has information redundancy at its own level. The SAN provides consolidated high-volume storage accessible by the computational devices over the network.

Personal area network (PAN). The coverage of the network ranges from only a few centimeters to a few meters and is capable of connecting various devices used as personal assistants to individuals or located near to their area of presence. The personal area network can be wired and supported by USB or Firewire. It can be wireless and supported by ZigBee, 6LoWPAN, Bluetooth, or Z-Wave.

In addition to these common types of network, a number of other networks have also been introduced, based on scale of extent and purpose. Some examples of these special and new types of network are the near field network (NFC), Internalnet, the body area network (BAN), the near-me network (NAN), the home area network (HAN), and the interplanetary Internet.

Based on the services concept, the network architecture can be broadly grouped into the following two categories:

Client–server architecture. The system is decomposed into two entities classified as the client and the server. The client and the server may be the processors (computers) or the processes. A producer providing the appropriate resources and services is termed the 'server', and the consumer using the provided services is termed the 'client'. There exists a relationship between multiple clients and multiple servers. The client–server architecture model works in a 'tier' approach, separating the functionality of the tiers on the basis of the concept of the services provided.

Peer-to-peer architecture. In this model, all the network nodes are believed to have equivalent computational power and resources and have equal capabilities and responsibilities in terms of service provision.

There can be instances where a combination of peer-to-peer architecture is embedded in client–server architecture. The client–server architecture distributes the system in tiers, and each tier can further have a peer-to-peer architecture running within it.

The network architecture can also be designed with an attempt to separate application from data so as to enhance security as well as accessibility. Such architecture has three tiers [6] – the Web tier (referred as the demilitarized zone), the application tier, and the data tier. The description of each of these tiers is as follows:

Demilitarized zone (DMZ). The demilitarized zone is the topmost level of the application in the network's hierarchy and is also known as the Web tier. It provides an interface to the external network for accessing data and utilizing the services of the applications and resources lying in the militarized zone without directly interacting with the internal system in a network. The functional implementation of the demilitarized zone is created using firewalls. A DMZ is categorized as the part of the network that is layered between a trusted internal network and an untrusted external network.

Application tier. The application tier is also known as the business logic layer. This is the middle layer between the demilitarized zone and the data tier. The application tier accesses the data tier to retrieve or modify data from the data tier and sends the processed data to the devices in the DMZ tier. Direct access to the application tier is not permissible to the users.

Data tier. The data tier is the innermost (i.e. core) tier of the network's architecture. This tier hosts the databases and database servers that store and access information of the systems. This tier is responsible for maintaining the neutrality and the independency of the data from application servers and business logic. Direct access to the data tier is not permissible to the users. This layer is also known as the database tier or intranet zone.

Setting up network standards facilitates the interoperability of network technologies and systems. A standard in the field of the networks can be proprietary, open, or de facto. The open standards generally emerge from the efforts of a consortium of industries, which are generally non-profit organizations. Some of the standards organizations in the field of networks are as follows:

- International Organization for Standardization (ISO),
- American National Standards Institute (ANSI),
- Institute of Electrical and Electronics Engineers (IEEE),
- Electronic Industries Alliance (EIA),
- Telecommunications Industry Association (TIA),
- International Telecommunication Union – Telecommunication Standardization Sector (ITU-T),
- European Telecommunications Standards Institute (ETSI).

Unlike international standards organizations, which generally work on open standards, there are also networking industry groups that work on creating, upgrading or promoting standards. However, they develop and promote specific standards that are generally product oriented. With the wide spread and usage of the Internet, a few Internet standards organizations [7] have also taken responsibility to develop policies, standards, and architecture related to the Internet. Some of these organizations are as follows:

- Internet Society (ISOC),
- Internet Architecture Board (IAB),
- Internet Engineering Task Force (IETF),
- Internet Research Task Force (IRTF),
- Internet Engineering Steering Group (IESG),
- Internet Research Steering Group (IRSG).

Table 1.2 A few major networks standardized by IEEE.

IEEE standard	Network/system standardized
802.1	Procedures for bridging and managing network
802.1 Q	Virtual LANs (VLANs) over Ethernet network
802.3	Physical layer and media access control layer of wired Ethernet
802.4	Token-passing bus
802.5	Token ring
802.6	Distributed queue dual bus (DQDB)
802.7	Broadband LAN
802.11	Wireless networking and Wi-Fi certification
802.15	Wireless PAN
802.20	Mobile broadband wireless access

Many networking standards set by IEEE are in common use, and the major ones are listed in Table 1.2.

The 802.11 standard includes 802.11a, 802.11b, and 802.11 g, the details of which are as follows:

802.11a. The 802.11a standard [8] sets the protocols in the data link layer and an orthogonal frequency-division multiplexing (OFDM)-based physical layer. It operates in a bandwidth spectrum of 5 GHz with a maximum data flow rate of 54 Mb/s and includes error correction procedures.

802.11b. The 802.11b standard uses the media access method and has a maximum data flow rate of 11 Mb/s. The 802.11b standard extends the modulation technique of the 802.11 standard directly.

802.11 g. The 802.11 g standard works on a bandwidth of 2.4 GHz and uses the OFDM-based physical layer scheme for data transmission. It has a maximum data flow rate of 54 Mb/s, exclusive of error correction codes.

Networking standards set by EIA: EIA-485 is a standard that defines the characteristics and electrical properties of drivers and receivers that are to be used in a balanced digital multipoint circumvented system. Digital communication networks possessing the EIA-485 standard can be used for long-range networks that can work effectively in an environment with electrical interference or other noise.

Networking standards set by ITU-T: ITU-T is a standard that defines the characteristics of an optical transport network (OTN), and hence this standard has played a characteristic role in transforming the Internet's bandwidth and spectrum capabilities.

Networking standards set by the International Organization for Standardization: In the networking area, ISO sets the standards for the implementation of an OSI reference model. This model defines and lays the networking framework for network protocol implementation in the seven distributed layers of the network. Control is passed in the network via these seven layers, starting from the topmost to the bottom-most layers, which are as follows:

Table 1.3 Layer-wise protocols in the ISO OSI model.

Layer	Protocols
Application (layer 7)	COPS, FANP, FTP, HTTP, IPDC, IMAP4, IRC, ISAKMP, NTP, POP3, RLOGIN, RTSP, SCTP, SLP, SMTP, SNMP, TELNET, WCCP
Presentation (layer 6)	BGP4, EGP, HSRP, EIGRP, TGRP, NARP, NHRP
Session (layer 5)	BGMP, DIS, DNS, ISAKMP/IKE, ISCSI, LDAP, MZAP, NetBIOS/IP
Transport (layer 4)	ISTP, Mobile IP, RUDP, TALI, TCP, UDP, Van Jacobson, XOT
Network (layer 3)	DVMRP, ICMP, IGMP, IP, IPv6, MARS, PIM, RIP2, RSVP, VRRP
Data link (layer 2)	ARP/RARP, MPLS, PPP, FDDI, SLIP
Physical (layer 1)	ATMP, L2F, L2TP, PPTP

Application (layer 7). This layer is application specific and provides services such as file transfers, email, and other network services and supports end-user processes.

Presentation (layer 6). This layer provides the representation of data by application to network format translation, and vice versa. This layer is also responsible for encryption and decryption.

Session (layer 5). This layer is responsible for establishing, managing, and terminating the network connections between applications.

Transport (layer 4). This layer provides the effective and fault-free transfer of data between end systems and is responsible for error recovery and control of flow so as to ensure complete data transfer.

Network (layer 3). This layer provides the switching and routing techniques to transmit the data from source node to destination node in a network.

Data link (layer 2). At this layer the encoding and decoding of data packets into bits take place.

Physical (layer 1). The bit stream, in the form of electrical impulse, light or radio waves, is transmitted through the network. It provides a hardware means of sending and receiving data on a carrier.

A list of selected protocols in the various layers of the ISO OSI model is given in Table 1.3.

1.3 Glimpse at the Network Layer

The network layer is also referred as the internet layer or the internetworking layer. It is designed for data delivery across the nodes in the network. It also performs device addressing, packet sequencing, path determination, congestion control, and error handling. The network layer operates over the physical layer and the data link layer. While the lower layers provide a mechanism for data transfer within the same network, the network layer has protocols for support, with due consideration for QoS, and data transfer across various interconnected networks or the internetwork and even the

Internet. The main network equipment operating in this layer is the router, which is used for routing packets and message forwarding within an internetwork. A few common protocols of the network layer are as follows:

- Internet Protocol (IPv4/IPv6),
- Internet Control Message Protocol (ICMP),
- Internet Group Management Protocol (IGMP),
- Internet Protocol security (IPsec),
- Internetwork Packet Exchange (IPX),
- Routing Information Protocol (RIP).

The need for the network layer [4] can be explained with a simple example of an internetwork connecting a few local area networks (LANs). When the source and destination are within the same LAN interconnected by a switch, the delivery of the packet can be achieved using the physical layer and the data link layer, employing the media access control (MAC) addresses of the source and destination available in the frame. However, to achieve data delivery across LANs connected by a router, as shown in Figure 1.5, the lower layers cannot provide information regarding forwarding of the frames to the appropriate interface on the router, as they lack the routing information. The network layer provides information regarding the link and interface on which a packet should be forwarded for delivery to the destination.

The network layer is responsible for host-to-host delivery and performs necessary functionality at the source, router, and destination [4]. At the source, which is the transmission end, the network layer creates packets from the data received from higher layers, such as the transport layer or any other appropriate layer in the protocol stack being used. For delivery of the packet across various links, a logical address of the destination is required, which is handled by the network layer. As logical addressing is handled at the network layer, the necessary information for routing, which also includes the source and destination addresses, is put in the network layer header to assist routing. Other information available in the header includes packet checksum

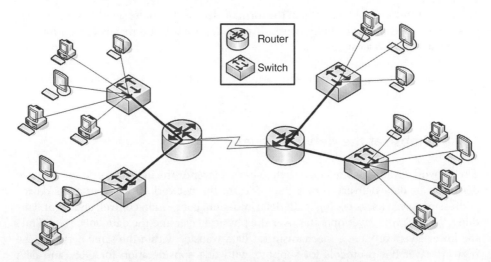

Figure 1.5 A network with two routers and two LANs connected to each router.

information for facilitating error detection, an identifier for the higher-level protocol being used, and some administrative data and some optional padded-out fields. Thereafter it adds the network layer header to the data packet and fragments the data packet if the size is larger than the maximum transmission unit (MTU). The network layer is responsible for maintaining the routing table, which contains information related to the interfaces through which a packet should be forwarded for a particular destination. The network layer refers to this routing table available at the source and determines the interface at the source through which the data packet should be forwarded out of the source. With this exit interface information, the data packet is sent to the lower layers for delivery.

Each protocol has an MTU, which is the maximum packet size that it can handle and transfer. When the network receives a data packet that is larger than its MTU, it has to fragment the data packet for transmission and then reassemble the data packet from the fragments at the destination. The format of a fragment, as well as that of a much larger data packet, is the same so as to support the reassembly. The header of the fragment contains information for identification of the fragment, i.e. the identification field to locate the packet to which the fragment belongs, as well as that for its reassembly, i.e. the fragmentation offset to identify the location of the fragment in the packet [9]. A 'More Fragments' flag field acts as a counter to indicate the assembly of all the fragments of the packet.

When the data packet reaches the router through any of its interfaces, it looks at the network header to read the destination address. If the packet is not meant for the router, then it looks up its routing table to search for a suitable exit interface for the destination address of the arrived packet and thereafter forwards the packet through that interface for framing at the lower layer and further delivery. During this process, necessary modifications are made in the network header to support a number of other functionalities of the network layer. The actual delivery of the packet to its destination is based on the MAC address. The network layer facilitates mapping of the MAC address with the logical address generally at the time of network connection. If a suitable entry corresponding to the destination address cannot be located in the routing table, then the packet is dropped.

At the destination, the network layer ensures that the delivered packed is meant for it. The identification is based on the destination address in the data packet header. As the network layer may not be connection oriented, this may lead to the arrival of fragments through different links and in a different sequence to that in which they were transmitted. The network layer assembles all the fragmented packets and arranges them sequentially. The reassembled fragments are then sent to the upper protocol layer.

There are two types of packet in the network layer – data packets and route update packets. Data packets are used to transmit the data received from the higher layers for transmission across the internetwork. These packets belong to any of the protocols, such as IP or IPX, that support routing of the packets across the routers. Such types of protocol are known as 'routed protocols'. The route update packets are generated by the routers to make the network aware of its interfaces and connected networks and enable the recipient routers to build, maintain, and update the routing table. The packets may be transmitted only to neighbors or to the entire internetwork, depending on the routing protocol in use. Protocols such as RIP and OSPF that transmit these route update packets are known as 'routing protocols'.

Connection-oriented vs connectionless network service. A network layer service can be connection oriented or connectionless, based on the specific protocol being used, the supported application, the type of network, and a few other related parameters. In a connection-oriented network service, a virtual path [4] is established between the source and the destination before initiation of the packet transmission, and so this is also known as the virtual circuit approach of packet switching. Once the path is established, all the packets are transmitted from the source to the destination using the same path, and the packets are transmitted in sequence. The transmission is reliable, and all the packets follow the same path. They arrive at the destination in the same sequence in which they were transmitted. The connection is terminated only after the entire message has been transmitted. Asynchronous transfer mode (ATM) and frame relay use the virtual circuit approach for packet switching.

In a connectionless network service, the path of each data packet is determined separately. The network header of each packet is read for its destination address along with the other network layer parameters. The destination address is then looked up in the routing table to determine the exit interface for forwarding the packet. As the routing table may be dynamic in nature and can change with time owing to change in the network parameters, this can lead to a different exit interface for the packets to the same destination address. Thus, the packets from a source to the same destination can travel across different links and reach the destination in a different sequence from the one in which they were transmitted. This type of service, which is also known as packet switching using the datagram approach, is commonly used on the Internet.

1.4 Addressing in TCP/IP Networks

TCP/IP is the basic communication language or protocol that enables computers to communicate over the network. Created by an agency of the United States Department of Defense, DAPRA, TCP/IP is an industry standard suite of protocols describing a set of guidelines and specifications to provide communication in a heterogeneous environment. It provides a routable, enterprise networking protocol and access to the Internet and its resources. In TCP/IP, a session is established between the sender and receiver before transmission of the actual data, thus making it a connection-oriented protocol. The end-to-end connection is established using the port numbers at the endpoints. The reliability of the delivery of data to the destination is ensured by using the sequence number in the data packets and acknowledgement from the destination.

The TCP specifications were first laid down in 1974, and the IP standard was published in 1981 in the form of RFC-791. TCP/IP is a bilayer standard. In the higher layer is the Transmission Control Protocol, which manages the assembling and reassembling of the packets that are transmitted over the network for communication. The received packets are converted into the original message. Thus, a TCP header primarily comprises the source port, the destination port, the sequence number, the acknowledgement number, the data offset, the flag bits, the window size, and the checksum. In the lower layer the Internet Protocol manages the address of each packet so that it goes to the right destination. It follows a point-to-point communication in which the message is transmitted from a point in the source computer to the destination computer. For the delivery of every TCP segment there is a TCP port, and the commonly used port numbers are: 20 for File Transfer Protocol (FTP) (data channel), 21 for FTP (control channel), 23 for Telnet, 25 for Simple Mail Transfer Protocol (SMTP) (for email), 80 for Hypertext Transfer Protocol (HTTP) used for the WWW, and 139 for the Net BIOS Session Service. The IP header mainly comprises time to live (TTL), the protocol, the header checksum, the source IP address, and the destination IP address.

The network nodes, such as computers, servers, routers, and other network-enabled devices, have a unique IP address. Internet Protocol Version 4 (IPv4) uses a 32 bit addressing scheme, limiting the number of uniquely addressable devices to 2^{32}, i.e. 4 294 967 296. The 32 bit IP address is represented in the form of four octets (8 bit field). Each octet, being 8 bit, represents a decimal number in the range 0–255. This format of representing the IP address as four decimal numbers in the range 0–255, each separated by a dot, is called dotted decimal notation. For example, an IPv4 address represented in binary form can be 10000010.01101111.00000010.00001100, the dotted decimal notation of which is 130.111.2.12

Classful Addressing

Originally, IP addresses were divided into two parts, namely the network ID and the host ID. The former used the first octet of the address, and the latter occupied the remainder of the address. This led to the development of only 256 networks. Later on, classes of network were created of higher-order octet. This type of class-based IP address was known as classful networking [7,10]. For the purpose of network identification, the classes A, B, and C had different bit lengths for both network ID and host ID, as shown in Table 1.4.

Subnetting

A network may be small and does not require the entire available host IDs to address its nodes. Alternatively, the organization may be willing to divide its network into a number of smaller networks based on the applications or physical distribution and would be interested in restricting the traffic flow within the smaller network unless specified. A subnet divides the host address space into smaller groups for preservation of address space, reduction in traffic congestion in the network, and enhancement of security. A subnet reduces the network traffic by limiting the broadcast domain within the subnet. A subnet address is created using the initial bits of the host ID, thus reducing the number of hosts that can be accommodated in the subnet.

Table 1.4 IP address class.

Class	Bits in network number	Bits in host number	Initial bits
Class A	8	24	0
Class B	16	16	10
Class C	24	8	110
Extended addressing			111
Class D	Not defined	Not defined	1110
Class E	Not defined	Not defined	1111

In a classful address [3], the host computer knows which initial bits of the address represent the network ID, and the remaining bits represent the host ID. However, in the case of subnet addressing, the same information is sent to the host computer using subnet masking, as it is not predefined. Like an IP address, the subnet mask is a 32 bit address with the initial bits as 1 and the trailing bits as 0. The 1 s in the subnet mask indicate the bits that will represent the subnet or the network ID in the IP address. The default subnet mask for class A, B, and C addresses is 255.0.0.0, 255.255.0.0, and 255.255.255.0 respectively. The number of bits available in class A, B, and C addresses for subnetting is 24, 16, and 8 respectively. Thus, in a class C address, the subnet masks in the fourth octet can be 10000000 (128), 11000000 (192), 11100000 (224), 11110000 (240), 11111000 (248), 11111100 (252), and 11111110 (254), which are created by increasing the masking by 1 bit successively. The subnet mask of 11111110 (254) cannot address any host computer, as all 0 s in the host ID represent the network or the subnet and all 1 s are reserved for broadcast. However, RFC 3021 uses a 31 bit subnet mask for a point-to-point link where network and broadcast addresses are not necessary. In a similar manner, the subnet mask of class B and A addresses can vary from 1000000.00000000 to 11111111.11111110 and from 1000000.00000000.000000 to 11111111.11111111.11111110 respectively. Consider a class C address with a subnet mask as 11000000 (192). Here, the two bits with value 1 represent the subnet and the six bits with value 0 represent the host. Thus, the subnet can have the values 00, 01, 10, and 11. For the subnet 01, the first host ID will be 000000, i.e. the network address (**01**000000), the next host ID will be 000001, i.e. the first valid host (**01**000001), up to the last valid host (**01**111110), followed by the broadcast ID (**01**1111111).

Classless interdomain routing (CIDR) is used to assign blocks of IP addresses with consecutive addresses within an address bit boundary. CIDR is represented using an IP address followed by a decimal number with a slash (/) in-between. The decimal number when converted into binary gives the sequence of leading 1 s, which represent the initial bits that cannot be changed in the address chunk. The trailing 0 s represent the bits that can be changed to obtain a sequence of IP addresses in the address block. Thus, a block with an IP address represented as 192.168.13.32/28 represents a subnet mask of 255.255.266.240. The default subnet mask in a class C address is 255.255.255.0, and a class C address, e.g. 192.168.13.70, with a default subnet mask can be represented in CIDR slash notation as 192.168.13.70/24 because 24 bits of 1 s followed by 0 s translate to 11111111 11111111 11111111 00000000, i.e. 255.255.255.0.

Variable-length subnet mask (VLSM). The conventional classful addressing scheme was a two-level addressing scheme with a network ID and a host ID. Subnetting converted this addressing to a three-level scheme, adding an extra level of subnet giving the network designer the liberty to divide a huge network or a large address space into smaller equisized addressing chunks. However, the equisized address blocks sometimes prove to be a bottleneck, as the subnet is generally determined by the size of the biggest chunk, thereby wasting address space in the smaller subnetworks.

The subnet masking will be inefficient when a network is divided into two or more parts with a huge variation in the number of host computers in the two networks. This can be explained with an example of a network with, say, 200 nodes that can be easily addressed by class C addressing. Now, if this network is to be divided into five subnetworks accommodating 6, 14, 30, 50, and 100 nodes each, it requires 3 bits in the subnet for the creation of five subnets. Now, with 5 bits left for the host ID, it will not be enough to address the nodes in the networks with 50 and 100 hosts, even though the addresses will be unused in the subnets with 6, 14, and 30 nodes. Using the conventional classful method, this would be resolved by taking one more block of class C address and dividing it into only two subnets, each accommodating 100 and 50 nodes respectively. This would effectively mean using an address space of 255 + 255, i.e. 510 nodes to address 200 nodes. The reason for the wastage is that subnetting creates only equisized subnetworks. Variable-length subnet masking [7] enables the creation of subnetworks of varying size. This is achieved by splitting a subnet further into smaller subnets, which can further be split into even smaller subnets until a subnet of an appropriate size is achieved. Thus, to accommodate 200 nodes in the example above, a class C address space (/24 network) will be split into subnetworks accommodating 126 hosts (/25 network), 62 hosts (/26 network), 30 hosts (/27 network), 14 hosts (/28 network), and six hosts (/29 network).

IPv6. IPv4 has an address space of 2^{32}, i.e. 4.29 billion, and every node on the Internet has to be given a unique address to enable its communication with others. Network address translation (NAT) is a technique to map a public IP with a number of private IPs, thus reducing the requirement of the number of unique IP addresses over the Internet. Still, as reported by the Asia-Pacific Network Information Center (APNIC) [11], 'the primary supply of unallocated IPv4 addresses was exhausted' by the Internet Assigned Numbers Authority (IANA) in February 2011. To cope with the shortage of IPv4 address space envisaged by IETF, it introduced IPv6 as the new protocol for Internet communication.

The protocol has an address length of 128 bits and has enhanced features of security, mobility, QoS, end-to-end connectivity, scalability, and autoconfiguration over IPv4. IPv6 has provided expansion of the available network addresses, and it has catered for the various demands of technological enhancements of IPv4, which has been in use for more than two decades. This is one of the reasons for NAT not being able to handle the address space problem and magnify the requirement of IPv6. As the existing networks are on IPv4, and the upcoming networks would be using IPv6, compatibility among IPv4 and IPv6 networks would be achieved using dual-stack translation or tunneling.

IPv6 is 128 bits (16 bytes) long and hence can address 2^{128} nodes. The address is represented as $b_1:b_2:b_3:b_4:b_5:b_6:b_7:b_8$, where b_i is a 16 bit binary number represented in hexadecimal form and thus requires only four digits. For example, 2012:0000:0000:9876:0000:0000:0000:9ABC:1234 is an IPv6 address. Some of the addressing rules [12,13] in IPv6 are as follows:

- Leading zeros in the address are optional
  2012:0000:0000:0076:0000:0000:0000:9ABC:1234
  →2012:0:0:076:0:0:0:9ABC:1234
- The address is case insensitive
  2012:0000:0000:0076:0000:0000:0000:9ABC:1234
  →2012:0000:0000:0076:0000:0000:0000:9abc:1234
- Once in an address, fields with successive 0 s can be represented as ":.:"
  2012:0000:0000:0076:0000:0000:0000:9ABC:1234
  →2012:0:0:076:0:0:0:9ABC:1234
  →2012:0:0:076::9ABC:1234
- In an URL, it is enclosed in a bracket
  http://[2012:0:0:076::9ABC:1234]:8080/index.html
- It should use a fully qualified domain name (FQDN)

In IPv6 address structuring, Internet service providers (ISPs) are assigned a /32 IPv6 address, customer sites are assigned a /48 address, /64 is used for subnets, and /128 is used for devices. Unlike IPv4, which had a unicast and broadcast address, IPv6 supports unicast, multicast, and anycast (one-to-one-to-one-..., leading to many by delivery to the nearest).

1.5   Overview of Routing

The routing calculates a route between the source and destination node so as to enable proficient utilization of the intermediate network connecting the source and destination. The parameters for proficiency vary with the algorithm used for routing and may be associated with the bandwidth utilization, the time delay, the number of hops across the intermediate routers in the network, the congestion in the network, or a combination of these. The simplest performance criterion to evaluate the proficiency of the routing algorithm is the hop count, which gives an idea of the number of intermediate links traversed by the packet. The overhead associated with counting the hop is also less, as each intermediate node has to change the hop counter by 1 and does not involve any detailed calculations, leading to minimum processing time and minimum resource consumption. However, hop count is an effective performance criterion if the links are similar in terms of bandwidth. The number of hops determines the 'cost' of the route and the routing algorithm attempts to achieve the minimum-cost route. If the links are of different bandwidth, the least hop count may not be the most efficient performance criterion as it may not give the least time delay path between the source and the destination. So, to improve the routing efficiency of the algorithm, the cost may depend on other parameters and may be directly or inversely proportional to the parameter. The routing algorithms are designed on the least-cost approach, and the costing parameters vary with the algorithm or the network parameters. The cost parameters may either be predefined in the algorithm or may be defined at the time of configuring the routing algorithm in the network to make it suite the network environment and parameters.

The routing decisions can be differentiated on the basis of the time when the decision is taken and the place where the decision is taken. The routing decision can be taken before initiation of the actual data transfer between the nodes or during the actual data transfer. In the case of a datagram network where packet switching takes place, the routing decision is taken throughout the path in which the packet is in transit, as each node decides on the next-hop node. In contrast to this, in virtual-circuit-based networks the virtual circuit is established before data transmission, and hence the entire path through which the data packets will traverse is decided at the time of establishment of the virtual circuit. Thus, in the case of packet switching networks, the routing decisions are taken throughout the duration when the packet is in transit, while in virtual-circuit-based networks, the routing decision is taken before the transmission of the packet. However, in advanced routing methodologies, a virtual circuit may be established before the transmission of the packets, but the virtual circuit may reconfigure itself during the process of data transmission. The reconfiguration of the virtual path depends on link congestion, link failure, or the introduction of new and better links. So, in such routing decisions, the time of decision is prior to as well as during the network transmission.

The 'decision place' of routing is a variable that tells the location of decision-making for routing a packet. The most commonly used, but relatively complex, strategy is *distributed decision-making*, where each intermediate node decides on the next link to which the packet should be forwarded. Each decision-making node should have complete or partial information about the network. The failure of a few intermediate nodes does not drastically affect the performance of the network, as the packet is forwarded through some alternative route. An alternative to this is the *centralized routing decision*, in which there is a centralized control node that takes all routing decisions. The control node has a view of the entire network topology and controls the routing of the packet through the network. The drawback of such a routing design is failure of the network routing in the case of failure of the control node. The control node is computationally overloaded, as it has to administer the entire network and may become a computational bottleneck for the network, and also leads to congestion of the links connecting the control node. A minor design improvement can be achieved by designating a few nodes as the network controllers instead of a central node.

The third policy of routing based on place of decision is *source routing*. In source routing the path of packet forwarding is decided by the source node and intimated to the network for data delivery. This routing strategy helps in deciding the route by the source on the basis of the parameters that the source may feel to be significant for that data delivery.

1.6   Delivery, Forwarding, Routing, and Switching

Delivery [4] refers to the handling of a packet under the supervision of the network layer to ensure it reaches the destination. The delivery of the packet is elaborated under two different concepts – connection type and the method of delivery. The connection types are handled by the data link layer in terms of two different types of switching – packet switching and circuit switching. The methods of delivery are direct delivery and indirect delivery.

Direct delivery can be achieved when two nodes are connected point-to-point on the same physical network and the packet can reach from the source to the destination in a single hop without any routers in-between. If the source and destination are connected through a wide area network and there are a number of routers in the path between the source and destination, leading to a multihop delivery, the last hop in which the packet is transmitted from the last/peripheral router to the destination is also known as direct delivery. The source node, looking at the network address of the destination, can determine before data transmission whether the packet will reach the destination by direct delivery, because if they have the same network address it will be sent using direct delivery.

When the source and the destination nodes are not on the same network, the packet has to pass through one or more routers in-between. The method of delivery from the source node to the router or between two intermediate routers is known as indirect delivery. Thus, a delivery will always involve at least one direct delivery and zero or more indirect deliveries, which is explained in Figure 1.6.

A message to be transmitted over the network is broken down into packets. Each packet contains some user data and the packet header, which has the control information such as source address and destination address. Each switch follows a

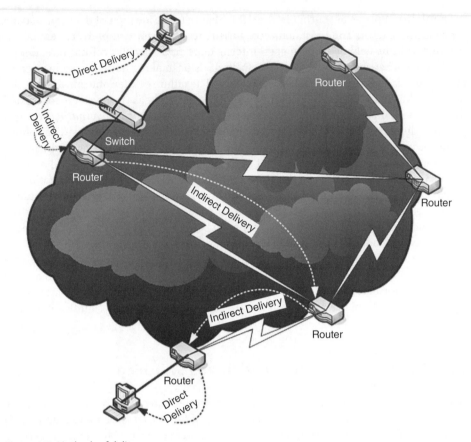

Figure 1.6 Methods of delivery.

store-and-forward technique, where it initially receives the packet which it may store in its buffer and then pass to the next network node. According to the type of switching, the network can be classified into two categories – the packet switching network and the circuit switching network.

A packet switching network is also known as the datagram approach, where each packet is treated independently. For each packet, the route through which it will travel in the network is decided independently and is not dependent on the preceding or the succeeding packet. Each packet can follow a different network route. Thus, the packets may become out of transmission order, and to reassemble the packet in proper sequence, a sequence number is used in the packet header as control information. In contrast to packet switching, in circuit switching a virtual circuit is established between the source and the destination before the exchange of data. As all the packets are transmitted over this virtual circuit, they reach the destination in the same sequence in which they were transmitted. Circuit switching does not require routing decisions for the packets, and thus it is fast. However, it takes time to establish a circuit, and once a circuit has been established between the source and the destination, the packets cannot take an alternative route if the channel is congested or fails, and no other packet meant for some other destination can be transmitted using the virtual circuit even if the circuit in underutilized.

Forwarding [4] is the process of determining the next hop for a packet in a network and placing it in the appropriate link for further transmission and delivery. Forwarding is done with the help of a routing table. The source node as well as all the intermediate routers perform the task of forwarding. The node consults the routing table before taking a decision concerning the link through which it should be forwarded. However, with an increasing number of nodes in the network and the Internet, a routing table cannot have entries of all the network nodes, and thus the routing table cannot be used for direct mapping of the address of the destination node and the related forwarding path. To cater for this requirement, various techniques are used to support forwarding with a minimal size of the routing table.

It is generally believed that a routing table should contain the entire route, i.e. details of all the intermediate nodes through which a packet should be forwarded to reach any destination from the forwarding node. This requires entry of all the host computers and routers in the routing table, along with the exact path that should be followed to reach the destination node from the node referring to its routing table. In the next hop method, the routing table maintains the information only about the next hop instead of the entire route through which the packet should be forwarded in order to enable the data to reach its destination network. In the network-specific method, the routing table contains route information for the networks and not for specific hosts. Thus, to forward a packet, the network ID of the destination is retrieved from the destination IP, and that network ID is searched in the routing table for a match to determine the next hop to which the packet should be forwarded.

The default forwarding method is used to forward the packets to a default link when the router is aware that the destination cannot be reached through any of its other available links. The default route is used for all those packets for which the related entries cannot be found in the routing table, and the default route link is connected to a much larger network, which may help in determining the exact path to the destination.

Routing aims to find the minimum cost path within the network so as to efficiently transmit data packets from the source node to the destination node. The parameters for cost and efficiency vary with the type of routing algorithm used. The cost and efficiency may be measured in terms of time taken in the transition, number of intermediate routers used (hops), time taken by the router to decide the next hop, time taken by the router to build the routing table on booting and rebuild the same in case of change of network topology.

1.7 Routing Taxonomy

Routing algorithms can be classified on the basis of two different criteria:

1) Global information vs decentralized information with the routing nodes.
2) Static information vs dynamic information in routing tables.

Algorithms that possess global information are aware of the topology of the entire network and the associated link costs. The routing node has information about all the links and the neighbors connected to the routing node. It also has information about all other nodes that are available in the network and indirectly connected to this routing node through one or more intermediate routers. It may also possess information about nodes that were once connected to the network but may not be available in the network presently. Algorithms with decentralized information have partial information about the network. They possess information about themselves and all their neighbors to which they are directly connected, along with associated link cost.

Algorithms with static information are those in which the routing table once created in the routing node is rarely changed. Algorithms with static information are suitable for those networks in which the topology does not change frequently, links are reliable, and the type of links between all the nodes is known. Static algorithms require less computational power at the routing nodes, as they do not require calculation of the network topology or frequent building of the routing table. They simply have to look up the static routing table for a suitable entry and forward the packet to the appropriate link. It is easy to implement static routing in small networks, but difficult in huge networks. However, once a static routing table has been built for a network, small or big, static routing is faster than dynamic routing.

Dynamic routing is more suitable for networks in which there are links and nodes that often fail and where frequent rebuilding of the routes is required for successful transmission of the packets from source to destination or between two routing nodes. In algorithms with dynamic information, the nodes regularly share with their neighbors information about all the links connected to them. By accumulating all such information, a routing node can have a fair idea about the entire network. In the case of a link or node failure, all nodes become aware of it from updates received directly or indirectly from neighbors concerning failed nodes. Algorithms with dynamic information have to rebuild the routing table if they receive from a neighbor an update that is different from its previous update.

Thus, in the case of a huge network, with links frequently going up and down, the dynamic algorithm may keep the routing node computationally busy. However, as this algorithm has updated information about the network topology, it can avoid transmitting packets on routes with congested or failed links.

The routing algorithms can also be classified into the following categories:

a) adaptive routing,
b) non-adaptive routing,
c) multipath routing,
d) hierarchical routing.

a. Adaptive routing. Adaptive routing algorithms are dynamic in nature and they recalculate the routing tables whenever there is a change in network topology in terms of available links and variation in link costs. The routing node may receive information about topological changes either from its neighbors or from any other routing node connected to the network. An adaptive routing algorithm can be centralized, isolated, or distributed.

In a centralized mode of operation, there can be one or more centralized nodes in the network that get information from all the routing nodes about their link states, link condition, and cost. Based on this input, the centralized node updates a routing table. The routing table either may be shared by the centralized node with all other routing nodes or the routing nodes may refer to the centralized node to make routing decisions without getting a copy of the routing table. The advantage of centralized adaptive routing is that there is only a central node that has to do all the computation related to routing. As all the nodes transmit their routing table with neighbor information only to a centralized node instead of flooding the network to send it to all other nodes, the network traffic is reduced. However, in the case of failure of the central node, the entire network is down.

An isolated routing algorithm does not require the routing nodes to share any information with other nodes. Every routing node takes a forwarding decision either based on the condition of its links or based on the information about the network it collects from the incoming packets using backward learning. The decision to forward a packet to a particular link may be based on the fact that the packet is forwarded to the link with the highest bandwidth, lowest congestion, or minimum queue length. Alternatively, in backward learning, the routing node uses backward learning on the incoming packets to gain an insight into the path followed by it, rebuilding the information about the available routing nodes in the network and the number of hops those routing nodes are away from this routing node. The routing node keeps updating this information based on the information retrieved from each incoming packet.

The distributed adaptive routing algorithm is the most commonly used routing technique, where every node receives information from its neighbors about the links connected to them. Based on this information, the routing node updates its routing table. Route table updates are exchanged between neighbors either in the case of change in the state of any of its links or at a regular frequency. The advantage of the algorithm is that it can take an optimized routing decision; its disadvantage is that the algorithm is computationally expensive and generates network traffic for exchange of routing tables.

b. Non-adaptive routing. Non-adaptive routing algorithms are static routing algorithms that do not calculate the route based on any of the changing characteristics of the network or traffic patterns such as link failures and congestion. The simplest implementation of the algorithm is by 'flooding'. A routing node receiving a packet on any of its links forwards it to all other links. Flooding may lead to packets going in infinite loops in the network, leading to congestion. Looping of the packets is avoided by using a sequence number, hop count, or spanning tree. A packet is assigned a unique sequence number by the source.

Each intermediate node maintains a list of source ID and the sequence number of all the packets forwarded by it. When the routing node receives a packet, it checks the source ID and sequence number from its list of forwarded packets and drops the packet if it has already forwarded it before. In the hop count method, the packet may be assigned an optimum hop count based on the number of hops required by the packet to travel from source to destination, which may be separated by the maximum number of hops. The hop count assigned to the packet by the source node is reduced by 1 by every routing node from which the packet is forwarded. If the hop count becomes 0, the packet is dropped. A spanning tree can be used in non-adaptive routing if the intermediate nodes between source and destination are aware of the entire network topology. This helps to create a spanning tree from source to destination without loops. Instead of flooding, the algorithm may forward the packet 'randomly' to any of its links except the one from which it received the packet. The outgoing link may also be selected on the basis of certain criteria, such as the link with the maximum available bandwidth or minimum queue length.

c. Multipath routing. There may be more than one route between the source and the destination. All the routes may be of equal cost or with variation in cost. In case of cost variation, each route between a source–destination pair is assigned a weight based on its relative cost. The routing table maintains information not only on a single route between the source and the destination but also on a number of other routes with equal cost or with varying costs, along with the associated weight. A path is selected based on a random number generator, which uses the weights of the routes as probabilities. The routing table with weights assigned to the alternative routes is generally static and computed manually.

d. Hierarchical routing. Hierarchal routing is best suited for networks with a tree topology. Although it can be used for wired networks, this is a common routing algorithm used in wireless sensor networks and mobile networks where a cluster of nodes elect a cluster head and this cluster head in turn communicates with the upper level. Thus, in hierarchical routing, the network is divided into hierarchical clusters. The information can be transmitted only through the intermediate cluster heads enabling a node to transmit and receive from a node just one level above or below it in the hierarchy.

1.8 Host Mobility and Routing

Host mobility has become an area of research and technological development to support wireless and mobile networking. The host may not only change its location between two consecutive communication sessions but also change its location during an established session involving transfer of data. The host mobility approaches have to consider factors such as scalability, hand-off, rate of movement, computational requirements, traffic generation, connection blackouts, and byzantine failures. There are a number of protocols that support host mobility. Mobile IP, I-TCP, end-to-end solution, and cellular IP are some of the protocols that handle host mobility.

Mobile IP keeps the mobility of the source and the receiver transparent with minimum overhead. The host is assigned a permanent IP address known as the home address, and this IP is known to the home agent of the host. The home agent is deployed in the home network of the mobile host, and all the data packets meant for the mobile host are sent through its home agent. When the mobile agent changes its network, it gets a new temporary IP from the foreign agent after the mobile host has performed agent discovery for the foreign host. The foreign agent resides in the network to which the mobile host has currently shifted. The address assigned by the foreign agent to the mobile host is known as the care-of-address. The foreign agent then updates the home agent of the mobile host about its presence in the foreign network. Thus, the home agent is able to keep a record of the location of the mobile agent. When a packet reaches the home agent for delivery to the mobile host, the home agent forwards it to the mobile node using the care-of-address. The forwarding to care-of-address is done by encapsulating the packet with a new IP header, which is the care-of-address of the mobile host. As the mobile host keeps moving, changing its locations, it discovers new foreign agents and gets a new care-of-address, and the home agent is kept updated about its latest location, which it duly acknowledges to the mobile host through its foreign agent.

In indirect TCP (I-TCP), an intermediate mobile support router (MSR) is used to support the communication between the mobile host and the fixed host. I-TCP divides the communication between the mobile and the fixed host into two parts, the first between the mobile host on the mobile network and the second between the MSR and the fixed host in the wired network. Thus, the problem related to host mobility and routing in a wireless network is confined only to one part of the network, and the other part uses fast and reliable connectivity with the TCP/IP network. I-TCP uses a variant of the mobile IP in the wireless network for communication between the mobile host and MSR. As the mobile agent moves, its MSR is changed. MSRs coordinate with each other to hand over the mobile host among themselves. They also keep the mobility of the mobile host transparent to the fixed host.

The end-to-end host mobility [14] approach does not provide transparency to the mobility, unlike mobile IP and I-TCP. The approach uses the Domain Name System (DNS) to update the new IP address of the mobile host as it leaves its home network and enters a foreign network, acquiring a new address. As the mobile node keeps changing its location, it keeps the DNS updated of its latest IP address. The migration from one network to another when a connection is already established and with data transfer in progress is supported by a new migrate TCP option by changing the method of handling sync packets.

Cellular IP uses a combination of the technology of mobile IP and cellular communication. Mobile IP is used for transmission of data packets to the mobile host. However, unlike mobile IP, the foreign agent from the care-of-address of the mobile host does not directly send the data packet to the home network of the mobile agent. The foreign agent sends the packet to its base station using a wireless access network. The foreign agent's base station then forwards it to the base station of the mobile host's home network. The base stations have the routing information for path determination. The mobility of the node from the cellular area of one base station to another base station is supported by the hand-off between the base stations.

For routing, mobile IP, I-TCP, and cellular IP use intermediate devices, namely foreign/home agents, MSRs, and base stations respectively. This leads to a delay in the routing and failure of the mobile hosts to communicate with each other in the region in the case of failure of the intermediate device. Two mobile nodes in close proximity cannot communicate directly and have to use the services of an intermediate device, which may be placed far away. There are improvements in all the protocols to overcome the drawbacks. In mobile IP, the sender is also informed of the care-of-address of the mobile host so that it can directly communicate with the mobile host. The packet is sent to the home agent only if the sender does not have the latest care-of-address of the mobile host. The end-to-end host mobility supports direct communication between the source and the destination. A separate mobile routing algorithm is not required for the same, as the sender looks for the latest address of the destination in the DNS and establishes a connection with the mobile host. As the connectivity is point-to-point, it is secured and fast. The method requires optimization only when the mobile host has just moved and the source has received its old IP address from the DNS before it could be updated with the new one.

References

1 W. Stallings. *Data and Computer Communications*. Prentice Hall of India Publication, 8th edition, 2007.
2 W. Odom. *Computer Networking First Step*. Cisco Press, 2004.

3 T. Lammle. *Cisco Certified Network Associate Study Guide*. BPB Publishers, 4th edition, 2003.

4 B. A. Forouzan. *Data Communications and Networking*. McGraw-Hill Publication, 4th edition, 2006.

5 Description of system area networks, Microsoft. http://support.microsoft.com/kb/260176.

6 Network architecture standard. http://www.servicecatalog.dts.ca.gov/docs/3117_Network_Architecture_Standard.pdf.

7 C. M. Kozierok. *TCP/IP Guide, Volume 3.0*. No Starch Press, 2005.

8 IEEE 802.11. http://standards.ieee.org/about/get/802/802.11.html.

9 C. Hunt. *TCP/IP Network Administration*. O'Reilly & Associates, Inc., 2nd edition, 1998.

10 Internet Protocol, RFC 791. http://www.ietf.org/rfc/rfc791.txt, September 1981.

11 Asia Pacific Network Information Center (APNIC). http://labs.apnic.net/blabs/.

12 S. Deering and R. Hinden. Internet Protocol, version 6 (IPv6) specification, RFC 1883. https://tools.ietf.org/html/rfc1883, 1995.

13 S. Deering and R. Hinden. Internet Protocol, version 6 (IPv6) specification, RFC 2460. https://tools.ietf.org/html/rfc2460, 1998.

14 A. C. Snoeren and H. Balakrishnan. An end-to-end approach to host mobility. 6th ACM MOBICOM, 2000.

Abbreviations/Terminologies

ANSI	American National Standards Institute
APNIC	Asia-Pacific Network Information Center
ARP	Address Resolution Protocol
ARPANET	Advanced Research Projects Agency Network
ATM	Asynchronous Transfer Mode
ATMP	Ascend Tunnel Management Protocol
BAN	Body Area Network
BGMP	Border Gateway Multicast Protocol
CAN	Campus Area Network
CIDR	Classless Interdomain Routing
COPS	Common Open Policy Service
DIS	Distributed Interactive Simulation
DMZ	Demilitarized Zone
DNS	Domain Name System
DoD	Department of Defense
DQDB	Distributed Queue Dual Bus
DVMRP	Distance Vector Multicast Routing Protocol
EGP	Exterior Gateway Protocol
EIA	Electronic Industries Alliance
EIGRP	Enhanced Interior Gateway Routing Protocol
ETSI	European Telecommunications Standards Institute
FANP	Flow Attribute Notification Protocol
FDDI	Fiber Distributed Data Interface

FQDN	Fully Qualified Domain Name
FTP	File Transfer Protocol
HAN	Home Area Network
HSRP	Hot Standby Router Protocol
HTTP	Hypertext Transfer Protocol
IAB	Internet Architecture Board
IANA	Internet Assigned Numbers Authority
ICMP	Internet Control Message Protocol
IEEE	Institute of Electrical and Electronics Engineers
IESG	Internet Engineering Steering Group
IETF	Internet Engineering Task Force
IGMP	Internet Group Management Protocol
IMAP	Internet Message Access Protocol
IPDC	Internet Protocol Device Control
IPsec	Internet Protocol security
IPX	Internetwork Packet Exchange
IRC	Internet Relay Chat
IRTF	Internet Research Task Force
IRSG	Internet Research Steering Group
ISAKMP	Internet Security Association and Key Management Protocol
ISCSI	Internet Small Computer Systems Interface
ISO	International Organization for Standardization
ISOC	Internet Society
ISP	Internet Service Provider
ISTP	Internet Signaling Transport Protocol
I-TCP	Indirect TCP
ITU-T	International Telecommunication Union – Telecommunication Standardization Sector
L2F	Layer 2 Forwarding
L2TP	Layer 2 Tunneling Protocol
LAN	Local Area Network
LDAP	Lightweight Directory Access Protocol
MAC	Media Access Control
MAN	Metropolitan Area Network
MARS	Multicast Address Resolution Server
MPLS	Multiprotocol Label Switching
MSR	Mobile Support Router
MTBF	Mean Time Between Failures
MTU	Maximum Transmission Unit
MZAP	Multicast-Scope Zone Announcement Protocol
NAN	Near-me Network
NARP	NBMA Address Resolution Protocol
NAT	Network Address Translation
NFN	Near Field Network
NHRP	Next Hop Resolution Protocol
NTP	Network Time Protocol

OFDM	Orthogonal Frequency-Division Multiplexing
OSI	Open System Interconnection
OSPF	Open Shortest Path First
OTN	Optical Transport Network
PAN	Personal Area Network
PIM	Protocol Independent Multicast
POP	Post Office Protocol
PPP	Point-to-Point Protocol
PPTP	Point-to-Point Tunneling Protocol
QoS	Quality of Service
RIP	Routing Information Protocol
RSVP	Resource Reservation Protocol
RTSP	Real-Time Streaming Protocol
RUDP	Reliable User Datagram Protocol
SAN	Storage Area Network
SCTP	Stream Control Transmission Protocol
SLP	Service Location Protocol
SLIP	Serial Line Internet Protocol
SMTP	Simple Mail Transfer Protocol
SNMP	Simple Network Management Protocol
SyAN	System Area Network
TALI	Transport Adapter Layer Interface
TCP/IP	Transmission Control Protocol/Internet Protocol
TGRP	Trunk Group Routing Protocol
TIA	Telecommunications Industry Association
TTL	Time To Live
UDP	User Datagram Protocol
VLAN	Virtual LAN
VLSM	Variable-Length Subnet Mask
VRRP	Virtual Router Redundancy Protocol
WAN	Wide Area Network
WCCP	Web Cache Communication Protocol
XOT	X.25 over TCP

Questions

1 Why are standards required for networking? What were the difficulties being faced with proprietary products and protocols?

2 State the difference between reliability and availability.

3 Name and then draw at least four network topologies that can be created from the basic network topologies, i.e. star, bus, ring, and mesh. Also mention the advantages and disadvantages of the network topologies drawn.

4 Explain the requirement of network address translation in IPv4.

5 What problems are faced in using a static routing algorithm for huge and scaling networks with regular link failures?

6 What are the disadvantages of a distributed routing algorithm compared with centralized routing?

7 How does a three-tier architecture with a demilitarized zone at the first layer enhance the security of the data and application?

8 Study from the Internet about the special and new types of network, such as the near field network (NFC), Internalnet, near-me network (NAN), home area network (HAN), and interplanetary Internet. Mention the basic features in terms of bandwidth, distance covered, and application of each of these.

9 Explain the following:
   A personal area network,
   B client–server architecture,
   C ISO OSI network layer,
   D classless interdomain routing,
   E variable-length subnet mask.

10 Differentiate between the following:
   A storage area network and system area network,
   B ISO OSI presentation layer and session layer,
   C routing protocols and routed protocols,
   D IPv4 and IPv6,
   E direct delivery and indirect delivery.

11 State whether the following statements are true or false and give reasons for the answer:
   A Mobile IP, I-TCP, end-to-end solution, and cellular IP are the protocols that support host mobility.
   B The end-to-end host mobility approach uses the care-of-address.
   C In multipath routing, the various routes between the source and the destination may be of different cost.
   D Static routing uses flooding to get information about the network nodes and the link status.
   E SNMP is an ISO OSI layer 4 protocol.
   F IEEE 802.15 defines standards for wireless personal area networks.
   G TCP/IP is known as the 'DoD model' as well as the 'public networking model'.
   H Non-adaptive routing algorithms calculate the route based on the changing characteristics of the network or traffic pattern.
   I 10.192.172.13 is a class A address.

Exercises

1 Assume a simple network with three nodes: A, B, and C. A is connected to B, B is connected to C, and C is connected to A. Node A has to transfer 800 GB of data to node B, and node B has to send 1 TB of data to node A [B = bytes, b = bits, K = kilo (1000), M = mega (1000 000), G = giga (1000 000 000)].

    i Assume that the network between node A and node B is on fiber (separate channels for transmit and receive) and has a full duplex communication capability, where node A can communicate with node B at an effective bandwidth (after removing bandwidth utilization for overhead processing) of 0.8 Gb/s, and at the same time node B can communicate with node A at 0.8 Gb/s. How much time will be required to transfer the total data between the two nodes.

    ii Now assume that the network between node A and node B is on copper wire with an effective bandwidth of 0.8 Gb/s and the transmission can only be one way at a time, with negligible time for switchover from one direction to the other direction of transmission. How much time will now be required to transfer the total data between the two nodes?

    iii What is the time required for the transmission over the copper wire if transmission of only 50 GB is permissible towards one direction and then the transmission starts in the other direction if some data is waiting in the other direction to be transmitted. The time required for pre-emption and processing for change in direction is 2 ms.

    iv What will be the total time required if the network allows only a simplex mode of communication from node A to node B and from node B to node C and from node C to node A. Node C has a processing time of 10 µs for receiving and forwarding each GB of data.

2 Hashing/checksum is a technique that can be used to check any change in values in a file. Consider a text file having a few hundred characters in it. Write a function to generate a hash value/checksum of blocks of 100 characters and insert the values just after each block or in a separate file. Also write a function to validate using the hash value/checksum if any character in the document has been changed. Can the exact location also be detected where the character has been changed?

3 A network router 'Alpha' had been operating for 1 year. During the year it failed only once, and it took just 1 day to repair and put back into operation. Another network router 'Bravo', which is installed in the shaft of the building, gets switched off once every weekend owing to removal of power cables caused by movement of rats. It takes 5 min to detect the failure through NMS and make it operational again by plugging in the power cable. Which network router has more availability and by how much percentage?

4 An organization has bought 50 network switches and it is open to implement any type of network topology for connecting these switches. Before selection of the topology, it wants to see the requirement of cables and connectors for each topology only in terms of numbers and irrespective of the type of cable or connecter. Calculate the number of cable and connectors required if the 50 network switches are connected in:

    i bus,

    ii single star with a separate core-switch in the center,

    iii ring,

    iv complete mesh,

    v partial mesh where each switch is connected to two other switches,

    vi binary tree.

If the number of nodes is denoted by n, can any formula be derived in terms of n to calculate the number of cables or connectors for the different topologies?

5 The railway station of a big city has an area of 500 m × 500 m. Wireless network accessibility has to be provided throughout the station. One wireless router has a range of 100 m.

    i A minimum of how many routers will be required to cover the entire area of the station under wireless communication so that no area is left without wireless signal coverage?

    ii If it is told that there should be no area with overlapping signals from two different wireless routers, even if certain areas in the station are left without the wireless signals, what is the maximum number of routers that can be deployed and what is the minimum percentage area of the station that will exist without a network signal?

6 A network has to be set up in a building. There are ten floors in the building, each floor has an area of 30 m × 20 m, and there will be 20 computers installed across each floor. Design the network architecture for the building along with the subnet masking scheme. The designed network will fall under which topology? What changes will have to be incorporated in the network and the subnet mask scheme if the number of computers in each floor has to be scaled up to 150?

7 Router 'Alpha' is connected to router 'Bravo' through router 'Charlie', 'Delta', and 'Echo' in-between. There are 20 hosts connected to router 'Alpha' and 15 computers connected to router 'Bravo'. Each of the 35 hosts sends one data packet to two other hosts in the same network and to any two hosts connected to the other network. What will be the total number of direct delivery hops and indirect delivery hops during this complete process of data transfer.

8 A travel agency has one office in the capital cities of 15 different countries. Each city office has seven computers. Design the network for the travel agency. What should be the preferred IP addresses for the computers on the LAN? What subnet mask should be preferred over the LAN and over the WAN? Which type of routing would be most preferred in this network?

9 The backup of SAN has to be taken from the data center to the disaster recovery data center, which are connected to each other on a 2 Mb/s link. The SAN has 35 Tb of data stored in it. Assume that 15% of the bandwidth is consumed for overheads, headers, and call set-up, and only 85% of the channel capacity can be used for actual data transmission. How much time will it take to take the backup of the SAN? What will be the reduction in time if the bandwidth is increased to 34 Mb/s or to 155 Mb/s?

10 Your organizational network was established a few years back. The network has catered for the expansion of the organization over the past few years by enhancing the capacity with additional network devices and is now not further scalable owing to difficulties in managing the infrastructure and lower bandwidth. Better security solutions also cannot be implemented over the network as it lacks proper design after the expansion. Design a new network for your organization to provide connectivity to all the existing devices and with a scalability of 25%. What is the topology of this newly designed network? Explain the subnetting that is being planned for the network and the type of routing that will be most suitable for this network?

2

Basic Routing Algorithms

2.1 Introduction to Routing Algorithms

Communication patterns can be classified on the basis of a number of source and destination nodes participating in the communication process. The two major communication models are point-to-point communication and collective communication [1]. In point-to-point communication there is only one source and one destination for each message. In a network, as per the information handling pattern, there can be only one source–destination pair at a point in time or many such one-to-one message passing source–destination pairs can communicate simultaneously. A special case of point-to-point communication, referred to as 'permutation routing', can also exist where a node can be source of one message and destination for another message at any time instance. In collective communication there can be multiple sources as well as multiple destinations.

As there can be various combinations of single source or multiple sources with single destination or multiple destinations, it leads to one-to-all, all-to-one, and all-to-all communication modes in collective communication. The one-to-one combination is not considered, as this becomes point-to-point communication. A one-to-many communication pattern can be either in multicast mode or broadcast mode. In broadcast, the message is sent from the source node to all other available nodes in the network, while in multicast, the message is sent from the source node to a specific set of destination nodes. A special kind of one-to-many communication is known as 'one-to-all scatter' or

Network Routing: Fundamentals, Applications, and Emerging Technologies, First Edition.
Sudip Misra and Sumit Goswami.
© 2017 John Wiley & Sons Ltd. Published 2017 by John Wiley & Sons Ltd.
Companion website: www.wiley.com/go/misra2204

'one-to-all personalized communication', wherein the source node sends a different message to all the destination nodes simultaneously.

A routing algorithm creates a complete or partial digital map of the network. The digital map is generally in the form of a graph. The partial digital map contains at least the source node and its neighbors. With the digital map of the network available, the routing algorithm plans for the path from the source to the destination on the basis of one or more attributes: delivery time, distance between nodes, number of hops, available bandwidth of the links, and congestion. The cost of the path is calculated as a weighted sum of one or more of these attributes. The routing algorithm attempts to find the minimum cost path from the source to the destination. In the process of determining the minimum cost path from source to destination, the routing algorithm is run on each intermediate routing node so as to decide which interface should be used for sending the packet out of the router for each of the incoming packets received.

The routing algorithms can be categorized [1] on the basis of different criteria: location of routing decision, implementation of routing algorithm, adaptivity, minimality, and progressiveness of routing.

When a packet moves from source to destination in the network, it passes through a number of intermediate nodes. The sequence of intermediate nodes through which the packet will pass can either be predecided before transmission of the packet or can be decided during the transmission, and the decision can be taken by the source node or the intermediate node at which the packet has arrived [1]. The decision may even be taken by a central routing controller node, which dictates the path by which the packet should travel, as the controller node has the complete digital map of the network and thus can calculate the best path. Based on the various permutations of where the routing decision can be taken, the routing algorithms can be categorized, on the basis of location of routing decision, as distributed routing, source routing, hybrid routing, and centralized routing.

In distributed routing, when any intermediate routing node receives a packet from the source for sending towards the destination, the intermediate node decides on its own the next neighboring node to which the packet should be forwarded so as to enable the data packet to reach nearer to the destination. The source node also transmits the packet to its neighbor as decided by the routing algorithm running at the source node. In the case of a distributed algorithm, the packet should contain only the information about the destination address. A forwarding decision taken by a node may be based on the node's awareness only of its nearest neighbors or its awareness about the partial or complete network. Generally, in order to decide the neighbor to which the packet should be forwarded by a node and thus placed in one of its links is not dependent on the routing done by previous nodes in the network through which the packet has passed, i.e. it is independent of the route followed up to that instance.

In source routing, the source node computes the entire path that the packet should take to reach its destination, and the information about this route is attached to the header of the packet. Any intermediate node receiving the packet decides on the next hop based on the route information available in the header of the packet. The source node may embed the entire route in the header of the packet. Each intermediate node after receiving the packet reads the header for the next-hop information, deletes that portion of the header, and forwards it to the next intermediate node in the path. Such a form of source routing is known as strict source and record routing (SSRR). The other form of source routing is known as loose source and record routing (LSRR), where abstract path information in terms of one or more intermediate nodes is specified by the source through which the packet must travel to reach the destination. The LSRR is also loosely referred to as hybrid routing, as it is a combination of source routing and distributed routing because the source node precalculates information about some of the intermediate nodes and adds it to the header of the packet, while the exact path between these nodes is determined by the intermediate nodes using distributed routing.

Centralized routing is done by a network controller using a centralized database. The network controller has a complete view of the network and the topological changes therein, if any. However, centralized routing is less scalable than distributed routing and cannot be used for huge networks such as the Internet. Although it relieves the intermediate nodes from the burden of calculating the next hop, it also leads to slower detection of link failures and restoration of links as compared with a distributed network in which the immediate neighbor detects the failure at a much faster rate and thus can use an alternative route immediately.

Implementation of the routing algorithm can be based on a lookup table or a finite state machine. In the case of a lookup table, each routing node maintains a table, which assists the node to decide the interface through which the router should forward the packet for a particular destination. The table may be static or dynamic. Dynamic routing tables are regularly modified and updated on the basis of the changing network topology and condition, while static routing tables are stored in the routing nodes and are generally not changed on the basis of any ongoing changes in the topology or traffic pattern of the network. The routing algorithm can also be implemented using a finite state machine. The finite state machine can detect any flaw in the protocol or message sequence. The state machine can be divided into three parts – input state machine, selection state machine, and output state machine, connected sequentially and interacting with each other through the events triggered by each of them. The finite state routing machine may be in the idle state, connect state, active state, open state, or established state, depending on the exact protocol using the finite state machine for its implementation.

Routing algorithms vary in the process of adaptivity, i.e. their capability to determine a better path in the network. The ideal way to adaptivity is to select the best possible path between the source and the destination for every packet being transmitted. To achieve the best path routing, each node should have complete information about the network, which requires time to build, and good computational capability to find the best path from source to destination from the knowledge of the complete network. This can lead to computational delays and bandwidth consumption due to broadcast of routing information to neighbors. So as to determine the next-hop node in a short time in a distributed manner at each node, the adaptivity of a routing algorithm [2] can be of the following types:

a) **Oblivious routing algorithm.** In these types of algorithm the set of routes for every source–destination pair is known to the routing node in advance. There may be a single or multiple routes between a source–destination pair. In the case of multiple routes, one of the routes is selected out of the possible routes randomly or in a cyclic order. The route information between the source–destination pair is available with the routing node before receiving the packet and the route determination does not take traffic into consideration for deciding the route. In the case of a continuous or burst of traffic between a particular source and destination pair, it avoids congestion as it forwards the packet by various available optimal routes from the set of available optimal routes between the source and the destination pair instead of the single optimum route. This process is known as 'static load balancing'.

b) **Deterministic routing algorithm.** Deterministic routing is a special case of oblivious routing. In deterministic routing, the single 'shortest path' is generated between the source and the destination pair, and all the packets between the source–destination pair are transmitted by this path. The traffic information is not considered for path determination, and even if a number of optimal paths exist between the source–destination pair, only a unique path is selected among the optimal paths, based on the deterministic shortest path algorithm. Such algorithms are simple and fast with in-order delivery of packets. The algorithm performs well in a network with uniform traffic and high reliability of links and nodes. In the case of failure of the node, the algorithm has to update its routing table and recalculate the shortest path between source–destination pairs. The routing also leads to delays in request–reply type of traffic, as the packets may use the same path.

c) **Adaptive routing algorithm.** The adaptive routing algorithm considers the network topology, its connectivity status, link status, and resource utilization before deciding the path between a source–destination pair. The connectivity status is based on factors such as path cost, edge cost, and congestion. The resource utilization and availability can be based on factors such as computational resource available with the nodes, buffer capacity of the nodes, and current available space in the buffer. The packets between a source–destination pair may be transmitted by different paths, depending on the condition of the network at that instance of time. Thus, the packets may reach the destination of an adaptive routing out of order.

The advantage of an adaptive routing is that it can adapt to conditions such as link or node failure at a much faster rate compared with oblivious routing. Adaptive routing may also have a congestion control mechanism, which may forward a packet by a path providing the best possible bandwidth at the instance of transmission. However, in adaptive routing, as the routing nodes have to calculate the best path for each packet in terms of distance as well as network condition, this may lead to delays in forwarding the packets by the routing nodes. It also consumes additional network bandwidth to share information between the routing nodes about the condition of the network. Adaptive routing is also known as 'load-based routing'. An ideal situation would be the load information of the entire network available at each node for calculation of the path, but this would be too computation intensive and data intensive, leading to delays. Thus, the routing nodes generally use neighborhood traffic information for local load balancing.

Based on the concept of minimizing the path from source to destination, termed 'minimality', the algorithms are classified into 'greedy' and 'non-greedy' [1]. In greedy routing algorithms, the greedy approach is followed, i.e. at every intermediate node where the routing decision is made, the algorithm tries to bring the packet closer to the destination and also at that point in time the path already traversed by the packet would be one of the shortest paths between the source and the intermediate node. Deterministic and oblivious routing algorithms belong to this category of routing, which is also referred to as minimal, direct, shortest-path, or profitable routing.

In the non-greedy routing approach, the routing algorithm may take a decision to forward the packet on a path that may lead to taking the packet away from the destination for that routing decision, leading to an increase in the path length or cost. Such decisions may be taken by the intermediate node to perform load balancing of the network or to avoid sending the packet on a congested link, which will lead to further delays and congestion. Forwarding the packet to a region away from the destination may lead to a faster delivery from the intermediate node to the destination through the detour, as the detour route may be congestion free with a higher bandwidth. For these reasons, the non-greedy approach of routing is also referred to as non-minimal, indirect, non-profitable, or optimistic routing or 'misrouting'.

The routing algorithms can be classified, on the basis of their progressiveness, as 'progressive routing algorithms' and 'backtracking routing algorithms' [3, 4]. The progressive routing algorithm searches for the next-hop node as a step towards the destination. The next-hop intermediate node may be nearer to the destination, as in the case of a greedy algorithm, or relatively farther from the destination node, as in case of a non-greedy algorithm. With an adaptive protocol, the progressive routing algorithm moves the packet to the next non-faulty profitable link on the shortest path, or detours it in a non-greedy protocol. With each execution of the progressive algorithm, the path traversed, the number of intermediate nodes visited, and the associated cost increase. Greedy algorithms and deterministic routing algorithms are always progressive.

A backtracking routing protocol calculates the next-hop path on the basis of the information of the network available with it. In the case of congested links or failure of links, the routing algorithm may backtrack for further connectivity. The backtracking algorithm also ensures that a path is traversed only once, and it helps to avoid loops and deadlocks. So as to support backtracking, the history of the path traversed by the packet has to be stored, which leads to storage requirement and message overheads for intimating about the path traversed. A backtracking routing algorithm is suitable for a network with various types of fault and frequent failures, and is used in fault-tolerant routing.

The routing algorithms have also been classified into three generations [5] based on the stages through which it has evolved in ARPANET. The first-generation routing algorithms were the distributed adaptive algorithms, which evolved in 1969 as a version of the Bellman–Ford algorithm. To overcome the shortcomings of the first-generation algorithms, such as non-consideration of line speed, low accuracy, slow response to congestion, and increase in delay, the second generation of algorithms was adopted in 1979. The second-generation algorithms were also distributed adaptive algorithms, but they used time stamping to calculate delay in the network, which was also the new criterion to decide the route. Dijkstra's algorithm was used to compute the routing tables. However, algorithms based on link delays led to problems of swing in routed links, as all the nodes avoided the congested links and forwarded the packets to the least congested link at any instance of time, leading to congestion in the link and a slow decongestion of the prior link, which again prompted the router to swing back from the recently congested link to the recently decongested link. To overcome this problem, the third-generation routing algorithms introduced in 1987 are based on an attempt to search for an average route to the destination rather than the best route for all destinations, as was the case with the second-generation algorithms. This helped to dampen the routing swings and reduce the routing overheads.

2.2 Routing Strategies

A packet-switched network is based on delivery of data packets from the source to the destination through certain routing nodes forming the route in the intermediate network. At any point in time, one or more than one route may exist between the particular source and destination, among which one is selected for packet forwarding based on a routing function. The routing function should generally possess the following characteristics [5]:

Correctness. The function should be able to determine an optimum path between the source and the destination. The destination should not be treated as unreachable if a path exists to it. In a similar manner, it should be able to route the packet to the destination and not land in an infinite loop making the destination unreachable or increasing the network congestion.

Simplicity. The function should be simple to implement. This will ensure its correctness during implementation. A simple function will lead to less computational overhead. In the case of a distributed routing decision, the function will be running at all the intermediate routing nodes. If the function is not simple and calls for extensive computation, it will lead to computational delay of the packet at each intermediate node.

Robustness. The routing function should be able to deliver the packet to the destination even in the case of link failure or congestion. It should attempt to route the packets by an alternative route without packet drop or looping in the case of a packet switching network. In the case of circuit-switched networks, the function should be robust enough to continue routing without breaking the virtual circuit.

Stability. The function should lead to a stable routing strategy and not have frequent changes in routes. If a routing function detects link congestion, it attempts to route all the packets by some alternative route, leading to decongestion of the previous route and hence reducing the cost of packet forwarding by the previous route if congestion is a parameter in route cost. So, the function will again try to route back the packet through that 'now decongested' link and change the routing path. A similar situation may arise if a link frequently fails and comes up immediately thereafter. This leads to instability in routing in the case of frequent route change and may lead to looping up of packets in traversal. A routing function should attempt to increase the stability but at the same time maintain robustness of the algorithm at an optimum level.

Fairness. The fairness of the routing function should ensure that no node should 'starve' of bandwidth allocation for the transmission of its packets. It should attempt to provide the maximum possible bandwidth for each node. The packets with nearby source–destination should have equal probability to packets with far source–destination with a number of intermediate hops. The bandwidth of the intermediate path should also not act as a deterrent to avoid transmission between certain source–destination pairs to maintain a minimum cost or maximum throughput [6]. Fairness is generally considered to be contradictory to optimality.

Optimality. A routing function should optimize the parameters that measure the routing efficiency of a network. The function should be optimum in terms of minimum packet delay, maximum network throughput, and minimum number of hops that a packet has to traverse between source and destination. However, optimality may go against fairness in its attempt to maximize or minimize a parameter. By the optimality principle, a source–destination pair with fewer intermediate hops would be preferred to a pair with large intermediate hops, as the former would indicate better performance because the packet transmission was successful with a fewer number of total hops.

Efficiency. Each routing function adds some computational overhead as well as transmission overhead to the routing protocol. The computational overhead is caused by processing requirements in the routing nodes. The router has to decide on the next hop to forward the packet, based on certain parameters. These parameters should be calculated either once after a certain period or in real time, depending on the protocol requirement. In a similar manner, the routing function requires the routing nodes to communicate among each other or with the source or the network controller to have details about the network or the path by which the data packet should be forwarded. All these communications are done through the exchange of information across the network, which leads to network overheads associated with routing. Although a routing function should be stable and robust, at the same time these factors should be optimized with the efficiency of the routing function also in terms of communication and computational overheads.

Adaptivity. The routing function should be aware of the nodes and the paths. From the knowledge of the paths to each node, it should be aware of the redundant or alternative paths between any two nodes. With this awareness, the routing function should be able to reroute the packets by the redundant path in the case of link or node failure. It should also be able to reroute packets by another path if congestion arises in the existing path or regions. Still, the path by which the packet has been rerouted should be an optimum path at that instance of time.

Fault tolerance. The nodes or links may fail in a network. In typical networks, there can even be misbehaving nodes and faulty nodes, which may randomly perform in an improper or a faulty manner. The failed nodes and links may remain as such or recover with time. To continue routing packets in such a scenario, there should be alternative paths between source and destination, of which the best path is selected at that instance of time. Fault tolerance enhances the 'reliability and performance' of a network routing. The fault tolerance feature of the network may comprise one or more of the following features – fault detection, fault diagnosis, and fault accommodation. The fault detection feature is of prime importance in fault tolerance, as it leads to knowledge of the existence of a fault in the network and triggers fault diagnosis or fault accommodation. A fault can be detected in a network by various methods, such as periodic exchange of information between neighbors about the active links and system check of the entire network by each node independently. Fault diagnosis is a more complex process than fault detection, as it has computational, storage, and resource overheads, and thus fault detection and fault accommodation [7, 8] are preferred until the network deteriorates to a minimum threshold.

Deadlock freedom. The routing function should be deadlock free with minimum delay and maximum throughput, even in a huge network with changing topology and frequent faults. The routing function has to optimize between routing performance and deadlock avoidance [9, 10]. Distributive adaptive routing is prone to deadlocks, as the routing node decides on the path by which a packet should be forwarded based on the local condition of the network without global awareness of the network. This may lead to forwarding of the packet in a loop. Adaptive routing with deadlock avoidance therefore either removes the cyclic dependency in the channel dependency graph or introduces an escape path in the channel.

The routing strategy is generally based on parameters such as knowledge of the network topology, type and status of the intermediate links in the network, congestion, time delays being faced in the network, and the data load. Still, there are routing strategies that do not require any information about the network and can successfully route the packets to the destination by flooding or some other routing strategies. In the case of a centralized route controller, it receives knowledge of the network from each node in the network. However, in the case of a distributed routing scenario, various options are possible. In the case of distributed routing, each routing node may use the information locally available to it, i.e. the information related to the links directly connected to it to forward the packet to the appropriate link. Alternatively, the node may obtain information from its immediate neighbor and also gain information about all the links connected to its neighbor, and this adds to its knowledge of the link status and congestion in all the links associated with its neighbors. Still further, each routing node may exchange information related to its link to the nodes beyond its immediate neighbor. This may finally lead to a complete or a partially complete knowledge of the network topology and affect the routing decision.

The routing strategy also governs the frequency of updates received by the node to decide on the routing path. In the case of fixed routing, in which a path is predecided between the source and the destination, it does not require any update on the status of the network from any node in it. Similarly, if the routing is based on flooding, it does not require any knowledge about the network. In the case of distributed routing, where each node uses the knowledge available with it, i.e. the information of the links directly connected to it, it gets information on a real-time basis and is always updated. In the case of distributed routing, where the routing nodes receive information from neighboring nodes or other nodes of the network, the routing strategy governs the frequency of updates that a node should receive from the other routing nodes. Adaptive routing, in which the routing path is decided on the basis of the network condition, is entirely dependent on the route updates received by the routing nodes.

Routing Metrics

A metric is a value assigned to each route by the routers for comparing the routes and finding the best among the routes. Different routing protocols use different, and sometimes multiple (hybrid) metrics: RIP uses hop count, IGRP and EIGRP use bandwidth, delay, load, and reliability, and IS-IS and OSPF use cost and bandwidth. If there are multiple routes to the same network, cost is used for finding the shortest or the best path. Common metrics [11] used in routing protocols are:

Hop count. This is a metric that specifies the number of hops a packet must take on its route from the source to the destination.

Reliability. This is a metric defining the dependability of each network link. Some network links are more prone to errors than others, and certain links might be repaired more quickly than other links. It is usually described in terms of bit error rate or may be calculated on the basis of previous failures, leading to assigning a probability of non-interrupted operation on each link.

Routing delay. This is the time taken for the packet to reach from source to destination. Delay can be calculated on the basis of a number of factors, such as bandwidth, router latency, queues at each router, and congestion in the network.

Load. This is a measure of the amount of traffic on the path. A heavily loaded path has busy routers. The load on a network depends on the CPU utilization and the number of packets processed. It is a dynamically changing metric. The path with the lowest load is preferred.

Bandwidth. The throughput of a network is directly proportional to the bandwidth offered by the links, if the network is not loaded. Routes with higher bandwidth are preferred.

Cost. This metric is defined by the network administrator to give preference to certain routes over others. Cost may be defined by any policy of the service provider or may depend on one of the above metrics or a hybrid of them.

2.2.1 Non-Adaptive Algorithms

A non-adaptive routing algorithm uses a static table to decide the route a packet should take to reach the destination. In the case of a source routing, the partial or entire route might be decided by the source node before transmission of the packet, while in distributed routing each intermediate node will decide on the next hop to forward the packet, based on the static routing table accessible to the intermediate node. Static routing is also known as 'directory routing', as it searches the fixed table or directory for the destination address and gets the next link from the directory. In the routing directory of any node, each destination address or a group of destination addresses are generally mapped to a unique link on the router to inform the router to forward the packet through that link. The route is determined on the basis only of the current location of the packet in the network and its intended destination. The decision regarding the route is independent of the network condition. The packet takes the same route every time if it has to go from a particular node to a destination.

When a node or a link fails in static routing, the packets forwarded to travel by that route as per the static route entries of the intermediate routers have to wait in the buffer of the intermediate nodes until the route is repaired, or the packet will be dropped in the case of limited buffer capacity or timeout. In the case of failure of any link or node in the network, the entries in the static routing table have to be deleted, added, or modified manually, which leads to delays and enhanced administrative cost. Modifying a static routing table may be possible for small networks, but making such changes for huge networks with complex topologies is generally not feasible as it leads to errors and delays.

Non-adaptive routing is simple to implement and performs well in networks with consistent topology and traffic. The static routing table is available to the routing node at the time of boot-up. The performance of non-adaptive routing is poor in the case of node or link failures, as the routing is unable to decide on alternative routes. The performance also deteriorates in the case of drastic changes in the traffic volume, because a burst of traffic in one of the links may lead to traffic congestion and all the packets intending to use the path have to wait even though an alternative free path might exist. In order to find the shortest path, non-adaptive routing creates a spanning tree from its knowledge of the entire network to decide the optimum path. Each link and node can also be assigned weights before calculating the shortest path.

Although flooding is also sometimes referred to as non-adaptive routing, it is generally studied as a separate class of routing from non-adaptive and adaptive routing. Still, non-adaptive routing may be classified into two categories – flooding and random walk. Flooding can be described as a kind of non-adaptive routing that uses sequence number, hop count, or the spanning tree to avoid looping or forwarding of similar copies of the packet. Random walk is used in a robust network with a high degree of connectivity. The packet is forwarded randomly to any of the neighboring nodes and the process is continued until the packet reaches the destination. Multipath routing may also be used which has high reliability, as multiple routes are possible from source to destination and the packet is expected to reach the destination by one of the alternative routes even if there are link failures or congestion in other routes.

2.2.2 Adaptive Algorithms

The adaptive routing technique is most commonly used in various routing algorithms for packet-switched networks. The technique helps to locate an alternative route if the routing node detects that it has become difficult, owing to changed network conditions, to route the packet by the previously calculated path. The two factors that lead to the requirement of an alternative path are link failure and link congestion. As the technique adapts to changing network conditions, it is referred to as the adaptive algorithm.

So as to enable the network to adapt to changing network conditions, the basic requirement is information about the state of the network at all times. In the case of distributed routing, knowledge of the network is possessed by each routing node, while in the case of centralized routing it is possessed by the central network controller. If knowledge of the network is required on a real-time basis, there has to be a continuous exchange of packets between the nodes in distributed routing and between the nodes and the network controller in centralized routing. Real-time knowledge of the network helps to make better routing decisions, but it leads to a high exchange of information related to packets through the network and computation by the nodes. So, it is computation as well as communication intensive. In order to avoid extensive computation and communication requirements, the information is exchanged after an optimum time period so as to keep the communication and computation requirements at an optimum level and achieve a tradeoff between freshness of the information and computation and communication overheads. Another disadvantage of near-real-time network knowledge with the nodes is that it leads to instability of the network, as it results in frequent path changes. As soon as a routing node detects congestion in a link, it starts forwarding the packets through the relatively less congested links. This leads to congestion of the later link and decongestion of the previous link. This information would again affect the routing pattern of the node, as it will now change its packet forwarding to the initial link, which has now been decongested. Such swings lead to unstable routing and to packet loops and losses.

2.2.3 Flooding

Flooding is a routing technique in which all the routing nodes participate in routing of all the packets, irrespective of source and destination. The routing nodes neither require any information about the network topology nor share link information with their neighbors or any other node in the network. As there is no exchange of routing information, this does not lead to communication overhead. Still, this routing technique has maximum utilization of the network resources, as it floods the network with data packets. The technique uses a wave-like transmission of data packets from source towards

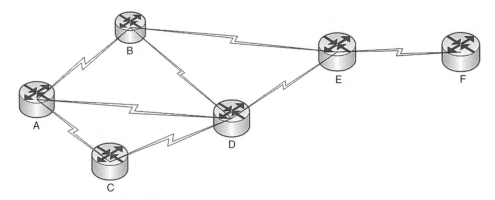

Figure 2.1 Sample network to demonstrate flooding.

destination across all links. The source node sends the packet across all its interfaces to its neighbors. Each neighbor, on receiving the data packet, forwards it on to all the links except the one from which it has received the data packet. Thereafter, each intermediate node, on receipt of a data packet that is not destined for it, forwards it on to all its interfaces except the one through which it has received the data packets. For example, in Figure 2.1, if node D wants to transmit a data packet to node F, then node D sends the packet to nodes A, B, C, and E, which are its neighbors. On receiving the packet, node A forwards it to nodes B and C, as they are its neighbors, but does not send the packet to node D, as it received the packet from node D; node B forwards it to nodes A and E; node C forwards it to node A; node E forwards it to nodes B and F, and again a similar forwarding starts from each node receiving a data packet not destined for it. Thus, node F receives the packet, but even now many packets destined for it are still in transit and will keep on reaching it subsequently.

To avoid receiving multiple packets at the destination and processing each of them, each node should only accept the first data packet received by it and then discard all subsequent packets received by it from other nodes. This can be achieved by attaching a unique packet identifier to each packet. As there is no central node to generate the unique identifier in the routing, the identifier is generated by the source node using a combination of the source node ID and a sequence number being maintained by it. In the case of a virtual circuit, the node identifier is generated at the source from the circuit ID and a sequence number.

In the routing technique based on flooding, multiple copies of the data packet are made by each routing node and are transmitted in the network. Each intermediate node receives one data packet from its neighbor, but transmits $(n-1)$ (where n is the number of neighbors of the routing node) data packets corresponding to the received data packet. This leads to the creation of a huge number of duplicate data packets moving in the network and can lead to congestion if not controlled. There are two commonly used mechanisms by which the infinite replication and transmission of packets are controlled in flooding. In the first method, each routing node maintains a list of unique packets, based on its packet identifier, forwarded by it to its neighbors. If the node receives a packet from any of its neighbors, it checks for the packet identifier in the list of packet identifiers already forwarded by it to its neighbors. If the identifier of the packet received by it matches any of the identifiers in the list, the data packet is immediately dropped or else it is forwarded to all its neighbors and an entry of it is made in the list of packets already forwarded by the node. Thus, by this technique each node forwards the packets only once along all its links. This reduces the looping of data packets as well as an uncontrolled replication of the data packets.

Another mechanism to prevent explosive increase in the number of data packets and looping of these packets is based on a hop counter. The hop count is set to a predefined value to ensure assured delivery of at least one copy of the data packet from source to destination. The diameter of the network is one such parameter to help to decide the maximum value of the hop count. In a network, the diameter is calculated by taking into account all the minimum hop counts among all the source destination pairs. The maximum value of all the minimum hop counts is known as the diameter of the network. It indicates the number of hops that would be sufficient to reach from any node in the network (source) to any other node in the network (destination) if an optimum path is taken. A hop counter is selected in the flooding technique, the value of which is set to the diameter of the network. The source adds the hop count to the packet before flooding it in the network.

Each routing node that receives the packet checks the value of the hop counter. If the value of the hop counter is 0 and the packet is not destined for it, it drops the packet or else it decrements the hop counter by one and transmits the packet to all other links. This technique ensures that the destination will definitely receive at least one data packet from the source node even if the source and destination may be the farthest separated nodes in the network. At the same time, the technique ensures that all data packets are discarded after a predefined number of hops in the network. Thus, it will stop indefinite loops as well as reduce congestion by regularly dropping packets after they reach end of life based on hop count.

The major disadvantage of flooding is the large number of data packets generated by a node running the flooding algorithm. This leads to congestion of the network. Each routing node receiving the packet has to put in computational resources for checking whether the packet is destined for it or else replicate and forward the packet to the other links. It also involves processing the packet to reduce the hop count if the technique uses the latter. The technique has computation and memory requirements to check the packet identifier and maintain the list of packet identifiers forwarded by it if applicable. The amount of generated packets increases with increase in the size of the network in terms of intermediate routing nodes or links. Thus, the congestion increases with increase in the size of the network. However, there are a few critical applications of the flooding technique that make it highly useful, robust, and reliable. Some of the advantages of the routing technique using flooding are:

Reliable delivery. The technique is very robust as it forwards the packets from source to destination by every possible route between them. Thus, even if some links are disconnected or blocked by congestion, the packet is assured to be forwarded through at least some other links. This makes it a useful protocol in critical networks like those in defense or emergency applications, which require a high reliability and robustness even if a major portion of the network in damaged or captured. An associated drawback would be that, if a portion of the network is compromised, the adversary will receive all the data packets, as every routing node receives all the data packets being transmitted through the network.

Discover best route. The technique can be used to determine the minimum hop route between the source and destination and can be used as a precursor of a fixed or adaptive routing or to establish a virtual circuit after determination of the shortest route using flooding. As the packets are replicated and forwarded through all the intermediate nodes, at least one of the packets will be the first to reach the destination. The route of this packet, first to reach the destination, can be traced back to find the minimum hop route.

Widespread communication. The technique generates a wave of data packet transmission from source in such a manner that it crosses all the routing nodes in the network. Thus, in a way it communicates with all the nodes through these packets, and this can be used to transmit various routing control information among the nodes. Flooding is used in a few routing protocols to share with all other nodes in the network the information about the links connected to a node. On receiving this information from all the other nodes in a network, every node is able to gain complete knowledge about the topology of the entire network and can thereafter decide on the least-cost route for forwarding the data packets.

2.3 Static Shortest Path Routing Algorithms

The routing table entries are made manually in static routing. The administrator should have a detailed understanding of the network, discover the routes with respect to each router, and then enter them in the individual routers. The entries made in the static routing tables by non-automated means are known as 'static routes.' In static routing, the routers do not communicate with each other for exchange of network status, thus making them 'blind' about the changes in the network. A router, whenever it receives a packet, consults the routing table to get the next-hop address and forwards it on to the exit link of the next-hop address without any consideration for traffic, congestion, or link failure beyond the next hop [12, 13]. As static routing does not change its routing pattern based on the existing network environment, it is also referred to as 'non-adaptive routing.'

Static routing is one of the simplest routing techniques and has minimal overheads in terms of processing power, memory, and bandwidth. However, it is not suitable for networks with inconsistent traffic pattern or bursty traffic as the protocol cannot adjust to the network environment. The routing lacks any central controller as well as an information-sharing mechanism between the nodes, making it impossible to implement any fault detection techniques in the protocol. As there is no automated fault rectification mechanism, the emergence of any fault cannot be detected by the system, and performance deteriorates.

A static routing protocol comprises a routing table and a routing algorithm. The routing table is also referred to as the routing information base (RIB). The common parameters that are entered in a routing table are the IP address of the destination network along with its subnet mask or the network ID, the next-hop address or exit interface, and the administrative distance. The next-hop address is the IP address of the next router in the network to which the packet should be forwarded so as to enable the packet to reach the destination. Alternatively, the interface of the router to which the next-hop router is connected can be mentioned so as to enable the router to forward the packet to the next-hop router through this interface. The administrative distance is used to assign weight, which is generally 0 for a directly connected next-hop router or 1 otherwise. Default routing is used to send the packets having destination addresses for which a corresponding entry is not available in the routing table. Default routing can be done on stub networks, i.e. those having only one exit route.

Based on the principle of optimality, the routing algorithm decides the exit route for an incoming packet. A routing algorithm builds the sink tree for the router and uses it to discover the shortest path. Sink trees are built by deciding optimal routes from a tree rooted at destination. In a static shortest path routing algorithm, to send a packet from one node to another, we find the shortest path between the pair of nodes. This decision of determining the shortest path is based on the routing algorithm. One algorithm that is mainly used for this purpose is Dijkstra's algorithm.

Static routing has its associated advantages and disadvantages. The advantages of static shortest path routing are as follows:

- As the router is already informed of the routes at the time of booting, the router has no computational requirement for calculating routes, which leads to minimal CPU processing.
- As there is no exchange of information related to the neighboring links and nodes between the routing nodes, there is no messaging overhead.
- Static shortest path routing adds a layer of security to the network, as any malicious router coming into the network will not get any network traffic because there are no routes mentioned for it in the static routing tables of the existing routers. Besides, the administrator also has the control to forward traffic by a desired route and to certain networks only on the basis of the security policy, if any.
- Static routing can be used to incorporate very complex routing policies on a relatively small network that cannot otherwise be implemented using a standard dynamic routing protocol.
- Static routing is easy to configure and understand. The administrator can easily analyze the path being followed by a data packet on the basis of static routing.
- Static routing is suitable for low-capacity WAN links or links based on dial-up connectivity, where the cost of the link is based on the time the link remains connected. It is suitable as it avoids regular exchange of routing information between nodes. A WAN link based on dial-up connectivity configured with dynamic routing protocol may end up paying for link connectivity for a major part of the day just for regular exchange of routing information among the routers, but may be transmitting data only once throughout the day.
- Default routes can be configured to forward the packet towards a network even when a specific route is not available to the destination.

The disadvantages of the static routing are as follows:

- The administrator should have detailed knowledge of the network topology, links, and bandwidth. This will be guiding the entire routing based on the configurations made in the routing table.
- The configuration and administration is time consuming. If a router, link, or network is added to or deleted from the network, or in the case of failure of the link or failure

of the routing node, corresponding entries have to be added, deleted, or modified manually by the administrator.

• Configuration is error prone because any fault in understanding the network or entering wrong entries may make a portion of the network unreachable. Configuration errors can also lead to flooding and congestion or send the packet in loops.

• Owing to manual management of the routing table, it becomes impossible to manage the routing table in static mode for huge networks and inconsistent links. Thus, static routing does not scale well with the growing size of the network and makes maintenance cumbersome.

• With static routing, configuring huge networks and networks with redundant links is difficult.

• In static routing, adaptation to changes in the network topology, link failures, and network scalability can neither be in near-real time nor in a dynamic manner.

Dijkstra's Algorithm

The algorithm was proposed by the Dutch computer scientist Edsger Dijkstra. The algorithm gives the single-source shortest path in a graph with non-negative edge cost. The algorithm creates a sink tree to give the shortest path from a node to every other node. The step-by-step working of Dijkstra's algorithm is given below:

1) Mark all nodes as UNVISITED. Set the start node as ACTIVE.
2) At the beginning of the algorithm, every node is assigned a 'distance value', which represents the distance of the node from the start node. Therefore, the distance value is set to zero for the start node and is set to infinity for all other nodes.
3) For the ACTIVE node, all its unvisited neighbors are considered and their distance from the start node is calculated. If this distance is less than the previously recorded distance (infinity in the beginning), the distance value is changed to the new, lesser value.
4) When all the neighbors of the ACTIVE node have been considered, they are marked as VISITED. The distance value of a VISITED node is final and minimal.
5) If all nodes have been VISITED, the algorithm stops. Otherwise, the UNVISITED node with the smallest distance from the start node is set as the next ACTIVE node and the algorithm is continued from step 3.

The step-by-step working of the algorithm can be explained on the basis of the graph shown in Figure 2.2:

1) Starting from the source, node A, it has three neighbors and thus three possible options to choose from – nodes 2, 3, or 6, but the neighbor with the minimum distance is node 2 with distance 8. Therefore, it is chosen, but at the same time values of 8, 10, and 15 are assigned to the three neighbors.
2) Next, node 2 is chosen from the queue because it has the smallest assigned value as of now. Now node 2 has four neighbors – nodes 1, 3, 4, and 7. Node 1 has already been covered, and among nodes 3, 4, and 7 the distance values are min(10, 8 + 11) for node 3, min(infinity, 8 + 16) for node 4, and min(infinity, 8 + 17) for node 7. Therefore, node 3 is chosen and is assigned the value 10.

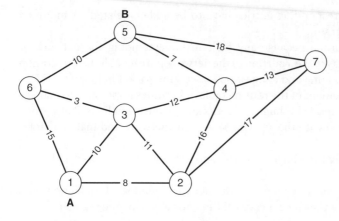

Figure 2.2 A graph to explain Dijkstra's algorithm.

3) Next, node 3 is chosen from the queue because it has the smallest assigned value as of now of the vertices left in the queue. Now node 3 has two neighbors – nodes 6 and 4. The values assigned to them are min (10 + 3, 15) for node 6 and min (10 + 12, 24) for node 4. Therefore, node 6 is chosen and is assigned a value 13.

4) Next, the neighbors of nodes 6 are nodes 5 and 3, out of which node 5 is chosen and assigned a value 23 on a similar basis.

Thus, we are able to find the shortest distance between nodes A and B on the basis of this algorithm.

2.4 Dynamic Shortest Path Routing Algorithms

Dynamic routing consists of two components: the routing protocol and the routing algorithm. The routing protocol is used for exchange of information among the routers about their current network status, and the routing algorithm is used for calculating the shortest path. The protocol allows continuous changes in routing decision to account for the current traffic and changing topologies. It allows routers to be continuously aware of the status of the remote networks and automatically add this knowledge to their routing tables, and allows the routing to act intelligently by avoiding network failures and congestion. In dynamic protocols, the routers exchange information whenever there is a failure or change in topologies. The routers are thus able to learn new and alternative routes and modify the shortest path in their routing tables after recalculations based on the latest network scenario. Thus, dynamic routing is more robust than static routing. Static routing allows routing tables in routers to be set up in a static manner where network routes for packets are preset. If a router on the route goes down, the destination becomes unreachable. In such cases, dynamic routing allows routing tables in the routers to change so as to enable the best alternative route available at any instance of time to be chosen as the possible routes change. Thus, the major activities and components [14] of a dynamic routing protocol are:

1) Having a method to discover other routers in the network.
2) Exchanging routing messages and maintaining up-to-date routing information about all other routers.

3) Determining optimal routes based on the information the router has, and recording this information in a route table.
4) Reacting to changing topologies and finding a new path if the current path fails.

The dynamic routing protocols have also been differentiated on the basis of various characteristics. The protocols have been classified [14] and described as:

- classful or classless protocols
- Interior Gateway or Exterior Gateway Protocol
- Distance Vector or Link State Routing Protocol

Classful vs Classless Routing Protocols

Classful routing protocols were used when network addresses were allocated on the basis of classes. In classful routing, the subnet masks are not sent in the periodic messages generated by the dynamic routing protocols as the subnets are predefined on the basis of the network address. Classful routing protocols exchange routes with subnetworks within the same network if the network administrator has configured all the subnetworks in the major network with the same routing mask. Examples of classful routing are Routing Information Protocol version 1 (RIPv1) and Interior Gateway Routing Protocol (IGRP). When routes are exchanged with foreign networks, subnetwork information from this network cannot be included because the subnet mask of the other network is not known. Classful routing protocols cannot be used when a network is subnetted using different subnet masks, i.e. classful routing protocols do not support variable-length subnet masks (VLSM), and thus IPv4 address utilization is poor.

The present-day networks are based on Classless Interdomain Routing (CIDR), a newer scheme of IPv4 addressing. In CIDR, the subnet mask cannot be determined from the network address. Classless routing protocols support CIDR. The protocol includes the subnet mask with the network address in routing updates. Classless routing protocols support VLSM and discontinuous networks and have better IPv4 address utilization. Some of the common classless routing protocols are RIP (v2, v3, v4), EIGRP, OSPF, IS-IS, and BGPv4.

Interior Gateway Protocol vs Exterior Gateway Protocol

Interior Gateway Protocols are used within a single autonomous system or a routing domain, which may comprise individual networks. They are used for networks under a single network administration. Each individual network also uses an IGP for shortest path determination within its own routing domains. As shown in Figure 2.3, IGPs can be further classified into distance vector and link state protocols. A major difference between distance vector and link state protocols is that routing information is stored in table form for distance vector routing and in a graph or database form for link state routing. The common Interior Gateway routing protocols are RIP, IGRP, EIGRP, OSPF, and IS-IS.

Exterior Gateway Protocols are used among different autonomous systems having independent administrative entities. EGPs do not always search for the shortest or the cheapest path between autonomous systems. EGPs propagate 'reachability indications' in terms of many different attributes to measure routes and not the true metrics. The routing table contains a list of known routers, the addresses they can reach, and the metric, called the 'distance', associated with the path to each router. The best available route is chosen depending on this distance. A router exchanges 'Hello' and 'I heard you'

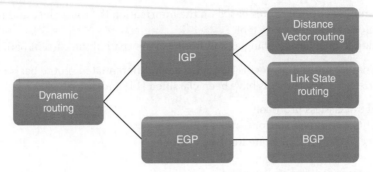

Figure 2.3 Classification of dynamic routing.

messages with another router to establish a connection to enable exchange of routing information. Each router then polls its neighbor at periodic intervals, and the neighbor responds by sending its complete routing table. A path vector protocol, Border Gateway Protocol (BGP), is the typically used EGP.

Distance Vector vs Link State Routing Protocols

Distance vector routing protocols such as Bellman–Ford or Ford–Fulkerson were among the earliest dynamic routing protocols. The name 'distance vector' is used because the routers exchange vectors containing distance and direction information. Distance is defined in terms of a simple metric such as hop count, load, delay, bandwidth, and reliability, or a complex metric derived from a combination of the simple metrics. Direction is the next-hop intermediate neighbor router or exit interface. The basic operating features in distance vector routing protocol are:

- Routers only know their immediate neighbors and the distance or cost to reach them.
- Periodically, this information is sent to all connected neighbors.
- The neighboring routers compare this information with the already known information and make changes if any shorter routes are available.
- After all the routers have exchanged their vectors, each router will have the best distance to each destination.

Distance vector routing protocols have a slow convergence, and hence they are used in simple and flat networks where convergence time is not a concern. A few examples of distance vector routing protocols are RIP, IGRP, and EIGRP.

Link state routing protocols are also known as shortest path first or distributed database protocols. In link state routing, each routing node makes a connectivity graph for the nodes in the network and independently calculates its shortest path to every other destination in the network. The information about the shortest paths is stored in the respective routing tables. Link state routing works best for hierarchical routing design and in networks where fast convergence is crucial. Examples of link state routing protocols are OSPF and IS-IS. The basic operations [15] in link state routing protocols are:

1) Each node discovers all the neighboring nodes.
2) Each router thereafter generates information about itself, its directly connected links along with their state, and any directly connected neighbors.

3) This message is then flooded throughout the network, and each router uses this information.
4) The link state database thus generated at each router is used by Dijkstra's algorithm to compute a graph of the network indicating the shortest path to each router.
5) Whenever there is a link failure or change in topology, link state messages are recomputed by the connected node and then flooded throughout the network.

Dynamic routing has its associated advantages and disadvantages [14]. The advantages of dynamic routing are as follows:

- As the entries in the routing table are automated with protocol for discovery of nodes and links, a dynamically routed network is scalable and does not pose any problem if it grows quickly.
- The protocol has a high degree of adaptability and it incorporates necessary changes in the routing path to adapt dynamically to variation in network topology with time.
- Configuration complexity is independent of the network size.
- Addition and deletion of nodes is easier for the administrator as it is not to be done manually as in static routing, but is taken care of by the protocol itself.
- The protocol has the ability to perform load balance between multiple links.
- Routing nodes can determine alternative paths in the case of failure of links or nodes.

The disadvantages of dynamic routing are as follows:

- Routing information updates are exchanged between routers, which consumes a lot of bandwidth.
- These protocols require more memory in the routing nodes to store their own view of the network.
- The protocols put a greater load on the CPU for route determination compared with the static routing algorithm.
- The route that a packet will follow cannot be known *a priori*. The decision concerning the shortest route is not in the hands of the network administrator as it is decided by the protocol.
- The administrator should have advanced knowledge about configuration, verification, and troubleshooting.

2.5 Stochastic Routing Algorithms

In static routing, the static routing table defines the hop-by-hop path to be followed by a packet for a particular source–destination path. In the case of link or node failure in the path, no other alternative link can be used. As the routing is not aware of the environment of the network based on traffic, alternative routes cannot be detected even in the case of congestion. A deviation from this is dynamic routing, in which the router is aware of the network environment, link and node status, latency, bandwidth, and congestion in the network. Based on one or more of these parameters, the algorithm can route its packet by alternative routes. However, in the case of a particular link going down and coming up frequently or regular congestion and decongestion along a particular route, the algorithm leads to swings in the routing pattern. However, in both cases, static as well as dynamic routing, the packets are forwarded from source to destination only by one route, which is discovered to be the best or the shortest possible route even though there may be many other alternative routes.

These routes may not be the shortest possible route at a particular instance of time, but these might be optimal routes between the source–destination pair. The throughput of a network will increase if the packets are forwarded by all available routes on a weighted approach, i.e. more packets are forwarded by the least-cost route and fewer packets are forwarded by the optimum-cost routes. The route with maximum bandwidth can be used to pump in the maximum number of packets from source to destination, but at the same time the other underutilized routes between the routing node and the destination can also be used to pump in packets from the node to the destination in proportion to their bandwidth so as to reduce the overall latency of message transmission between the source–destination pair by using multiple routes.

There are several techniques for distributing the next-hop packet into various available routes for load balancing. The simplest approach is to forward the packets by all the available routes to the destination in a round robin fashion. Alternatively, they can be forwarded in proportion to the transmission cost. However, in both cases, the routing decision is based only on the destination node and is not dependent on the source node or on the intermediate routers through which the packet has already travelled. Such a source-invariant approach of distributing packets over a set of multiple routes has been called 'multipath datagram routing'.

An alternative approach to multipath datagram routing is 'per flow routing', which also performs dynamic load balancing by routing all the packets of the same flow on the same route. Multipath datagram routing takes a decision on a per-packet basis, and thus the packets may reach the destination out of order, while the 'flow' reaches the destination in order, saving the time of reassembly. Alternative routes may be selected for different flows from the routing node to the same destination. The algorithm requires regular monitoring of loads on all ports for efficient load balancing of flows using near-equal paths based on near-equal cost. Flow routing also leads to faster error recovery, as all the alternative routes from a node to the destination are determined before distribution of flow, and thus any link failure will lead to rerouting of the flow through an alternative route that the node is already aware of. The other major characteristics of flow routing are guaranteed bandwidth, guaranteed flow group, maximum rate traffic, TCP fairness, and high trunk utilization.

In stochastic routing [16], for any particular destination, each routing node may forward the packet to any of its neighboring nodes on the basis of a probability distribution maintained by the node. When a router receives a packet for a specific destination, it forwards the packet to any of the neighboring nodes on the basis of the probability distribution function for the destination node. Thus, all the outgoing links may receive the packets for the same destination from an intermediate node on a proportional basis. Stochastic routing is a destination-based routing scheme. As it randomly forwards packets to different neighbors, the forwarded packet may end up in a loop. Looping of the packet in stochastic routing can be avoided by adding the information of all the visited nodes in the header of the packet, which increases the packet size as well as the processing overhead of the packet. An alternative to ensure loop freedom is to assign cost to the neighboring nodes in such a manner that only a restricted set of neighbors can be used to route the packet for a particular destination.

References

1 P. Tvrdik. Topics in parallel computing – routing algorithms and switching techniques. http://pages.cs.wisc.edu/~tvrdik/7/html/Section7.html.

2 C. Scheideler. Theory of network communication. http://www.cs.jhu.edu/~scheideler/courses/600.348_F02/.

3 J. Duato, S. Yalamanchili, and N. Lionel. *Interconnection Networks: An Engineering Approach.* Morgan Kaufmann Publishers Inc., 2002.

4 M. Tsai and S. Wang. A fully adaptive routing algorithm for dynamically injured hypercubes, meshes and tori. *IEEE Transactions on Parallel and Distributed Systems,* **9**(2):163–178, 1998.

5 W. Stallings. *Data and Computer Communications.* Prentice Hall of India Publication, 8th edition, 2007.

6 J. Kleinberg, Y. Rabani, and E. Tardos. Fairness in routing and load balancing. *Journal of Computer and System Science,* **63**(1):2–20, 2001.

7 F. Xiushan and H. Chengde. A fault-tolerant routing scheme in dynamic networks. *Journal of Computer Science and Technology,* **16**(4):371–380, 2001.

8 M. Bhatia and A. Youssef. Performance analysis and fault tolerance of randomized routing on close networks. *Telecommunication Systems,* **10**(1–2):157–173, 1998.

9 J. C. Sancho, A. Robles, J. Flich, P. Lopez, and J. Duato. Effective methodology for deadlock-free minimal routing in infiniband networks. *IEEE International Conference on Parallel Processing,* pp. 409–418, 2002.

10 A. Jouraku, M. Koibuchi, and H. Amano. An effective design of deadlock-free routing algorithms based on 2D turn model for irregular networks. *IEEE Transactions on Parallel and Distributed Systems,* **18**(3):320–333, 2007.

11 J. Doyle and J. Carroll. *Routing TCP/IP, Volume 2.* CCIE Professional Development, CiscoPress, 2004.

12 T. Lammle. *Cisco Certified Network Associate Study Guide.* BPB Publishers, 4th edition, 2003.

13 L. Parziale, D. T. Britt, C. Davis, J. Forrester, W. Liu, C. Matthews, and N. Rosselot, *TCP/IP Tutorial and Technical Overview.* IBM Red Book, 2006.

14 R. Graziani and A. Johnson, *Routing Protocols and Concepts, CCNA Exploration Companion Guide.* Cisco Press, 1st edition, 2012.

15 J. Doyle and J. Carroll, *Routing TCP/IP, Volume 1.* CCIE Professional Development, Cisco Press, 2nd edition, 2010.

16 S. Vutukury and J. J. Garcia-Luna-Aceves, A simple approximation to minimum-delay routing. *ACM,* **29**(4):227–238, 1999.

Abbreviations/Terminologies

BGP   Border Gateway Protocol
CPU   Central Processing Unit
CIDR   Classless Interdomain Routing
EIGRP   Enhanced Interior Gateway Routing Protocol
IGP   Interior Gateway Protocol
IGRP   Interior Gateway Routing Protocol

IS-IS   Intermediate System to Intermediate System (routing)
LSRR   Loose Source and Record Routing
OSPF   Open Shortest Path First
RIB   Routing Information Base
RIP   Routing Information Protocol
SSRR   Strict Source and Record Routing
TCP   Transmission Control Protocol
VLSM   Variable-Length Subnet Mask
WAN   Wide Area Network

Questions

1   State the difference between a routing algorithm and a routing protocol.

2   How is point-to-point communication different from collective communication?

3   Explain the disadvantages of a greedy routing algorithm and draw a network diagram to show failure of greedy routing in it.

4   Describe the three generations of routing algorithms.

5   State the common metrics used in routing protocols.

6   Describe the various mechanisms that can be used to reduce the explosive increase in packets during flooding.

7   Mention the advantages and disadvantages of static routing and dynamic routing.

8   Explain the following:
    i source routing,
    ii oblivious routing algorithm,
    iii Dijkstra's algorithm,
    iv Interior Gateway Protocols,
    v stochastic routing algorithm.

9   Differentiate between the following
    i centralized routing vs distributed routing,
    ii adaptive vs non-adaptive routing algorithms,
    iii distance vector vs link state routing.

10   State whether the following statements are true or false and give reasons for the answer:
    i Multipath datagram routing and per-flow routing perform dynamic load balancing.

ii Interior Gateway Protocol works within the same autonomous system, while Exterior Gateway Protocol is used across various autonomous systems.

iii Dijkstra's algorithm can be used for shortest path computation in a graph even with negative edge cost.

iv Static routing requires a central controller to build the static routes for the routers.

v Adaptive routing algorithms are not of much use in a circuit-switched network.

vi In the case of static routing, if a link fails, the packet waits in the intermediate node till the intermediate router gets information about the failed link and then the router changes its routing table and forwards the packet by the new determined path.

vii Adaptive algorithms have to deal with the problem of path swings.

viii An organization has ten routers in the network. It has special routing policies where certain packets have to follow some special predefined paths. The links and routers are stable, and there is rarely a change in the topology. The routing should ensure security. Static routing suits best in such a scenario.

ix Centralized routing is best suited for networks such as the Internet.

x Strict source and record routing can be used in EGP.

Exercises

1 Please consider the graph in Figure 2.1, wherein a data packet has to be sent from node A to node F. Write a program/algorithm for forwarding the packet strictly using point-to-point communication. Similarly, write a program/algorithm for the network that uses a plain broadcast mode of communication.

2 Prepare a partial digital map either in the form of tables or in the form of subgraphs for each node given in Figure 2.2.

3 Assume that each node in a network is aware of the complete network. Use Figure 2.1 and prepare a lookup table for each node, assuming that hop count is used as a metric and the routing attempts to minimize the hop count.

4 Consider a network that uses flooding for routing of packets, and the diameter of the network is used as the maximum hop count. What should the maximum permissible hop count be for such a network, shown in Figure 2.1, and for the network shown in Figure 2.2.

5 Give a step-by-step working of Dijkastra's algorithm to find the shortest path between node 6 and node 7 in Figure 2.2.

6 Consider the network indicated in the figure below:

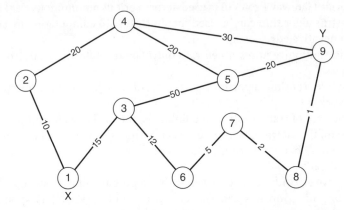

Give a step-by-step working of Dijkastra's algorithm to find the shortest path from node X to all other nodes. Is the path discovered between node X and node Y using this algorithm actually the shortest path seen visually?

7 Assume that a network distributes the next-hop packet into various available routes for load balancing in proportion to the transmission cost. Consider the network indicated in Figure 2.2 where the values shown in the links are the transmission costs. 100 data packets are to be sent from node A to node B. Calculate the number of data packets that will move from each of the links in the network. Please do a similar calculation for the network given in exercise 6 above for transmission of 100 data packets from node X to node Y.

8 Consider the network indicated in Figure 2.2, which uses simple flooding for data communication, where a node sends out the packet to all the links except the one from which it was received. How many copies of each data packet will be generated for transmitting a packet from node 6 to node 7. How many copies of each data packet will be generated for transmitting a packet from node 7 to node 6. Now, under the condition that each data packet is uniquely identified by an identifier and the node forwards the data packet along all its links only once, how many copies of each data packet will be generated for transmitting a packet from node 6 to node 7, and how many copies of each data packet will be generated for transmitting a packet from node 7 to node 6. Is the number of data packet copies generated equal to the number of links in the network?

9 Consider the network given in Figure 2.1, where 100 packets are to be transmitted from node A to node F. The two paths ABEF and ADEF are used, and equal numbers of packets are sent by the two paths. Assume that each hop takes a unit time, and the buffer of each router can accommodate 100 packets. The processing time at each router is infinitesimally small and can be assumed to be 0. Calculate the time for the first packet and the last packet to reach node F. Calculate the average waiting time in the buffer for this transmission.

10 Write a program/algorithm to detect routing swings in a network with adaptive routing.

3

Fundamental Routing Protocols

3.1 Routing Protocols

Routing is the major functionality of layer 3, i.e. the network layer in the OSI model. In the TCP/IP model, routing is performed at the Internet layer of the protocol stack. Routing helps to interconnect small networks so as to form a huge network. The routing protocol controls the flow of data between these smaller networks to avoid traffic congestion in the links. The router acts as the point of data entry and exit from the connected neighboring networks in the smaller networks. The prime functionality of the router is to find the best path between two or more nodes that may belong to different networks connected together. The path can be best in terms of the minimum transit time, least hop counts, least congestion, minimum cost, high reliability, or any other metrics as used in the specific protocol for routing.

The router has a routing table in its memory, and the entries of the routing table and the size of the table depend on the routing protocol being used in the network. A routing table contains information such as the routing nodes directly connected to it, the routing nodes that are not directly connected to it but that can be reached through its

Network Routing: Fundamentals, Applications, and Emerging Technologies, First Edition.
Sudip Misra and Sumit Goswami.
© 2017 John Wiley & Sons Ltd. Published 2017 by John Wiley & Sons Ltd.
Companion website: www.wiley.com/go/misra2204

neighbors, and finally complete or partial knowledge about the network topology. The router also has awareness of the links connected to it and the addressing scheme and the subnet mask being used in the network.

The routing path can be selected before the nodes start to exchange data. It will therefore not take into consideration the real-time situation of the network. This type of routing decision is known as static routing. Alternatively, the path can be selected and changed while the nodes are communicating with each other in the network. This category of routing decision is called dynamic routing. In static routing, the routes are predefined in the router by the network administrator and the router has to route the traffic on the basis of the static routing policies available with it. However, in dynamic routing, the router is responsible for understanding the network topology and its level of reachability and then performs route discovery and route optimization before sending the packets through the network. Moreover, the router also has to keep itself aware of the changing scenarios in the network and adjust its routing schemes and path on the basis of the changes taking place in the network with time. The router may get information from its neighboring routers or from other routers in the network. Based on this information, the router can build, modify, and update its routing table. However, the routing decision is taken by the router itself, based on its routing table.

The routing decision taken at a node is rarely a collective effort of two or more routers, even though the other routers may provide data to support decision-making. Any dynamic routing protocol has three major steps: (1) network discovery and filling the routing table; (2) updating its knowledge about the network status and changing the routing table entries accordingly; (3) selecting the best path for forwarding the data packet to the destination. The dynamic routing protocols can be classified into three categories:

- distance vector protocol,
- link state protocol,
- path vector protocol.

Apart from these three categories, a routing protocol may also be classified either as an Interior Gateway Protocol (IGP) or an Exterior Gateway Protocol (EGP). IGP is generally used for intradomain routing, i.e. routing within the autonomous system. EGP is generally used for interdomain routing, i.e. routing among autonomous systems. The IGP uses distance vector routing or link state routing. Exterior Gateway Protocols are very few and are restricted generally to path vector routing.

Distance vector routing shares its entire routing table with its neighbors at regular intervals. The routing table entries comprise the cost and exit paths to reach the node, known as the vector of distance. The router receiving such routing table advertisements from all its neighbors consolidates all the information and modifies its own routing table so as to maintain the best path to each destination, and thereafter transmits this updated routing table to its neighbors. Similar to distance vector routing, the path vector routing also shares its routing table with its neighbors, but the routing table does not contain any distance or cost metrics. It contains the entire path to reach the destination, and these paths are shared and updated. In link state routing, the routers exchange information related to changes in link status, based on which each routing node generates a topology of the network.

3.2   Distance Vector Routing

Distance vector routing is one of the simplest routing algorithms of a packet-switched network implemented in a node running a routing protocol. It was the routing protocol used in ARPANET way back in the early 1970s and is also known as the first-generation routing algorithm. The protocol runs by using frequent topological information exchange among the neighboring routers. Two routers are termed neighbors if they are directly connected by a link and can be reached in a single hop. A hop is the path traversed by a data packet between two adjacent nodes connected by a physical link.

The information is exchanged on a regular basis as well as on occurrence of a change. 'Distance' and 'vector' are two important items of information stored in a routing table in this protocol. Distance refers to the cost of reaching the node. The 'cost' can be the number of hops, effective bandwidth, link congestion, actual cost of using a link, or a combination of these or any other factors. Cisco Inc. has introduced the term 'administrative distance' based on the reliability of the intermediate protocol used. Administrative distance helps to decide between alternative paths on the basis of routing protocols used in the intermediate path to the destination. Vector refers to the direction in which the packet should be forwarded by the node. As a router has a number of interfaces, it gives the direction in terms of the interface through which the packet should be forwarded.

Each node initially builds up a routing table that has information about the link cost to its immediate neighbor and the interface through which it is connected. Thereafter it updates the table with the reachability information of other nodes based on the routing information received from its neighbor. The nodes cannot possess topological information of the network by personal discovery of the entire network, but have to rely on hearsay information received from its neighbors. As a node believes the information sent by its neighbor and manipulates its own routing table based on it, the protocol is referred as 'routing by rumor'.

As distance vector routing regularly exchanges data with its neighbor, it transmits a considerable amount of routing data through the network. However, as the exchange of information is only between the neighbors, it requires a considerable amount of time for information propagation across the network for events such as link failure or introduction of a cheaper path. A few commonly used routing protocols based on distance vector routing are Routing Information Protocol (RIP), Interior Gateway Routing Protocol (IGRP), and Enhanced Interior Gateway Routing Protocol (EIGRP).

3.2.1   Working of the Protocol

1) **Neighbor discovery.** When a routing node starts up, it is able to discover the neighbor nodes connected on its interfaces and the associated direct cost to reach the neighbors. This is done by sending a message to its immediate neighbor nodes and calculating the distance between itself and its neighbor on the basis of the reply received from the neighbor. Based on this information, it builds the distance vector table, which has a listing of the neighbor nodes, the associated cost to reach the neighbors directly, and the interface through which the neighbors are connected. The outgoing interface information related to the neighbor gives the vector information, and the other two items, i.e. the list of neighbors and the cost, give the distance

information. If the cost is based only on hop count, the cost will be 1 for all the initial entries in the distance vector table, as presently it has entries related only to its neighbors. Alternatively, if the cost is based on some other metrics, the related values would be entered.

2) **Initial sharing.** The router then shares this routing table with its neighbors on a regular basis. The exchange of information generally happens every 30 s in a typical routing environment. Any specified interrouter protocol may be used for such information exchange. The vector information in the table is not passed on to the neighbors, as the node receiving the distance table from its neighbor and updating the table for a less-cost path based on it would add the entry for this neighbor in the exit interface to this better path. As the information is shared only with the neighbors, the entire network is not yet aware of this node or any optimum path through it.

3) **Appending.** On receiving new information from the neighbor node, the router reconstructs its distance vector table. The router adds all those nodes and the associated cost to its distance vector table for which it has received information from the immediate neighbor and which did not exist in its table earlier.

4) **Updating.** The router compares the cost involved in reaching a network, based on the existing entry in its distance vector table, and that of reaching it through the neighbor from which the update has been received. If the former is costlier, the values of cost and interface for reaching the network are changed to the cost that has been received from the neighbor with the addition of the cost to reach the neighbor from this node. The entry for the interface is changed to the interface available on the router to reach this neighbor. If the existing entry of the router is of less or equal cost than that received from the neighbor, no change is made in the table.

5) **Convergence.** The modified routing table is then transmitted to all the immediate neighbors. In this way, each node generates awareness about the nodes in the network connected to its immediate neighbor. As neighbor discovery keeps propagating through distance table exchange between nodes, the routers gain knowledge about the entire network topology and are said to converge.

In addition to the regular update of the distance vector table, triggered updates can also occur in the network. An update is triggered when a node detects a link or node failure in its neighbor or when it receives a distance table from its neighbor, which leads to changes in its existing distance vector table.

3.2.2 Convergence of Distance Vector Table

Convergence of a distance vector table is shown with a sample network presented in Figure 3.1.

Assuming the network has come up at time T0, routers exchange and rebuild their distance vector table at each instant of time T0, T1, T2 ....Tables 3.1 to 3.4 show the distance vector table of the network at each of these instants of time.

Time T0 (neighbor discovery). At T0 it is assumed that all the routing nodes have started up and become aware of their links to their immediate neighbors. Based on the information of its neighbors only, each router fills the distance vector table as shown in Table 3.1 and has entries related only to directly connected neighbors.

Time T1. The routing nodes exchange the distance parts (node, cost) of their distance vector table with their neighbors. The nodes add information about any new nodes

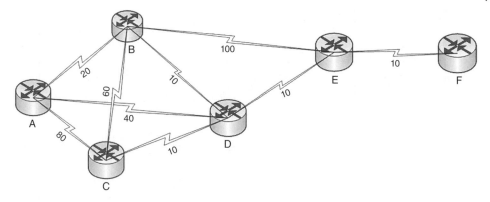

Figure 3.1 A sample network of interconnected routers.

Table 3.1 Distance vector table at time T0.

								Distance Vector Table									
Node A			Node B			Node C			Node D			Node E			Node F		
Node	Cost	Int	Node	Cost	Int	Node	Cost	Int	Node	Cost	Int	Node	Cost	Int	Node	Cost	Int
B	20	B	A	20	A	A	80	A	A	40	A	B	100	B	E	10	E
C	80	C	C	60	C	B	60	B	B	10	B	D	10	D			
D	40	D	D	10	D	D	10	D	C	10	C	F	10	F			
			E	100	E				E	10	E						

received through this distance table obtained from their neighbors. The cost value and interface entries in the existing table are also replaced with those for a node to which a less-cost entry has been provided by the shared neighbor distance table. The routing table at time T1 after appending and updating is presented in Table 3.2. The multiple paths to remote nodes are eliminated by removing the higher cost paths from the table, which leads to entries in Table 3.3.

Time T2 (converged routing table). The distance vector table accommodates the information received from its neighbor at T1 and recalculates the distance vector table and then shares this new table with its neighbor at T2. For this particular example, it can be seen that at T2, all the nodes gain complete topological information about the network and recalculate the distance vector table to obtain the final table shown in Table 3.4. At this stage, the distance vector tables are said to have 'converged'.

3.2.3 Issues in Distance Vector Routing

Pinhole congestion. There may be multiple routes between two nodes. These routes may be connected on different types of link and running separate routing protocols. However, if the cost is based only on hop count, then the link characteristics and protocol are not considered for deciding the best route. As can be seen in Figure 3.2, there are two paths

Table 3.2 Intermediate distance vector table at time T1 after receiving tables for updates from neighboring nodes.

Distance Vector Table

Node A			Node B			Node C			Node D			Node E			Node F		
Node	Cost	Int	Node	Cost	Int	Node	Cost	Int	Node	Cost	Int	Node	Cost	Int	Node	Cost	Int
B√	20	B	A√	20	A	A	80	A	A	40	A	B	100	B	E√	10	E
C	80	C	C	60	C	B	60	B	B√	10	B	D√	10	D	B√	110	E
D	40	D	D√	10	D	D√	10	D	C√	10	C	F√	10	F	D√	20	E
			E	100	E	E√	10	E	E√	10	E						
C	80	B	C	100	B	B	100	A	B	60	A	A	120	B			
D√	30	B	D	60	B	D	120	A	C	120	A	C	160	B			
E	120	B										D	110	B			
B	140	C	A	140	C	A	80	B	A√	30	B	A√	50	D			
D	90	C	D	70	C	D	70	B	C	70	B	B√	20	D			
						E	160	B				C√	20	D			
B	50	D	A	50	D	A√	50	D	E	110	C						
C√	50	D	C√	20	D	B√	20	D	A	90	C						
E√	50	D	E√	20	D	E√	20	D	B	70	C						
			D	110	E				B	110	E						
			F√	110	E				D	20	E						
									F√	20	E						

√ – Selected entry

Table 3.3 The final distance vector table at time T1 after selecting lowest-cost entries for each available destination.

Distance Vector Table

Node A			Node B			Node C			Node D			Node E			Node F		
Node	Cost	Int	Node	Cost	Int	Node	Cost	Int	Node	Cost	Int	Node	Cost	Int	Node	Cost	Int
B	20	B	A	20	A	A	50	D	A	30	B	A	50	D	E	10	E
C	50	D	C	20	D	B	20	D	B	10	B	B	20	D	B	110	E
D	30	B	D	10	D	D	10	D	C	10	C	C	20	D	D	20	E
E	50	D	E	20	D	E	20	D	E	10	E	D	10	D			
			F	110	E				F	20	E	F	10	F			

Table 3.4 The distance vector table at time T2.

Distance Vector Table

Node A			Node B			Node C			Node D			Node E			Node F		
Node	Cost	Int	Node	Cost	Int	Node	Cost	Int	Node	Cost	Int	Node	Cost	Int	Node	Cost	Int
B	20	B	A	20	A	A	40	D	A	30	B	A	40	D	A	50	E
C	40	B	C	20	D	B	20	D	B	10	B	B	20	D	B	30	E
D	30	B	D	10	D	D	10	D	C	10	C	C	20	D	C	30	E
E	40	B	E	20	D	E	20	D	E	10	E	D	10	D	D	20	E
F	50	B	F	30	D	F	30	D	F	20	F	F	10	F	E	10	E

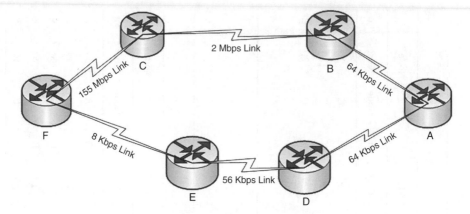

Figure 3.2 Example of a network that leads to pinhole congestion.

from A to F: one goes through B and C and the other goes through D and E. The link through B and C should be the preferred, as it has higher bandwidth compared with the link through D and E, but as we are considering only the hop count for metrics, the cost of the two paths remains the same for the routing protocol. Such an improper evaluation of the path metrics leads to pinhole congestion.

Routing loops. In distance vector routing protocol, the distance table is transmitted only to its immediate neighbors and is not broadcast throughout the network. The neighbors in turn forward their own tables after incorporating this information. This leads to delays in forwarding the information related to a node throughout the network, and all the nodes are not updated simultaneously. Such improper updating of information can lead to routing loops, which can be elaborated with an example of the network shown in Figure 3.1. The information about node F is propagated to the remaining network through node E. After the network has converged, all the nodes have information related to node F in them. Suppose the link between E and F fails. Node E propagates this information to nodes B and D, which update their distance vector tables. However, at the same time, nodes A and C send their distance tables to nodes B and D respectively. Based on this information, node B makes an entry for F through interface A, and node D makes an entry for F through interface C. Now, when a packet reaches node B for node F, it forwards the packet to node A, and node A in turn forwards the packet to node B and creates a routing loop. The routing loop is also referred to as 'count to infinity'. The process continues until the distance vector table entries are modified further.

Flapping. The route update information is transmitted at regular intervals. In addition to the regular updates, the information is also transmitted when a link comes up or goes down. When a node or the link to it is not stable and regularly goes up and down, it is known as flapping. As the distance table is shared only between neighbors and is not broadcast, it takes time for each node to gather topological information about the entire network and then process it for convergence. However, in the case of node flapping, this will lead to regular transmission of updated distance tables from its neighbors, which will initiate a process of recalculating the distance vector tables throughout the network. This would not allow the network to converge and lead to instability and inconsistent route paths.

3.2.4 Improvements in Distance Vector Routing

Maximum hop count. The routing loop leads to an infinite number of hops between the routing nodes possessing improper distance vector information. The problem can be reduced by adding a hop counter to the routing algorithm with a maximum permissible hop limit. If the algorithm enters an infinite routing loop, the hop count keeps increasing whenever the packet is forwarded. However, the maximum hop count incorporated in the algorithm limits the hop count to a predefined maximum. The packet will be dropped on reaching this limit of maximum hop count. This will also invalidate the entry related to the destination node that has led to the routing loop. RIP permits a maximum hop count of 15.

Split horizon. Routing node A passes its distance table only to its neighbor, say B, which in turn updates its distance vector table and then shares it back with node A. The distance table received by A from B may contain many entries that have been sent previously by A to B and returned back to it without any change. This leads to computational overheads and improper entries in the distance vector table. To reduce this problem, the router keeps itself aware of the routing node from which it has received routing information. The router will not transmit back distance entries that it has received from the neighbor and not changed. This technique is known as split horizon and acts as another solution for avoiding routing loop. The slit horizon strategy, however, has a drawback. Every node has a timer for updating node information. If no information about a node is received in the specified time limit, the entry for the node is deleted from the distance vector table. However, if a split horizon is implemented, a node is unable to understand the reason for its non-receipt of the node information – whether the node has ceased to exist or the information has not been forwarded to it by its neighbor on account of the split horizon. Route poisoning and poison reverse is used to solve this problem.

Route poisoning. Route poisoning is a technique to avoid infinite loops and inconsistencies in topological updates. If a routing node detects that the link to its neighbor has gone down, it puts the cost to this neighbor to a value greater than the one permitted in max loop count and sends this to its other immediate neighbors, thus poisoning the route. This high value of loop count in turn propagates the inaccessibility of this node. Each node receiving the poisoned route information, after transmitting the same to its neighbors, sends a reverse poison to the neighbor that has transmitted the poisoned route information to intimate that all its neighbors have received the route poison information.

Hold-down timer. The hold-down timer counters the effect of flapping nodes. It prevents frequent change in the distance vector table owing to an unstable link. It permits a prespecified time, programmed in its algorithm for the unstable node to stabilize or for the network to converge. When a node receives information from its neighbor about an inaccessible link, the node starts a hold-down timer. If any further updates are received by this node running the hold-down timer, the node decides to accept or reject this update on the basis of the cost value received therein. If an update has an equal or higher cost than the existing one, it is ignored, and if it is lower than the existing one, the hold-down timer is removed and the distance vector table is recalculated. Triggered updates are used in hold-down to create awareness in the neighbors that a change has occurred in the network and the distance vector table should be recalculated. The triggered updates are initiated in the following scenario:

- expiry of hold-down timer,
- receiving an update with better cost,
- deletion of a node from the distance vector table owing to non-receipt of any update on it in the specified time.

3.2.5   Advantages and Disadvantages

Distance vector routing is simple to implement, requires little management of the nodes, and is very effective in a small network where it stabilizes very quickly and requires much less CPU processing and link bandwidth.

However, some of the factors that prevent its usage in huge networks are the computation, storage, and transmission of large distance vector tables. With increase in the number of routing nodes in the network, the size of the distance vector table grows, and thus CPU utilization increases with the number of nodes as it has to calculate huge distance vector tables. This also increases the consumption of link bandwidth in transmitting the routing updates. The result is slow convergence of the distance-vector-routing-based network, leading to inconsistent routing tables and low scalability in huge networks.

3.3   Link State Routing

Link state routing protocol [1], also known as the shortest path first protocol or the distributed database protocol, attempts to make a router aware of the entire network topology. This makes it easier for the router to calculate the shortest path to any destination node. It is an improvement over distance vector routing as every node has knowledge about all other nodes and links in the network and builds the complete routing table by itself. The node also runs the shortest path algorithm independently to decide the exit interface for the next hop of the packet. These improvements made link state routing a second-generation routing algorithm. The protocol is more complex and computation and memory intensive. However, it is more reliable, easier to manage, consumes less bandwidth, is more scalable, and can be used for larger networks. A few examples of link state routing protocol are Open Shortest Path First (OSPF) and OSI's Intermediate System to Intermediate System (IS-IS), EIGRP, DEC's DNA Phase V, and Novel's NetWare Link State Protocol (NLSP).

3.3.1   Working of the Protocol

When a routing node comes up, it determines the active links connected to it and their associated cost. As in the case of distance vector routing, the cost can be based on hop count, bandwidth, and/or congestion. The routing node then broadcasts the information of these links to all the routing nodes in the network. Thus, every node receives actively connected interface information from all other nodes of the network, based on which the node creates a graph depicting the topology of the entire network. The topological information about the network also has information on the condition, type, and cost of the links. Based on this network graph, the node runs a shortest path algorithm to determine the optimum path to destination. The preferred shortest path algorithm used by this protocol is Dijkstra's algorithm. The shortest paths and the associated exit interface for the destination are stored to build the routing table by each node. The nodes keep a regular watch on any changes in its links. Whenever a link fails or a new link comes up, the node broadcasts the information about the link and associated cost throughout the network for topology reappropriation and rebuilding of the routing table. These broadcasts, which notify the other nodes in the network about the state of a link, are known as link state advertisements (LSAs) and are done using link state packets (LSPs).

1) **Neighbor discovery.** As soon as a routing node boots up, it sends a beacon message to its neighbors along with the node ID. Other information, some of which may be optional, that is carried by the beacon message includes the IP address of the interface of the router, the subnet mask, the interval of beacon transmission, and link details. The node continues to send beacons at regular intervals thereafter. The messages are sent at regular intervals so as to inform the neighbors that it is alive. A beacon message may be lost during delivery, and thus a regular transmission of the beacon message ensures its delivery. A node discovers all its neighbors by collecting all beacon messages, including discovering the nodes previously unheard of in its neighborhood. A neighbor is dropped from its neighbor list if the beacon from it is not received for a prespecified time period, which is generally sufficient to receive n messages. The value of n is usually 2–4, or as defined in the neighbor discovery protocol. This protocol is also known as the 'reachability protocol'.

Though neighbor discovery has a very simple and reliable process in link state routing, it generates network traffic at regular intervals. The drawback of this process is that it may discover asymmetric links and lead to flooding in the case of a dense network with a large number of neighbors, and link failure to a neighbor is discovered only after waiting for n beacons from the neighbor. In an asymmetric link, only one-way transmission is possible, and hence a neighbor may be able to send its beacon message through the asymmetric link to the routing node, based on which the node puts it in its neighbor list. But the node will not be able to forward packets to this neighbor directly through this asymmetric link. To overcome these drawbacks, an enhanced version of the neighbor discovery protocol is used in which the neighbor list is transmitted along with the beacon message. Although the process increases the size of the beacon message, this solves the problem of the asymmetric link, as the node receiving the hello packet can check whether it is present in the neighbor list forwarded with the beacon message to assure that the link is symmetric. This also gives a better overview of the network, helping in the building of a partial network map by the node, as it has two-hop information with it.

2) **Generation of link state packet.** The nodes periodically flood their link information across the network using LSPs. The minimum information that should be carried by an LSP is as follows:

- the node ID that created the LSP,
- a list of direct links connected to the node,
- the cost of the links to the directly connected node,
- the sequence number of the LSP, and
- the time to live (TTL) or the time stamp of generation for the LSP.

The node ID, list of direct links, and cost of the link are used to generate the network topology. The sequence number of the LSP helps to distinguish between old and new LSPs, and the TTL or time stamp facilitates discarding of expired LSPs. The sequence number is assigned in an incremental manner to the LSPs by the node and is reset to 0 on rebooting.

3) **Flooding LSPs.** Unlike the neighbor discovery beacon, which is transmitted only to the neighbors, the LSPs are flooded in the entire network and are meant for receipt by all active nodes. The node generating an LSP transmits it through all its interfaces. Any other node on receiving this LSP compares the sequence number of the LSP for that particular node ID with the sequence number of the LSP that it already has for that particular node ID. If the sequence number of the received LSP is greater than the one already stored, indicating that it is a newer one, the older LSP is replaced with this new LSP for the node ID, or else it discards the LSP. Thereafter, it transmits this LSP to all its neighbors through its interfaces, except the one from which it received the LSP. The TTL is decreased before transmitting the LSP to the neighbors. The LSP is flooded at regular intervals as well as on detection of any link failure. The frequency of transmitting regular LSPs is low compared with that in distance vector routing, and LSPs may be sent generally once every 1 or 2 h. As each node immediately forwards LSPs on all other links, this leads to a faster convergence of the network compared with distance vector routing, where each node has to compute the routing table on the basis of the updates and forward them thereafter.

4) **Formation of routing table.** After receiving the LSPs from all active nodes, each routing node has the topological information about the entire network in the form of a graph, which is also known as the topological database or the link state database. To create this graph, the node iterates over each LSP and creates the node in the graph and the edges connected to it based on the information received in its LSP. In the case of symmetric links, an edge, representing the link, is introduced into the graph only if the information about the link is received from the LSP of both the nodes at either end of the link. Thereafter, each node runs its shortest path algorithm, generally the Dijkstra algorithm, to generate the shortest path tree with itself at its root. The shortest path tree is used to construct the routing table, which has entries for all the remaining nodes of the network, the cost of reaching the node, and the interface on the routing node through which the packet should be forwarded for the destination node.

3.3.2 Routing Tables

In the link state routing protocol, each routing node builds three tables during its entire process of initialization, broadcasting, and calculation of the shortest path. These tables are used by the node to store the information and to build the final routing table. The first table has entries related to the links directly attached to the node. The second table is built with the help of broadcast interface information received from all other nodes and contains the topological information about the entire network, preferably in the form of graph representation. The third table is the actual routing table built by running the shortest path algorithm on the network graph. This results in the formation of a rooted tree with the routing node as its root, and the path to the destination node in the tree is the least-cost path. The tree is used to find the next best hop in the shortest route from the routing node to the destination node in the network.

3.4   Path Vector Routing

Interdomain routing is challenging in terms of scalability, security, and diversity. These challenges cannot be met well by link state routing or distance vector routing as they have been designed inherently for intradomain routing. In order to adjust to the interdomain routing environment, path vector routing does not implement the typical shortest path routing algorithm completely, but has a flexible implementation of it along with a loop detection mechanism. An interdomain routing differs from an intradomain routing in terms of the following challenges:

- The destination is represented in terms of a network address such as 202.54.15.0/24.
- The routing is between autonomous systems (ASs) and not end nodes.
- The internal topology of the ASs is generally not known.
- There are thousands of ASs on the Internet, and the number is growing rapidly.
- As the number of ASs is huge, the number of links and intermediate routing nodes between the ASs is also extremely high.
- The routing has to satisfy interdomain security and access policies. This leads to controlled forwarding of traffic over specific paths only.
- In the case of the Internet, examples of ASs are the Internet service provider (ISP) networks. The shortest path routing cannot be directly implemented as the ASs may have different business policies among them, leading to calculation of preferred routes based on these policies.

Link state routing cannot be used in interdomain routing because it floods the network with topology information. As the number of nodes in interdomain routing is very large, this leads to a high bandwidth consumption and requires huge storage space in each routing node to store the entire network topology locally. As each node calculates the shortest path from itself to every other node in the network, if the network size is large this will lead to a huge computational effort. Such computational overhead calls for high processing power at each of the intermediate routing nodes, which otherwise leads to processing delays in the intermediate routing nodes. Moreover, link state routing also propagates the topological information in the network, which is restrictive in interdomain routing. The entire link state routing is based on a common policy for routing and route cost calculation, which is not the case in interdomain routing. On account of these factors, link state routing can be used only within an AS and not for routing between ASs.

Distance vector routing, if amended to suit requirements, is a more suitable protocol for implementation in interdomain routing as it does not broadcast the network topology and thus keeps the topological details safe from sharing across AS. In addition to this, in distance vector routing the intermediate routing nodes run a less computation-intensive protocol to determine only the next-hop link to forward a packet towards the destination. Amendments are required in distance vector routing to make it suitable for interdomain routing for two major reasons. Firstly, the protocol has a slow convergence rate, and secondly it is susceptible to landing in a routing loop.

Path vector routing can be thought of as an extension of distance vector routing with routing loop avoidance and flexible routing policies. In contrast to distance vector routing, in which the node advertises the cost of the path to reach a destination, in path vector routing the entire path to reach a destination is advertised. This leads to the protocol being called path vector routing because the routers receive a path vector, which consists of the path in terms of intermediate nodes from a given node to the destination. As the entire path is advertised, any node can detect as well as avoid a routing loop in the path. If a node detects that its entry is already there in the path it has received as a broadcast, it discards the advertisement so as to avoid routing loop. As there may be multiple paths between two nodes, path vector routing allows flexibility to the node to select any of the paths based on its policy. Moreover, it also allows the node to selectively broadcast the paths that it wants to advertise.

Policy routing. The path vector provides the listing of AS that a packet has to traverse to reach the destination. This awareness about the path helps the router to implement routing based on policies. When a gateway router receives a path vector advertisement from the network, it verifies whether the intermediate paths in the vector are as per its policies. If the path vector does not comply with the policy of the AS in which the border router has received the path vector advertisement, then the border router in the AS does not amend its routing table based on this advertisement. In path vector routing, the router may decide an alternative path in order to pass most of the traffic through a particular AS that is in close association with the AS. Alternatively, the router may also avoid certain paths so as to prevent the datagram from traversing through some specific AS for privacy reasons or for lack of business collaboration with it. In framing the routing policy, a few other factors can also be used, such as the source address of the packet, the size of the packet, the protocol of the payload, or any other information available in the packet header or the payload. The routing can also be based on the performance-related information of a portion of the network between certain ASs or within ASs. The performance-related metrics [2] used in interdomain routing are bandwidth, congestion, quality of service, and number of intermediate ASs.

3.4.1 Working of the Protocol

There can be one or more border routers in an AS, which perform the interdomain routing for the AS. A routing table is maintained by each border router, which contains the path to other ASs and not the metrics associated with the path. This routing table is periodically advertised by the border router to other border routers in different ASs as well as to the nodes within the ASs. Unlike distance vector routing, which contains path metrics in the route broadcast by the router, in path vector routing, the route itself is advertised through the broadcast. The working of path vector routing [1] can be explained with the help of an example of a network of ASs, as shown in Figure 3.3.

The network comprises five ASs, namely AS1, AS2, AS3, AS4, and AS5, with interconnections. Each AS has one border router, which are designated as BRoAS1, BRoAS2, BRoAS3, BRoAS4, and BRoAS5 for AS1, AS2, AS3, AS4, and AS5 respectively. Within each AS, there are a few routing nodes that perform the intradomain routing. For example, there are three intradomain routing nodes in AS1, i.e. R1, R2, and R3, which we will refer to as AS1R1, AS1R2, and AS1R3 for unique naming of the routers across various ASs.

Startup. At the time of booting up, the border router is able to detect only the nodes present in its AS. With this information, it builds the path vector routing table as shown in Table 3.5. BRoAS1 is the border router of AS1. At the time of booting up, BRoAS1 creates the initial path vector table which has entries related only to the routers in AS1,

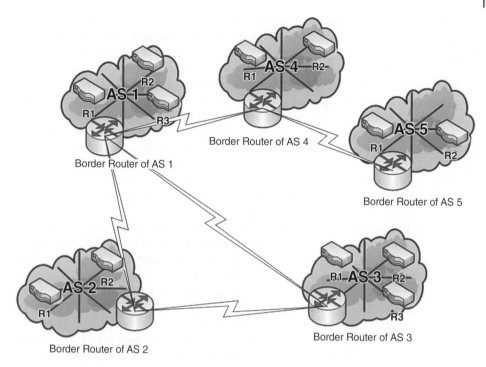

Figure 3.3 A sample network to explain path vector routing.

Table 3.5 Path vector routing table at startup.

Routing Table									
BRoAS 1		BRoAS 2		BRoAS 3		BRoAS 4		BRoAS 5	
Destination	Path	Destination	Path	Destination	Path	Destination	Path	Destination	Path
AS1R1	AS1	AS2R1	AS2	AS3R1	AS3	AS4R1	AS4	AS5R1	AS5
AS1R2	AS1	AS2R2	AS2	AS3R2	AS3	AS4R2	AS4	AS5R2	AS5
AS1R3	AS1			AS3R3	AS3				

which are AS1R1, AS1R2, and AS1R3 as they can be reached within the AS. Similar path vector routing tables are generated by BRoAS2, BRoAS3, BRoAS4, and BRoAS5.

Advertisement. The gateway routers share their routing tables with their neighbors. Thus, BRoAS1 sends its routing table to BRoAS2, BRoAS3, and BRoAS4. Similarly, routing tables are shared by BRoAS2, BRoAS3, BRoAS4, and BRoAS5 with their neighbors only. BRoAS2 sends its routing table to BRoAS1 and BRoAS3. BRoAS3 sends its routing table to BRoAS1 and BRoAS2. BRoAS4 sends its routing table to BRoAS1 and BRoAS5. BRoAS5 sends its routing table to BRoAS4.

Updating. When the border router receives a path vector table from its neighbor, it updates its routing table based on the entries received from the neighbor. It adds to its table those nodes that it can reach through its neighbor. Thus, the border router now adds the nodes that are reachable through its neighbor, and in the path to that node it puts the entry of itself and then the neighbor border router through which it can reach that node. The routing table at this stage has entries similar to those shown in Table 3.6.

Stabilization. With time, the updated path vector table is again shared among the neighboring nodes, and this gives the nodes greater awareness of the reachability to other gateway routers and the associated path. Slowly, the gateway routers become aware of almost all other gateway routers of the other ASs in the network and the path to reach any other AS that is connected to this AS. The stabilized table for the network in Figure 3.3 is shown in Table 3.7. In the table, the shortest path between two ASs has been selected; for example, the path from AS4 to AS2 has been shown as AS4-AS1-AS2. However, in path vector routing the shortest path may not be the optimum path and hence may not be the criterion for path selection.

In the path vector router, as no metrics are involved, the selection of the path AS2-AS1-AS4 is not based on the concept of minimum hop or least-cost path, but on the basis of any policy being implemented in the path vector routing. The path between AS4 to AS2 could also be AS4-AS1-AS3-AS2, depending on the policy implemented. Similarly, the path between AS3 and AS2 could be AS3-AS1-AS2, the path between AS1 and AS3 could be AS1-AS2-AS3, and the path between AS3 and AS4 could be AS3-AS2-AS1-AS4. Various ASs in the network may be running different protocols, and thus may define path optimality as per the protocol being used and the associated path metrics. A number of paths may exist between two ASs, and the optimum path is selected by each AS based on the protocol being followed by it and the policy implemented in the AS.

3.4.2 Advantages and Disadvantages

Path vector routing is used for interdomain routing [3] owing to the following advantages.

Low computational overhead. The protocol just has to read the routes received from the neighbors, check for looping, compare with the policy, and append it to its table. The path vector also does not contain any metrics for calculation or manipulation. The protocol also does not require running a shortest path algorithm or building a spanning tree to the destination, which all reduces the computational requirement at the routing nodes.

Support of heterogeneous policies. The path vector routing supports heterogeneous routing policies across ASs. It does not require all the ASs and gateway routers to share a common policy and implement it. Each AS can have its own routing policy, which can be changed on its own without affecting the other ASs. Change in the routing policy in an AS does not require any path recomputation by any other AS. In path vector routing, the route selection policy used by one AS may not be shared with other ASs for the purpose of privacy and security.

Loop prevention. The path vector contains the entire path to be followed to a destination. Any changes to the path in terms of adding intermediate routing nodes are first verified for looping before changing the path. Looping can be easily detected in the path and can be avoided by not allowing more than a single entry of a border router in the path vector. It does not involve any complex calculation.

Selective route information advertisement. The border router appends the route information it receives from its immediate neighbor to its path vector routing table.

Table 3.6 Path vector routing table after updating from advertisement from neighbor.

Routing Table									
BRoAS 1		BRoAS 2		BRoAS 3		BRoAS 4		BRoAS 5	
Destn	Path	Destn	Path	Destn	Path	Destn	Path	Destn	Path
AS1R1	BRoAS1	AS2R1	BRoAS2	AS3R1	BRoAS3	AS4R1	BRoAS4	AS5R1	BRoAS5
AS1R2	BRoAS1	AS2R2	BRoAS2	AS3R2	BRoAS3	AS4R2	BRoAS4	AS5R2	BRoAS5
AS1R3	BRoAS1			AS3R3	BRoAS3			BRoAS4	BRoAS5-4
		AS1R1	BRoAS2-1			AS1R1	BRoAS4-1	AS4R1	BRoAS5-4
AS2R1	BRoAS1-2	AS1R2	BRoAS2-1	AS1R1	BRoAS3-1	AS1R2	BRoAS4-1	AS4R2	BRoAS5-4
AS2R2	BRoAS1-2	AS1R3	BRoAS2-1	AS1R2	BRoAS3-1	AS1R3	BRoAS4-1		
BRoAS2	BRoAS1-2	BRoAS1	BRoAS2-1	AS1R3	BRoAS3-1	BRoAS1	BRoAS4-1		
AS3R1	BRoAS1-3	AS3R1	BRoAS2-3	BRoAS1	BRoAS3-1	AS5R1	BRoAS4-5		
AS3R2	BRoAS1-3	AS3R2	BRoAS2-3	AS2R1	BRoAS3-2	AS5R2	BRoAS4-5		
AS3R3	BRoAS1-3	AS3R3	BRoAS2-3	AS2R2	BRoAS3-2	BRoAS5	BRoAS4-5		
BRoAS3	BRoAS1-3	BRoAS3	BRoAS2-3	BRoAS2	BRoAS3-2				
AS4R1	BRoAS1-4								
AS4R2	BRoAS1-4								
BRoAS4	BRoAS1-4								

Table 3.7 Path vector routing table after stabilization.

					Routing Table				
BRoAS 1		BRoAS 2		BRoAS 3		BRoAS 4		BRoAS 5	
Destn	Path	Destn	Path	Destn	Path	Destn	Path	Destn	Path
AS1R1	BRoAS1	AS2R1	BRoAS2	AS3R1	BRoAS3	AS4R1	BRoAS4	AS5R1	BRoAS5
AS1R2	BRoAS1	AS2R2	BRoAS2	AS3R2	BRoAS3	AS4R2	BRoAS4	AS5R2	BRoAS5
AS1R3	BRoAS1	AS1R1	BRoAS2-1	AS3R3	BRoAS3	AS1R1	BRoAS4-1	AS4R1	BRoAS5-4
AS2R1	BRoAS1-2	AS1R2	BRoAS2-1	AS1R1	BRoAS3-1	AS1R2	BRoAS4-1	AS4R2	BRoAS5-4
AS2R2	BRoAS1-2	AS1R3	BRoAS2-1	AS1R2	BRoAS3-1	AS1R3	BRoAS4-1	BRoAS4	BRoAS5-4
BRoAS2	BRoAS1-2	BRoAS1	BRoAS2-1	AS1R3	BRoAS3-1	BRoAS1	BRoAS4-1	AS1R1	BRoAS5-4-1
AS3R1	BRoAS1-3	AS3R1	BRoAS2-3	BRoAS1	BRoAS3-1	AS5R1	BRoAS4-5	AS1R2	BRoAS5-4-1
AS3R2	BRoAS1-3	AS3R2	BRoAS2-3	AS2R1	BRoAS3-2	AS5R2	BRoAS4-5	AS1R3	BRoAS5-4-1
AS3R3	BRoAS1-3	AS3R3	BRoAS2-3	AS2R2	BRoAS3-2	BRoAS5	BRoAS4-5	BRoAS1	BRoAS5-4-1
BRoAS3	BRoAS1-3	BRoAS3	BRoAS2-3	BRoAS2	BRoAS3-2	AS2R1	BRoAS4-1-2	AS2R1	BRoAS5-4-1-2
AS4R1	BRoAS1-4	AS4R1	BRoAS2-1-4	AS4R1	BRoAS3-1-4	AS2R2	BRoAS4-1-2	AS2R2	BRoAS5-4-1-2
AS4R2	BRoAS1-4	AS4R2	BRoAS2-1-4	AS4R2	BRoAS3-1-4	BRoAS2	BRoAS4-1-2	BRoAS2	BRoAS5-4-1-2
BRoAS4	BRoAS1-4	BRoAS4	BRoAS2-1-4	BRoAS4	BRoAS3-1-4	AS3R1	BRoAS4-1-3	AS3R1	BRoAS5-4-1-3
AS5R1	BRoAS1-4-5	AS5R1	BRoAS2-1-4-5	AS5R1	BRoAS3-1-4-5	AS3R2	BRoAS4-1-3	AS3R2	BRoAS5-4-1-3
AS5R2	BRoAS1-4-5	AS5R2	BRoAS2-1-4-5	AS5R2	BRoAS3-1-4-5	AS3R3	BRoAS4-1-3	AS3R3	BRoAS5-4-1-3
BRoAS5	BRoAS1-4-5	BRoAS5	BRoAS2-1-4-5	BRoAS5	BRoAS3-1-4-5	BRoAS3	BRoAS4-1-3	BRoAS3	BRoAS5-4-1-3

There may be multiple possible paths to a destination, based on the information received from the neighbors. However, based on the neighborhood information, the border router selects and appends only those entries in its routing table that meet the policy requirements of the AS of the border router. The other route information violating the policies is discarded. Now the border router shares this appended table only with its neighbors. The table contains only that information which was found to be suitable by the border router in that particular AS and does not contain the path information discarded by it owing to policy violation or for some other reason. The border router also has the capability to selectively hide the information during advertisement to its neighbors.

The disadvantages of path vector routing are as follows.

Complex. Path vector routing can be very complex to configure in the network.

Lack of congestion control. The routing policies may be heterogeneous across the ASs. The network traffic or the link congestion may not be the criterion for path selection in the policies of any AS. Moreover, the path vector routing protocol converges very quickly and stabilizes. Thus, it may not be suitable for handling network congestion efficiently.

Inefficient load balancing. The basic path vector routing protocol does not support load balancing. The path vector table may contain alternative paths to a destination, but an alternative path is selected generally on the failure of an existing path. Although load balancing between the source and destination can be done by disseminating packets through each of the alternative paths, multiple paths for packet dissemination are not selected to support load balancing.

3.5   Unicast, Multicast, and Broadcast Routing

There are three different kinds of addressing scheme in a network – unicast, multicast, and broadcast [1, 4, 5]. Movement of the packet from source to destination in unicast, multicast, and broadcast is indicated in Figure 3.4 to 3.6 respectively. Broadcast is a one-to-all mode of communication. In broadcast addressing, the message is sent to all the nodes on the LAN or in the network. Thus, there is one source, and all the nodes in the network are the destination. The broadcast can be in layer 2 or in layer 3. When the broadcast is in layer 2, which is also known as hardware broadcast, then the message is sent to all the nodes in the LAN. The broadcast does not traverse beyond the LAN or across a router. The broadcast packet in layer 3 reaches all the nodes in all the LANs across all the networks, i.e. it covers the entire broadcast domain. It is also referred to as broadcast across 'all network and all hosts'. In order to broadcast a packet, each intermediate router makes copies of the broadcast packet and sends a packet through each of its exit links. Thus, an innumerable number of copies of the packet are generated during the time between generation of the packet by the source node and receipt of the packet in all the nodes in the network, each of which is a destination node. In the broadcast mode also, the source generates a single packet (if it is not a router but a host connected to a network). Address Resolution Protocol (ARP) uses this type of broadcast. Huge networks such as the Internet do not support broadcast, as it will lead to choking of the network by the tremendous amount of traffic generated.

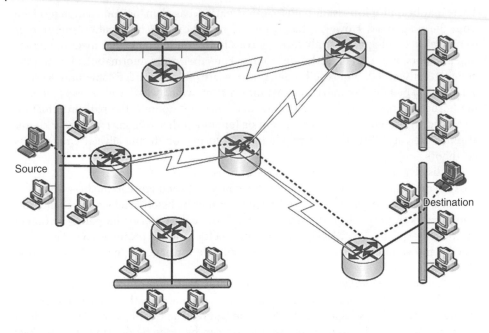

Figure 3.4 Movement of a packet from source to destination in unicast.

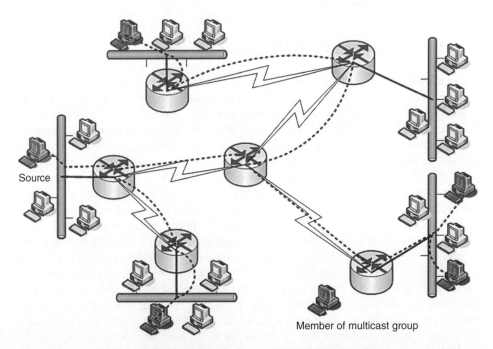

Figure 3.5 Movement of a packet and its copies from source to all members of the multicast group.

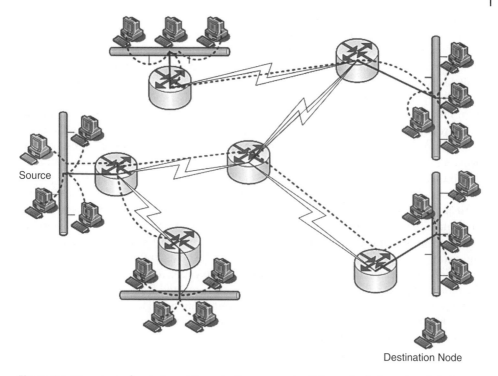

Source

Destination Node

Figure 3.6 Movement of packets and its copies from source to all the nodes in the network during broadcast.

Unicast is a one-to-one communication, which involves only one source and only one destination. In unicast, the message is forwarded only to a unique destination node. The router detects the node to which the packet is being sent by looking at the destination IP address in the packet header, and then puts the packet in its exit link that leads to the destination LAN. Any intermediate router does not make copies of the packet for transmission along different links or for broadcasting. At any point in time between the transmission of the packet from the source and its arrival at the destination, there is just the single copy of the packet present in the network – either in any of the intermediate nodes awaiting processing or in the connecting links. Based on the routing table entries, if an intermediate router is unable to detect the exit link by which it should forward the packet for the destination address, the router may drop the packet.

Multicast is a point-to-multipoint or one-to-many mode of communication. However, it is not a combination of unicast and broadcast. Multicast enables the forwarding of packets to various nodes across different LANs without flooding any of the LANs or the network. Multicast uses the IP multicast group address to send the message. When a router receives a multicast packet, it forwards a copy of the packet through all those links that have at least one of the destination nodes among those in the multicast group. This results in selective duplication of the packet by the intermediate routing node to be transmitted across the specific exit links that lead to the nodes of the multicast group. In the case of multicasting, all the duplicated packets have the same destination address, which is the multicast group address.

Multicast abstraction. A typical multicasting scenario relates to a single source of data transmission, but multiple destinations for receiving the same data. This can be implemented in three different ways [6]. Firstly, multiple unicasts in which the sender generates multiple replicas of the same packet, each meant for one of the destinations. This leads to multiple unicasts where there is a separate unicast from the source to each destination. Thus, multiple duplicate packets may traverse through some common links so as to reach from the source to one of the destinations. Secondly, host-based replication, in which the source sends the packets as a multiple unicast to a few hosts from the list of hosts in the multicast group. These few hosts, on receiving the packet, replicate the packet and transmit it to a few more hosts within the multicast group that have not yet received the packet. These hosts further replicate the packet and send it to some other members of the multicast group, and the process continues until all the members of the multicast group have received a copy of the packet. Each multiple unicasting host may have a restriction on the maximum number of hosts to which it can transmit the packet in multiple unicast mode after replicating the packet, and a multicast group controller keeps a record of the hosts in the multicast group that have already received the packet and those that are yet to receive the packet. Thirdly, the typical multicast in which a single packet is transmitted from the source and each network router replicates the packet and transmits it along all those links that contain a host belonging to the multicast group. Such a multicast requires support from the network layer and is the most efficient form of unicast, as only a single packet will traverse through each link.

For a network layer to support multicast, it has to resolve two issues related to multicast addressing and node identification. Each packet transmitted through the network carries the destination address. However, in the case of multicasting, the packet has multiple destinations, but it is not possible to have the IP address of all the destinations in the packet header. The network should also be aware of all the hosts belonging to a multicast group. Each router should know whether it has hosts that belong to the multicast group. If a router possesses hosts belonging to the multicast group, it should be able to identify such nodes. The network should also be aware of the routers possessing hosts belonging to the multicast group. It may happen that a packet lands up at a router that itself does not have any multicast hosts but that has exit links connected to some other routers that have hosts from the multicast group. Furthermore, these routers may be further connected to some other routers that have hosts from a particular multicast group. In such a scenario, the intermediate routers should ensure that they send a copy of the packet by the related links so as to reach the hosts of the multicast group through their respective routers. The addressing of a multicast group is done with the help of a class D multicast addressing scheme. All the multicast hosts forming a multicast group are associated with a unique group address [7]. Thus, a single group address uniquely identifies all the receivers of the group and sends a copy of the packet to all the multicast receivers in the group. This is known as address indirection. However, the IP address of the unicast host, which can be used for unicast, is separate from this multicast group identifier.

IGMP Protocol. Network layer multicast routing functionality has been divided into two different boundaries. The first is for a host that is a member of a multicast group and has a directly attached router, which we call its first hop router. The second is for routers to support multicasting. Multicast routing from a router to its directly

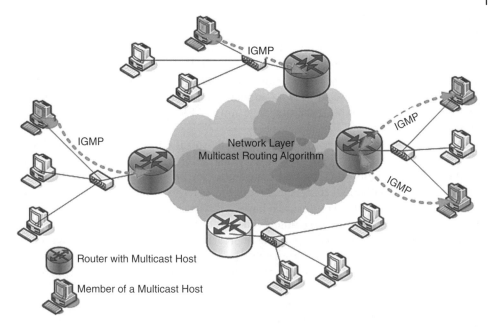

Figure 3.7 Scope of operation for protocols supporting multicasting – IGMP and the multicast routing algorithm.

connected host is supported by Internet Group Management Protocol (IGMP), as indicated in Figure 3.7, and multicast routing among routers is supported by a number of multicast routing algorithms such as DVMRP, MOSPF, and PIM. As IGMP operates locally, it has a very small domain of operation, i.e. it works between the host and its directly attached router and is used by the host to inform the router that the application running on it wants to be a member of a multicast group. There is no protocol to enable direct communication between all hosts of a multicast group together, or to enable a host to know about all other hosts connected to its multicast group.

IGMP uses three different types of message.

Membership query message. This is sent by the router to each host connected to it to enquire about all the multicast groups that have been joined by the host. Through this message, a router can also learn from the hosts whether they have joined any specific multicast group with a particular multicast address.

Membership report message. This is used by the host to respond to the router. In the membership report, the host informs its directly connected router about the multicast address of any multicast group it has joined. This message can also be directly generated by the host without a probe from the router whenever the host joins a multicast group. However, the membership report message goes to the router as well as to all other hosts connected to the router. A router does not require information about how many hosts connected to it are members of a multicast group and which particular hosts are the members. The router only needs to know that it has at least one host connected to a multicast group so as to enable the router to participate in the multicast routing algorithm along with other routers in the network. Thus, when a membership report message is sent by a host to all other hosts, the remaining hosts that are also members of the same multicast group do not send the membership report reply to the router.

Leave group message. This is used by the host to inform the router that it is no longer a member of the multicast group. However, the functionality of this message can also be performed by membership query and membership report messages, because if the host has ceased to be a member of a multicast group it will not respond to the periodic membership query message and will be removed from the router after timeout.

Multicast routing. In a network, there are only specific routers that have multicast group hosts connected to them. These routers may either receive traffic for these hosts, or the multicast group hosts may generate traffic for the other hosts in the multicast group connected with other routers to be sent through this router. The goal of multicast routing is to connect all the routers that have hosts belonging to a multicast group. This is done by forming a tree, which connects all such routers. However, in the formation of such a tree, it may not always be possible to prevent all routers that do not have a host belonging to the multicast group from being part of the tree. Once the multicast tree is formed, the multicast packets are sent along the path of this tree to reach all the routers with multicast group hosts.

In multicast routing, two different types of multicast tree may be formed by the source – a shared tree and a source-based tree [1, 6, 7]. In a shared tree there is only one tree that connects all the multicast-group-host-attached routers. In a source-based tree, there is a tree from each of the source routers to all other receiver routers. Thus, if there are N routers in the network that have multicast hosts connected to them, there will be N trees, each of which will be rooted at one of these routers and connecting all other routers having multicast group hosts. Each tree is a shortest path tree based on Dijkstra's algorithm and uses the reverse path forwarding approach. A few common multicast routing protocols are Distance Vector Multicast Routing Protocol (DVMRP), Multicast Open Shortest Path First Protocol (MOSPF), Protocol Independent Multicast (PIM), and Border Gateway Multicast Protocol (BGMP).

References

1 B. A. Forouzan. *Data Communication and Networking.* McGraw-Hill, 2007.

2 W. Stallings. *Data and Computer Communications.* Prentice Hall, 8th edition, 2007.

3 A. Rodriguez, J. Gatrell, J. Karas, and R. Peschke. *TCP/IP Tutorial and Technical Overview, IBM Redbook.* IBM Corporation, Prentice Hall, International Technical Support Organization, 7th edition, 2001.

4 T. Lammle. *CCNA IOS Commands Survival Guide.* SYBEX Inc., 2007.

5 S. Mueller, T. W. Ogletree, and M. E. Soper. *Upgrading and Repairing Networks.* Que Corp., 2005.

6 N. K. Singhal, L. H. Sahasrabuddhe, and B. Mukherjee. Optimal multicasting of multiple light-trees of different bandwidth granularities in a WDM mesh network with sparse splitting capabilities. *IEEE/ACM Transactions on Networking,* **14**(5):1104–1117, 2006.

7 J. F. Kurose and K. Ross. *Computer Networking: A Top-Down Approach Featuring the Internet.* Addison-Wesley Longman Publishing Co. Inc., 2nd edition, 2002.

Abbreviations/Terminologies

ARP	Address Resolution Protocol
ARPANET	Advanced Research Projects Agency Network
ASs	Autonomous Systems
BGMP	Border Gateway Multicast Protocol
CPU	Central Processing Unit
DVMRP	Distance Vector Multicast Routing Protocol
EGP	Exterior Gateway Protocol
EIGRP	Enhanced Interior Gateway Routing Protocol
IGP	Interior Gateway Protocol
IGMP	Internet Group Management Protocol
IGRP	Interior Gateway Routing Protocol
IS-IS	Intermediate System to Intermediate System
ISP	Internet Service Provider
LAN	Local Area Network
LSA	Link State Advertisement
LSP	Link State Packet
MOSPF	Multicast Open Shortest Path First
NLSP	Network Link State Protocol
OSI	Open System Interconnection
OSPF	Open Shortest Path First
PIM	Protocol Independent Multicast
RIP	Routing Information Protocol
TCP/IP	Transmission Control Protocol/Internet Protocol
TTL	Time To Live

Questions

1 Differentiate between static routing and dynamic routing.

2 Explain the working of the following protocols along with their advantages and disadvantages:
  A distance vector protocol,
  B link state protocol,
  C path vector protocol.

3 Explain the following terms:
  A flapping,
  B pinhole congestion,
  C multicast abstraction.

4 What can be the various techniques to avoid routing loops in any algorithm?

5 Explain how hold-down times reduce the effect of flapping.

6 What is an autonomous system, and why is path vector routing preferred for communication across ASs?

7 Explain the working of IGMP and show the exchange of different types of IGMP message by drawing a sample network.

8 Mention the contents of a link state packet used in link state routing.

9 Why does a gateway router in an AS advertise only selective route information?

10 State whether the following statements are true or false and give reasons for the answer:

A Distance vector and link state protocols are types of EGP.

B Distance and cost are the two important items of information stored in the routing table of a distance vector routing.

C Routing loops can be avoided by using split horizon.

D On reaching the maximum hop count, the packet is sent back to the source.

E Route poisoning is a technique used to avoid pinhole congestion.

F In link state routing, the router is aware of the entire network topology.

G Multiple unicasts can be done to perform multicast, and multiple unicasts and multicast have the same effect with the same amount of generated traffic.

H IGMP uses two different types of message.

I In a shared tree, there is only one tree that connects all the multicast-group-host-attached routers.

J In a source-based tree, there is a tree from each of the source routers to all other receiver routers.

Exercises

1 Please consider the network given below where the values along the link indicate the link cost:

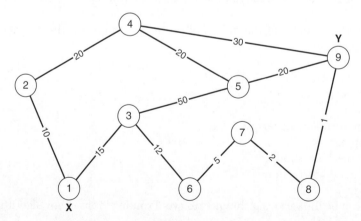

Make the table that each node will make initially and share with its neighbor in distance vector routing. Similarly, make the table that each node will make and share with its neighbor in path vector routing.

2 A network is indicated below with the link cost shown along the links:

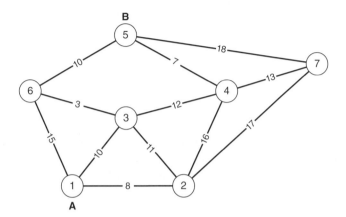

Prepare the initial distance vector table when the network boots up and then all the intermediate distance vector tables for all the nodes till the network converges.

3 Start with the converged routing table for all the nodes of the network at time = t1. Now assume that certain network policies are implemented and the same network is changed to the one indicated at time = t2. Draw all the intermediate stages of the routing table in the process of updating until convergence is achieved for the network at time t2. Assume that it takes one unit time to exchange and update the table once. How many units of time woul be required to obtain the new converged table?

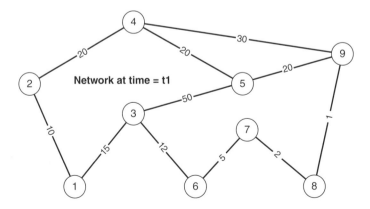

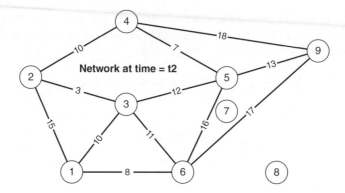

Network at time = t2

4 Consider the network given in exercise question 1. If only the hop count is considered as a metric, will there be pinhole congestion for data transmission between nodes X and Y? Similarly, if hop count is the only metric in the network indicated in exercise question 2, will there be pinhole congestion for data transmission between nodes A and B?

5 Consider the network at time = t2 given in exercise 3, without nodes 7 and 8. Attempt to explain the routing loop in the network by assuming that a certain number of network links go down. Is the assumed set of network links that have gone down the minimal set?

6 In link state routing, every node receives actively connected interface information from all other nodes of the network, based on which the node creates a graph depicting the topology of the entire network. The node then runs the shortest path algorithm independently to decide the exit interface for the next hop of the packet. Consider the network given in exercise 1 and draw the tables that each node has got before starting the shortest path algorithm. At node Y, run Dijkstra's algorithm to find the shortest path to all other nodes.

7 The network indicated at time t2 of exercise 3 runs link state routing. The network boots up at the same instant of time. How many beacon messages will be generated in the network after the network has been operational for exactly 1 h when a beacon message is generated every 4 s? How many beacon packets will be received by each node during the 1 h of operation? How many LSP packets will be received by each node during 1 h of operation if the frequency of transmission of LSP packets is also 4 s by each node?

8 Consider the network given in exercise 2 (ignore the values given along the links). Distance vector routing runs on the network with hop count as the metric. It is well understood that the maximum hop count will be equal to the diameter of the network. The node is uniquely identified by the node ID. Calculate the number of bits required for each entry in distance vector routing. Assume that the distance vector table is exchanged once in 15 s. How much bandwidth will be used by the exchange of distance vector tables in the network during 1 h of its operation?

9 In path vector routing, the routing table is periodically advertised by the border router to other border routers in different ASs as well as to the nodes within the AS. Consider Figure 3.3. How many copies of the routing table will be created for each border router during advertisement and the total number of advertisements generated in the network for all border routers at any instant of time? What will be the total number of advertised routing tables in the network when there is a redundant border router in each AS connected to the same internal routers and border routers in other ASs. The redundant border router is also connected to the primary border router and has the same type of connectivity with other ASs.

10 Consider the network with a redundant border router, as explained in the previous question. For this network, draw the path vector routing table in various stages, i.e. startup, updating, and stabilization.

Part II

Routing with Quality-of-Service and Traffic Engineering

4

Quality-of-Service Routing

4.1   Introduction

The setting up of LANs and WANs paved the way for Intranets. The Internet emerged as the largest WAN. Intranets were used for the client–server type of applications where data was sent from the client to the server over the network. The server processed the data and sent back the results to the client on the network. With the Internet came applications such as email (SMTP), file transfer (FTP), remote login (telnet), and web access (HTTP), which supported the web-based applications in which the client terminal was able to use the standard web browser to view or request applications. The initial applications planned for the networks were not bandwidth hungry.

The emergence of huge data crunching and analysis applications such as data mining, simulations, modeling, analysis, and visualization required tremendous processing capabilities supported by grid computing and supercomputing. Although these processor-hungry applications required a huge amount of data to be transmitted from the client to the servers for their processing, once the data was transmitted from the client to the grid or the cluster, the processor became busy processing it and then sent back the results to the clients. All these applications had different types of data transmission pattern over the network. The pattern varied from a small amount of data transmission occasionally over the network to a huge amount of bursty traffic over the network. Still, none of these was affected much by delay in data transmission. They were all dependent on accuracy of the data transmission without any significant consideration of the delay in transmission. These early-day applications were affected if there was a bit error or a packet loss that was to be detected and retransmitted. Moreover, these applications

Network Routing: Fundamentals, Applications, and Emerging Technologies, First Edition.
Sudip Misra and Sumit Goswami.
© 2017 John Wiley & Sons Ltd. Published 2017 by John Wiley & Sons Ltd.
Companion website: www.wiley.com/go/misra2204

were not affected if the packet reached the destination with a considerable amount of delay or if the packet did not reach the destination in the same sequence in which it was sent because the packet was reassembled back as per the sequence number at the destination. Characteristics such as these also led to categorization of this traffic type as 'elastic traffic' [1].

The emergence of real-time applications changed the requirement scenarios. Some of the real-time applications are Voiceover IP (VoIP), live webcasting, video conferencing, tele conferencing, and Internet telephony. The real-time applications were critical to the delay in data transmission and the sequence of arrival of the packets at the destination. However, they were tolerant to bit errors, packet losses, and unordered delivery of packets. Low bandwidth or fluctuating bandwidth affects real-time applications. Still, there are applications such as interactive computing and computer games that are delay sensitive as well as error sensitive. Although real-time applications may require transmission of a huge amount of data across the network, the transmission cannot be done in a bursty mode but has to be done at a constant rate over a period of time. Reduction below the minimum required bandwidth of the real-time application leads to call drop, freezing, jitter, and hangs. These requirements led to categorization of this traffic type as 'inelastic traffic'. As real-time applications are transmitted over the TCP/IP network, they are dependent on the underlying capabilities of the IP network for assured minimum bandwidth.

Quality of service (QoS) [2, 3] is a parameter for standardizing the performance of a network in terms of the assurance and level of the services it offers. The QoS is a parameter of significance in multimedia streaming applications. It guarantees a certain level of performance of the data flow over the network by using specific data transmission techniques and protocols that enable the network to deliver satisfactory prioritization, bandwidth, and performance. A network can be divided into certain paths that provide QoS as well as satisfying QoS criteria. Thus, in order to satisfy QoS constraints, it is essential for the underlying protocol to determine the path that satisfies the QoS parameters. QoS requires prioritization of the network traffic into various categories based on its priority. The network packets have to be prioritized according to the urgency of their delivery to the destination. Generally, all packets belonging to a certain application are given the same priority. QoS is of importance in those networks where the network bandwidth is not sufficient to cater for all applications simultaneously. If sufficient bandwidth was available with the network, each application would have received its share of bandwidth on an 'as per requirement' basis without affecting the performance of other applications using the bandwidth at the same time. But generally bandwidth is a constrained resource that is not sufficient to meet the performance parameters of all the applications together and has to be shared among applications. QoS prioritizes the usage of this bandwidth among various applications.

A network is said to support QoS if it has the capability of treating different packets differently. QoS technology has enabled service providers to support different levels of service for different customers, thereby capacitating them with the option to provide better levels of paid services to some customers than to others. For example, some groups of customers may be concerned with a service that guarantees packet delivery, even if that means paying a higher price for these services, while others may just as well be satisfied with relatively less reliable data transfer by paying less for their subscribed services. Networks that transport multimedia traffic, i.e. voice, data, and video, need differential treatments to different packets; while voice traffic is highly sensitive to time delay and the order of delivery of packets, data traffic is relatively less sensitive to these, and video conferencing traffic requires a dedicated connection for a fixed amount of time for the real-time, orderly delivery of packets [4].

Typical QoS-routing-based performance metrics [5] are bandwidth, delay, and throughput. While some applications require bandwidth guarantees, some others mandate the satisfiability of strict end-to-end delays, and others still require a high throughput, or a combination of both of these criteria.

Routing protocols in the pre-QoS era did not consider QoS requirements of connections (e.g. delay, bandwidth, and throughput). Furthermore, optimization of resource utilization was not a primary goal. As a result, while there were flow requests that were rejected because of non-availability of sufficient underlying resources, there were some other resources that remained available. To address such deficiencies with conventional routing protocols, QoS routing algorithms were devised that could locate network paths that satisfied QoS requirements, and that made better use of the network [2, 5]. Routing of QoS traffic requires stringent performance guarantees of the QoS metrics, i.e. delay, bandwidth, and throughput over the paths selected by the routing algorithms. Accordingly, whereas the traditional shortest path algorithms, e.g. Dijkstra's algorithm or the Bellman–Ford algorithm [6], indeed have the potential for selecting a feasible path for routing, QoS routing algorithms must consider multiple QoS and resource utilization constraints, typically making the problem intractable. Thus, QoS routing is different from the routing in traditional circuit-switched and packet-switched networks.

In ATM networks, for example, a connection is accepted or not by the connection admission control (CAC), depending on whether or not sufficient resources (e.g. bandwidth) are available. The CAC operates by taking into account factors such as the incoming QoS requests and the available resources in the network. The CAC operates the QoS routing algorithms, which identify whether the different possible candidate paths satisfy the QoS requirements or not. The network architects and the engineers want to choose a routing algorithm that will help the service providers to maximize the network resources (e.g. the amount of bandwidth to be routed) while satisfying the requirements from the customers. Therefore, the design of any good QoS routing algorithm takes into account factors such as satisfying the QoS requirements, optimizing the consumption of the network resources (e.g. buffer space, link bandwidth), and balancing the traffic load across different paths [7]. In addition to these, a good QoS routing algorithm should characterize itself by its ability to adapt to the periodic dynamic behavior of the network [4].

QoS routing can be divided into two entities – QoS routing protocol and QoS routing algorithm. The QoS routing protocol determines the condition of the network in terms of links, nodes, bandwidth, congestion, and delay. The dynamic routing protocols regularly analyze the underlying network to capture the condition of the network on a regular basis to help determine the behaviour of the network. The routing protocol distributes this collected information across the network to various routing nodes. The routing algorithm uses this information provided by the routing protocol to calculate the path by which it should forward the packet in order to satisfy certain parameter criteria to deliver the packet with assured constraints [8].

The QoS criteria may be based on a number of constraints such as bandwidth assurance, buffer usage, priority queuing, CPU usage, loss ratio, and jitter, in addition to the hop count or delay, which are the common criteria for operation of a shortest path routing algorithm. QoS routing is unavoidable for certain applications that are delay sensitive; for example, the packets of a network management system should be given priority over other data packets in a congested network in order to enable the NMS to display the congestion in the network. If the NMS packet is not given priority in the congested network, it will remain queued up in one of the nodes of the congested link, and this will lead to the NMS believing that the link has failed, even though it is operational, or the congestion information will reach the NMS after it has been resolved and this will bring down the reliability of the NMS owing to its non-real-time behaviour.

The prime objective of QoS routing is to find a path that has sufficient resources to satisfy the service level constraints imposed on the network by the QoS. The minimum amount of resources should be assured even in the case of some other applications using the network and its resources. Even if another application uses the path already selected by the QoS application, the residual resources should be able to serve the QoS service criteria of the application.

The QoS constraints can be broken down into two categories: (1) link constraints and (2) path constraints [9]. The link constraint refers to the quality of a particular link and its ability to route the traffic so as to satisfy the minimum service criteria. The link constraint is of more importance in a packet switching network where a link is selected for each individual packet by the routing algorithm. Thus, the QoS parameter of the link has to be assured before a packet is placed on it. The link constraint must ensure a minimum level of service assurance across its endpoints, e.g. transmission with an assured maximum delay or assured minimum available bandwidth across it. The path constraint requires minimum service guarantee across the entire path from source to destination. It is of greater significance in the case of point-to-point circuit switching networks than the packet switching network. The path has to be carefully selected with the assured QoS, as the same path will be used throughout the transmission and its performance should not degrade beyond the assured level of service during the entire period of connection. In the case of QoS in multicasting, the tree constraint also gains significance. The tree constraint can be assumed to be a modified form of the path constraint. In the case of a tree constraint, the data has to traverse the tree with a minimum assured level of service, e.g. the packet should traverse the entire tree within a specified period so as to enable the source node to broadcast the message within the permissible delay to all the members of the multicast group.

As the QoS is a user-oriented parameter, it has to be implemented end-to-end across a network. Thus, all the intermediate network components between the two endpoints, i.e. end nodes, switches, routers, and gateways, should have the capability of QoS implementation over them and should be capable of guaranteeing the QoS over the packet being processed or passing across it. The intermediate nodes should be capable of segregating the QoS packet from the normal packet before processing it. As the QoS packets may get priority of the network, QoS attacks are also possible by a network flow to get priority over the other packets. To prevent any QoS attack, 'traffic policing' [10] is one of the components of the QoS framework. The policing is done at the edge of the network to ensure that only the legitimate traffic is requesting QoS from the network.

The packets violating the traffic policy are dropped at the edge. Alternatively, instead of dropping the packet, the packet that is not conforming to the traffic policy may be accepted by the processing node and stored in its buffer for processing with a lesser priority than the QoS packet, and then when processed the undesirable characteristics of the packet are removed before it is sent back to the network. This, referred to as 'traffic shaping', can smooth out bursty traffic. Leaky bucket and token bucket are common traffic-shaping mechanisms. The QoS traffic should be managed by the intermediate nodes to provide fair competition among the flows with equal priority but with minimum assured level of service.

4.2 QoS Measures

QoS is a user-oriented parameter. While the connectivity and the level of services between the service providers is measured in terms of network performance (NP), QoS measures the level of satisfaction of the user with the performance of all the services provided to her by the network. NP depends on the interconnect network between the service providers and its associated devices and does not rely on the performance of the end terminals and user input/output. The QoS, however, is the delivery of the services at the assured level of satisfaction as offered to the customer or demanded by the customer [11, 12]. In order to measure the level of performance over the assured level of service, the QoS should be quantifiable. The user should be able to define the quality of service in some measurable parameters and not as linguistic adjectives such as excellent, very good, assured, high bandwidth, low delay, or jitter-free service. The QoS thus converts the quality into measurable parameters, and the parameters are such that they can be understood and measured by the user as well as by the service provider as they are related to user satisfaction.

Although QoS metrics quantify the measurable parameters, it is difficult to list a common set of QoS parameters for all types of application and network. The QoS metrics and their permissible values vary across applications. Thus, most of the QoS measures are defined on an application and network basis and not in general terms. In order to illustrate this, we present different sets of QoS parameters used by various organizations or mentioned in different literature sources. The European Telecommunications Standards Institute (ETSI) defines nine performance parameters to quantify the QoS. These parameters as mentioned by the ETSI are:

- access speed,
- access accuracy,
- access dependability,
- disengagement accuracy,
- disengagement dependability,
- disengagement speed,
- information transfer accuracy,
- information transfer dependability, and
- information transfer speed.

Dependability has been defined as the probability of successful completion of the activity irrespective of any assurance of speed or accuracy.

The critical factors [13] to measure the satisfaction of the user for a multimedia transmission over the Internet can be as follows:

- video rate,
- video smoothness,
- picture detail,
- picture color quality,
- audio quality, and
- audio/video synchronization.

The key indicators [14] of the QoS metrics from a user's perspective in respect of a mobile network can be as follows:

- *Coverage.* This indicates the geographical area over which the network can provide its services.
- *Accessibility.* This describes the ease with which two remote sites connected through the network can communicate with each other by accessing the network. It also indicates the capability of the network to establish communication between the interconnected mobile terminal and the fixed-line terminal.
- *Audio quality.* This indicates the ability of the network to maintain voice clarity over a certain period of time on the network with call disruption.

Thus, as the application changes, there is a change in the criteria for the measurement of its satisfiability. However, from a user point of view, a standard metric that can encompass some of the specific metrics across all applications can be defined as a guiding parameter for QoS measurement. A few common QoS measures are as follows:

- bit rate,
- delay,
- packet loss,
- jitters,
- response time,
- cross-talk and echo levels,
- signal-to-noise ratio, and
- grade of service (ratio of lost traffic to offered traffic).

The QoS requirements can be categorized into levels [13] based on user requirements: criticality, cost, security, and reliability. Criticality of the QoS depends on the application being considered, and the measurement parameter varies with application. However, as the criticality of the application varies with the users, so do the metrics for measuring the criticality of the QoS. In a video conferencing environment, a particular user may give more importance to the voice quality, while for another user the video resolution may be more important. Cost is considered as another criterion for measuring the QoS, as the user expects assured level of service in return for the service charges paid. When the service provider charges money for the service to the user, it also assures a minimum level of service. The user should be able to measure whether she is being provided the minimum assured level of service. There can be different charging criteria, such as per usage cost, per unit cost, or per unit of time cost. The QoS should also deal with ensuring the security of the data being delivered. The QoS should not be delivered at the cost of security. The minimum assured security parameters should be in place at all levels when a packet with QoS assurance moves across the network. However, to ensure faster delivery, the security encapsulation of the packet should not be reduced or removed. The delay should not be reduced by decreasing the packet size and processing requirement by stripping the security parameters.

The reliability of a system is the amount of time a system is expected to be operational before it fails. It is different from availability, which is the amount of time a system is operational out of the total specified time. A system going down very frequently for a fraction of a second every hour may be highly available but has very low reliability. Depending on the application, a system may require high availability or high reliability or both, e.g. an airline ticket reservation network requires high availability, while the autopilot of an aircraft requires high reliability. A network controlling a nuclear reactor requires high availability as well as high reliability. Some other reliability-based parameters that can be used to measure the QoS can be mean time to failure, mean time between failures, or mean time to repair.

The latest network-related technologies, applications, and products being launched in the market are coming with an assured QoS equivalent to or better than the legacy systems. Service providers have various levels of cost for the same product or service, based on the assured resource availability. It is challenging to make a new technology meet the customer service expectations in terms of quality, availability, and reliability and at the same time enable the network quickly to adapt to the new technology. QoS allows measurement of the changing dynamics of the network and prioritization of traffic flow through it. Thus, QoS is becoming a key factor in the interconnected networks across a large number of service providers to deliver interactive multimedia communication support.

4.3 Differentiated and Integrated Services

Previously, various mechanisms in place to incorporate QoS in a network were not point-to-point across the network, as this involved consensus among all intermediate service providers and the devices. Such collaboration could be achieved only with standardization of QoS over an IP network because an IP network does not directly support QoS, rather it offers a best-effort service. Moreover, QoS-aware routing protocols were also being designed. An application that requires QoS over the network can attempt to achieve the same in two different ways; it should be able to state its QoS requirements either before initiation of the traffic or on the fly during movement of the data packets in the network. In a similar approach, the International Engineering Task Force (IETF) has defined two models for implementation of QoS – differentiated services and integrated services.

Differentiated services. Differentiated service architecture [15] divides the data flow in the network on the basis of committed performance and provides a separate QoS to each group of the flow. Differentiated service is implementable in IPv4 as well as IPv6 networks, and it uses an 8 bit differentiated service field in the header to assign a priority to the packet and group it in one of the QoS classes in the network. The differentiated service header uses the same eight bits of an IPv4 packet header that were previously kept for type of service (ToS) in bit positions 8 to 15 of the header but were not used. From the eight bits assigned for the differentiated service header, two bits are presently unused. The remaining six bits form the differentiated service code point. Thus, when a packet with a differentiated service header reaches a router, the entries of unused bits are ignored and only the differentiated service code point entries are read to ascertain the level of service assured to the packet. The differentiated service code point occupies bit positions 8 to 13 in the IPv4 header.

The assignment of the QoS priority value in the header is based on the service level agreement (SLA) between the customer and the service provider before generation of QoS-assured traffic. The customer can be an individual, an organization, or another service provider that has its own differentiated service zone and would be exchanging traffic with this service provider. This enables the network of the service provider to be aware of the level of QoS assured to a particular customer before it starts receiving the data from the customer on its network. Once the service provider enters into an SLA with the customer, the customer marks its packets with its agreed priority value encoded in the form of a bit sequence in the differentiated service header. The header value may change for various traffic flows from the same customer, as the SLA might be different for different types of application from the same customer. As the priority level is decided on the basis of the value in the eight bits of the differentiated header, a number of customers or their specific applications may be assigned the same priority level. All the data packets that are assigned the same priority level are 'aggregated' together, even though they may belong to different customers or applications.

The differentiated service has to be implemented on each of the intermediate nodes in the network by configuring suitable ingress, egress, processing, and buffering policies, as each of them has to conform to the assured level of service. This also helps in distributing QoS assurance across the network, and each packet can be dealt with separately by the intermediate node with the assured level of service. However, as all the intermediate nodes in a huge network may not be under one administrative control, the differentiated services are generally restricted within a portion of the network that is under one administrative control or within even bigger portions of the network that may be under various ISPs each forming its own differentiated service domain, and the ISPs agree on providing QoS among the differentiated service domains of each other. The individual differentiated service domain of a single service provider or a combination of a number of differentiated service domains with mutual agreement among the service providers should be continuous in nature because, if the traffic has to go out of the domain, even if to a single routing node that is not configured for the differentiated services or does not recognize or accept the differentiated services of the service provider, it may hold or delay the packet beyond the service level agreed by the service provider and thus reduce the end-to-end QoS of the entire flow below acceptable limits.

An SLA may define the service parameters in qualitative terms or quantitative terms [15]. In the case of quantitatively defining the parameters, it is easy to measure the parameter and compare its compliance with the assured level. Some of the parameters defined in quantitative terms can be 2 Mbps bandwidth, 120 ms latency, 5% packet drop, <6 hops, or 99.8% availability. The parameters may be expressed qualitatively using linguistic expressions such as high bandwidth utilization, low jitter, and higher reliability. Although it is difficult to ensure the compliance of such an SLA because it is difficult to measure the service level, in such cases the service level of normal traffic, i.e. without any QoS based on SLAs but with a best-effort service, is taken as the base

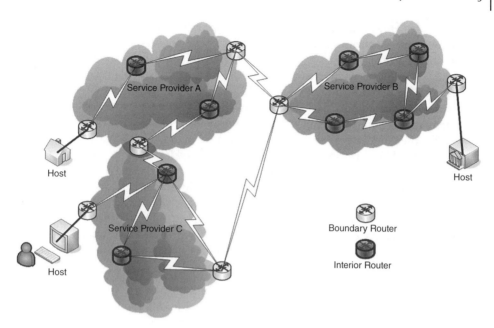

Figure 4.1 Interior router and boundary router in a differentiated service architecture.

reference to quantify high or low. An SLA can also have a mix of qualitative and quantitative representation of the different service levels.

There are two categories of routers in a service provider network – the interior routers and the boundary routers, as depicted in Figure 4.1.

Interior routers are connected only to those routers in the network that belong to the same service provider. When a service provider enters into an SLA with one or more customers, it has to configure all the routers in its network to make them aware of the QoS. In the case of an internal router, applying the QoS is simple as it receives packets that have come from some other router of the same service provider and thus have the assured level of service prior to reaching the entry point of this router. Also, once the router forwards the packet to another router within the same service provider network, that router too will ensure delivery of the assured level of service. Thus, this intermediate router just has to follow the differentiated service specifications across its input and output and treat the packet in accordance with its differentiated header value. This information available with the router on the type of treatment to be given to the data packet is also known as 'per hop behavior' (PHB). The internal router should process QoS packets with a priority over normal packets. This can be done by allowing them assured service along with empty or shorter queues. A shorter queue will reduce the latency, delay, jitter, and packet loss. This will ensure the assured level of service across the network of the service provider for the data packet.

Boundary routers are connected to at least one router that is not a part of the service provider network. These routers are responsible for regulating the traffic entering the differentiated service zone of the service provider. Traffic regulation consists of five processes [15]: classification, metering, marking, shaping, and dropping. During classification, the packet is distinguished into various classes based on the differentiated service header. However, the classifier may also take into consideration the payload and any other parameter from the packet header for classifying the traffic. Metering is the process of measuring at least those packet parameters that have an assured level of service guarantee so as to ensure compliance of SLA. Marking is the process of marking or changing the value of the differentiated service header. This is required when the performance parameter of a packet falls below the assured level or when a packet enters the differentiated service zone of one service provider from the differentiated service zone of another service provider. If the performance parameter of a packet falls below the assured level, the priority of the packet may be raised further to compensate for the loss or it may be changed to lowest-priority best-effort forwarding so as not to affect the performance of the other packets maintaining the performance parameter in their present flow. Shaping is done to make the packet conform to the traffic rate assured to the packet. It smooths out bursty traffic to a constant flow rate of traffic. Boundary routers may also drop those packets that do not meet the differentiated service specifications or are not of assured priority during congestion in link or buffer saturation.

Advantages of differentiated services are as follows:

- The routers can perform their primary job of routing without being affected by differentiated services, as the major processing related to differentiated services takes place at the boundary of the differentiated service zone.
- End-to-end connection negotiation for an assured level of service is not required.
- They are much easier to implement, and they work in a distributed manner.
- The QoS configurations are made in the router and are hence static. Thus, an application does not require any special configuration per flow to be sought for assured QoS.

Disadvantages of the differentiated services are as follows:

- As each router deals with the packet separately, it becomes difficult to predict the level of service that will be offered by the router.
- An initiator may tag a packet with high priority to get preference, even though it does not qualify for the same.

Integrated services. Integrated services provide end-to-end signaling to give a solution that meets most of the QoS requirements. They use Resource Reservation Protocol (RSVP) to reserve the resources to deliver the assured QoS to each 'flow' [15]. Flow is a unidirectional data stream from an application to one or more recipients. A flow is uniquely identified by five tuples: source IP address, source port number, destination IP address, destination port number, and transport protocol. A flow comprises a number of data packets generated from the same application, and all the packets in a flow have the same QoS requirement.

In a legacy IP network, routers attempt to provide the best-effort service based on the routing algorithm running in it and the buffer capacity of the router. The routing algorithm attempts to find the least-cost path from source to destination. A dynamic routing algorithm also finds alternative paths for forwarding data traffic in the case of congestion. However, no preferential treatment is given to a packet in the case of congestion. The router also drops any packet arriving at its interface if the buffer of the router is full. However, these attributes of IP routing do not suit implementation of QoS in the network. Integrated services manage routing and congestion control in a different manner to ensure assured level of service to the network flow. In an integrated service, all the routers collectively assure the availability of resources to process the packets in the flow within the assured level of service, and only thereafter is the flow accepted by the network. The routing algorithms consider QoS parameters in addition to the minimum cost to determine the next exit link to forward the packet. The buffering and packet drop policy of the routers are modified, as otherwise the FIFO buffering and processing policy will lead to treating all the packets with the same priority. Therefore, priority queues are created and assigned to QoS traffic, and a policy of dropping an existing non-QoS data packet from the buffer is used in order to admit a packet with an assured QoS in the buffer even when the buffer is full.

In a router, the flows are categorized into various classes [15] to decide on their forwarding mechanism. There can be a single flow categorized into a particular class, or there can be a number of flows from one or more applications or customers categorized into a particular class. The class of the flow is selected on the basis of the five tuples in the IP header. All the flows and its packets belonging to a particular class are given similar resource allocation and preferential queue allocation across the network. A packet scheduler in the router is assigned the activity of managing the queues. The router may have one or more queues. The packet scheduler, based on the class of the data packet, decides on the ordering of the data packets in the queue and their processing and forwarding. It is also responsible for deciding about dropping packets from the buffer, if required, to accommodate packets with assured level of service. When a new flow is to be accepted for transmission, the RSVP invokes the router to check whether sufficient resources are available to process the flow as per the assured QoS. The router has to assure availability of resources at the assured QoS to this new flow after taking into consideration the existing flows with the router as well as the other normal traffic being processed by the router. As RSVP gets similar assurance from all routers and end systems along the path of the flow, it reserves these assured resources for the flow.

There are two categories of integrated services that can be requested through RSVP: guaranteed service and controlled load service. There can be various levels of services within these two categories, and the specific amount of service requested within a category by a flow is based on certain parameter values that are known as 'traffic specifications'. Guaranteed service provides assurances such as a strict upper bound on the end-to-end delay, assured bandwidth for the traffic that conforms to the reserved specification, and assured queuing of packets in intermediate routers leading to zero packet loss due to buffer overflow. Thus, a guaranteed service is the requirement for a network supporting a hard real-time system as it has a strict deadline. The controlled load service is better than the best-effort service when the network is not under heavy load. The controlled load service does not provide an upper bound on the queuing delays or queuing loss. However, the service attempts to provide a high QoS by trying to minimize the packet drops and waiting time at buffers [1, 15]. It is better than a best-effort service because, during congestion, a best-effort service may fail to deliver the packet, but in the case of a controlled load service the network keeps aside a certain minimum amount of resources to process the flow even in the case of congestion. A controlled load type of service is generally used in the case of video-streaming-type applications, which are adaptive in nature and can tolerate a few packet drops and small delays.

Resource Reservation Protocol (RSVP)

The complexity and requirement of resource allocation for QoS are different in unicast and multicast data transmission. In the case of unicast transmission, the requirement is for exchange of data between two applications or hosts with assured QoS. The host can inform the network before transmission of the flow about its QoS requirement, and the network guarantees the same by reservation of resources along the transmission channel [15]. If the network is overloaded and it is not possible to assure QoS to the flow, the same may be informed back to the hosts. The host may either initiate the transmission at a reduced QoS level, which still will be better than the best-effort service, or back off and wait for some time before transmission so as to ensure release of the resources in the network.

Resource reservation for multicast transmission is more challenging, as it can produce a high volume of data either owing to the sheer volume of data generated by the application, as in the case of multilocation video conferencing, or owing to a huge number of hosts in the multicast group, leading to replication of data by the intermediate routing nodes. However, effective management in delivery of multicast data is required according to the actual requirements at the nodes of the multicast group because a node may be a member of a multicast group, which makes it eligible for receiving a flow, but it may not require the flow at the moment. A node of a multicast group may be a member of two different multicast groups and receive data from one group, but it may not require the flow from the other group. Some of the nodes of a multicast group may also not be able to handle the entire multicast data. A multicast flow of a video conference application may have audio and video components. A member of the multicast group with a low bandwidth may only be interested in the audio data.

RSVP is used in integrated services, and so it has to be implemented in all the routing nodes of the network that form a path for end-to-end QoS. The protocol runs over IP and UDP and can be used for resource reservation for a flow for unicast as well as multicast transmission. The challenges in the multicast QoS are resolved by RSVP because it is a one-way receiver-initiated protocol. The sender transmits an RSVP signal to the receiver, intimating it about the QoS from the sender. Based on this signaling, the receiver sends a reservation request to the sender for the flow at the required QoS level, which may be equal to or less than the QoS signaled by the sender. This enables the network to have different QoS levels between the sender and the members of the multicast group for the same flow [1]. As RSVP is a one-way protocol, to support QoS in interactive communication such as teleconference, videoconference, and interactive gaming, two separate RSVP sessions are initiated simultaneously, one in each direction.

Working of RSVP. In RSVP, the sender transmits a path message for resource allocation along the path by which it will send the flow. The path message is different from the flow and is used to reserve resources along the path of the flow. The path message is sent periodically by the sender along the path to enable the receiver to send a reservation request based on the path message. The path message uses the routing protocol to hop from one router to another until it reaches the destination. Routers in the path by which the path message flows keep a note of the routing node from which they have received the path message and the routing node to which they have sent the path message. This enables backward tracing of the path by the routers when a data packet comes from downstream from receiver to sender. As the path message contains the information about the QoS desired by the flow, all the intermediate routers through which the path message has been forwarded now become aware that they may be shortly requested for resource reservation along this path.

When a receiver wants to receive a message from the sender, the receiver sends a 'reservation' message back to the sender. This reservation message traverses through the same path from which the 'path message' was sent from the sender to the receiver but now in the opposite direction, i.e. from the receiver to the sender. When an intermediate router receives the reservation message, it checks the availability of the requested resources with it and the authorization of the application to reserve the resources at the requested QoS. If even one of these is not satisfied, a 'reservation error' message is sent back to the receiver. If both conditions are satisfied, the router reserves the resources for the flow and sends the reservation request message upstream towards the router in the direction of the sender. In the upstream router the same process is followed again until the reservation request reaches the sender. So, if a reservation request has reached the sender, this implies that the intermediate routers in the path have reserved resources for the requested QoS and the sender initiates the flow.

The RSVP reservations made in the intermediate routers along the path time out if they are not refreshed periodically by the RSVP using the 'refresh message'. This helps in tearing off the connection after transmission of the flow and selection of an alternative route, if found more suitable, by the dynamic routing algorithm for another flow. The sender, receiver, or the router can initiate a teardown request for the path or the resource reservation if a timeout is observed. The sender can tear off the path from it to the receiver, releasing all the reservations related to the path in the intermediate routers by sending a 'path teardown message'. The receiver can tear down the reservation of resources in the router of the path by sending a 'reservation teardown message'.

Disadvantages of the integrated services are as follows:

- The hosts as well as the networking devices have to be aware of RSVP. All the routers in the path should support RSVP. If even a single router in the path does not support RSVP, QoS cannot be assured for the flow.
- The reservation of resources along the QoS path for a flow is 'soft' and times out. It has to be periodically refreshed. These periodic 'refresh reservation' messages increase the traffic in the network.
- The state information of each reservation has to be maintained by each of the routing nodes. A node has to store the information of the previous node and the next node in the routing path for each flow. This increases the memory and computational requirements of the router.

4.4 QoS Routing Algorithms

Several QoS-based path selection algorithms [16], such as the shortest widest path (SWP) [17], widest shortest path [2], and utilization-based algorithms, have been proposed in the literature, and comparisons of their performance have also been made [5]. An overview of some of these algorithms is presented here.

SWP Algorithm

In distance vector routing [3], each node in the network is aware of the link costs of its immediate neighbors. When this information is exchanged by each of the nodes with its immediate neighbors, all the nodes are made aware of the distances to the rest of the nodes. When the algorithm stabilizes, upon updating the routing tables of each of the nodes, each node knows the costs and the next-hop nodes for the rest of the nodes in the network.

In the case of linkstate routing [3], each node is made aware of the link state information (i.e. whether the links to its neighbors are active or dead and how much the cost of each link is) by a procedure called *flooding*, so that each of the nodes can find the shortest paths to all the remaining nodes in the network. Upon stabilization of the algorithm, after flooding has disseminated all the link state information to all the nodes in the network, each node is capable of calculating the shortest path to the remaining nodes in the network. The algorithm maintains two lists for a node, called the *tentative* and the *confirmed* lists, each of which contains entries that store information about the destination node, the cost to reach the neighbor of the node and the next hop of that node. The entries labeled *tentative* are those that are being processed but are not yet confirmed to belong to the shortest path routes.

In simplified terms, the link-state-based SWP algorithm can be stated in the following manner:

1) Firstly, from the list of tentatively labeled nodes, find the nodes with the maximum width.
2) If there are several such nodes satisfying step 1, select the node with the minimum length path, and label it as *confirmed*.
3) Finally, update the maximum width (or the link residual capacity) and the length for the shortest widest path for the nodes labeled as *confirmed*.

The SWP algorithm [17] has the following properties:

1) The time complexity of the SWP algorithm is proportional to that of the corresponding shortest path algorithm.
2) The SWP algorithm can be considered as a general case of the shortest path algorithm when all links are considered to have different link capacities. In other words, if all link capacities are considered to be the same, the shortest path algorithm can be viewed as a particular case of the SWP algorithm.
3) A path's width is decided by the bottleneck link, i.e. the link in a path with the minimum residual bandwidth.
4) The paths computed using SWP algorithms do not form a loop. This property of SWP is analytically *appealing and proven*.

WSP Algorithm

The WSP algorithm was proposed by Guerin *et al.* [2]. Unlike the SWP algorithm, WSP first attempts to compute the shortest path. If there exists more than one alternative, the algorithm chooses the one with the largest residual bandwidth in the bottleneck link (i.e. the widest path) [4]. If there is still a tie with one or more such path(s), one of the prospective paths is randomly chosen [5].

This process can be described in four principal steps [2, 5, 16, 18] stated below:

Step 1. For a route request with b units of bandwidth, scan and discard all links, e, whose residual bandwidth $r_j(e) < b$.

Step 2. Apply a suitable shortest path algorithm (e.g. Dijkstra's), and compute the shortest path. If all paths have the same edge lengths (costs), then select the path with the least number of hops.

Step 3. If step 2 results in more than one choice, select the widest path with the largest residual bandwidth.

Step 4. If it turns out that there is more than one path with the same widest path, randomly select one path from among the available alternatives.

The major limitation of WSP lies principally in the fact that the algorithm is essentially a shortest path greedy algorithm. The notion of widest paths arises only when there are multiple shortest paths. On the other hand, as the SWP algorithm first selects the path that has the largest residual bandwidth, it can select very long paths to satisfy its requirement. The choice of the algorithm should depend on the network environment and requirements. In a meshed networking with intricately connected equal-length paths, WSP could be a better alternative than SWP. On the other hand, if (i) there is a general non-meshed (or even small) network, where propagation delays are small, or (ii) network resources are inexpensive, SWP could be a better alternative.

The difference between WSP and SWP clearly lies in the fact that, whereas WSP attributes higher priority to minimize the distance or the hop count, SWP places more emphasis on balancing network loads.

SELA Routing Algorithm

The stochastic estimator learning automata (SELA) routing scheme is a QoS-based scheme that was proposed by Vasilakos *et al.* [7] for use in CAC problems in ATM networks. It uses the reinforcement learning scheme called SELA for optimizing the network revenue while ensuring that the QoS requirements (e.g. *cell loss ratio*, *maximum cell transfer delay*, and *cell delay variation*) for the different connections are satisfied in the ATM networks.

In SELA routing, an automaton operates at each source node. The purpose of the automaton is to choose one of the different possible paths, represented as the different actions of the learning automata (LA) operating at the source node, for routing requests between pairs of source and destination nodes. The SELA algorithm maintains a table containing the current link utilizations for the entire network. This table helps to compute the feedback of SELA to the different possible actions. The algorithm reserves trunks for network conditions where uncontrolled alternative path routing can lead to unacceptably increased rejections of connection setup requests [7, 19]. As a result, the algorithm tries to improve its QoS performance by discouraging the loading of the congested links, unless the link happens to be in the shortest path (in terms of the number of hops) between the source and the destination nodes.

The algorithm [7] defines a parameter ρ_{TRT} that represents a predefined route utilization threshold that any route can have. The SELA algorithm routes a request through an overutilized path only if the maximum expected utilization that the links comprising a particular route can assume ($\rho_{\text{exp}}^{\text{route}}$) exceeds ρ_{TRT}, but that overutilized path happens to be a shortest distance path. If the path is not a shortest distance path, the algorithm looks for an alternative path.

The SELA algorithm utilizes a feedback function $b(t)$ for any link in a set of links comprising a route as follows [7]:

$$b(t) = 1.0 \times \text{MAX\_NO\_HOPS} - \sum_{i=0}^{n} \rho_i$$

where

$$\rho_i = \begin{cases} 1, & \text{if } \rho_{\text{exp}}^i > \rho_{\text{TRT}} \\ \rho_{\text{exp}}^i, & \text{otherwise} \end{cases}$$

ρ_{exp}^i is the expected utilization of the ith of the n available link segments comprising any route, and MAX_NO_HOPS is the maximum number of hops that is possible between any pair of source–destination nodes.

The structure of the function $b(t)$ demonstrates that increasing congestion would reduce its value. Similarly, the value of $b(t)$ increases with increasing number of hops in a route, thereby signifying that the algorithm favors minimum-hop routes [16].

4.5 QoS Unicast Routing Protocols

In a QoS unicast routing [9, 20], the data has to be sent from the source to only one destination through the feasible path that satisfies the QoS constraints. The feasible path is one that minimizes the routing cost across the network. The least cost can be based on any of the normal routing metrics such as hop count, delay, or bandwidth. The QoS constraints can be mapped in terms of link constraints and path constraints. Optimization of link constraints depends on availability of resources in the network, i.e. bandwidth, and availability of resources in the intermediate routing nodes, i.e. size of buffer, queue length, and availability of priority queue. The parameters of concern for the path metrics can be additive in nature along the path, e.g. delay, or can be multiplicative in nature along the path, e.g. packet loss or reliability.

The prime objective of QoS routing is to meet the service level parameter constraints across the network. This is achieved by trying to manage these constraints in each individual link and routing node along the path from source to destination to adhere effectively to the overall transmission constraint. QoS unicast routing may use source-based routing, distributed routing or hierarchical routing strategy. In source-based routing the sender computes, before transmitting the packet, the complete or outline path by which the packet should travel. In distributed routing, each intermediate node in the network between source and destination decides on the next link on which the packet should be forwarded. In a hierarchical network, the network is divided into a hierarchical structure forming groups, and the routing decision is made by the group leader or parent node of the tree recursively. A few [21] unicast source routing algorithms are the Wang–Crowcroft algorithm [22], the Ma–Steenkiste algorithm [23], the Chen–Nahrstedt algorithm [24], the Awerbuch *et al.* algorithm [25], and the Guerin–Orda algorithm [26], and a few unicast distributed routing algorithms are the Wang–Crowcroft algorithm [22], the Salama *et al.* algorithm [27], the Cidon *et al.* algorithm [28], and the Shin–Chou algorithm [29]. The commonly used hierarchical routing algorithm is the private network–network interface (PNNI). In all three cases the decision of the routing node is based on the QoS constraints as well as path optimization. When the routing decision is made, a single feasible path may be selected or multiple feasible paths satisfying the QoS constraints may be obtained, of which one is selected based on certain criteria as per the requirements. QoS unicast routing can also be categorized as distance vector routing, where the nodes have information only about their neighbors, or link state routing, where the routing node is aware of the complete network topology.

A QoS unicast routing protocol is broadly classified on the basis of the routing metrics into two categories: link-constrained routing and link optimization routing. A link optimization routing finds a path from source to destination in which the constraint metrics are satisfied, i.e. the path metrics of the QoS parameter are above the required values if the assured level of service has to be more than the constrained parameter, e.g. bandwidth, and the path metrics of the QoS parameter are below the required value if the assured level of service has to be less than the constrained parameter, e.g. delay. A link optimization routing finds the path that maximizes or minimizes the path constraint from the source to all possible destinations in the network. An example of path optimization routing is Delay-Constrained Least-Cost Routing Protocol, and an example of link-constrained routing is Delay-Constrained Unicast Routing Protocol.

Delay-Constrained Unicast Routing (DCUR) Protocol. DCUR is a distributed protocol where each routing node decides only the next-hop node. The protocol is initiated by the sender, and the path of packet traversal is decided node-wise as it moves from the source to the destination, i.e. it is a forward propagating protocol [27]. The protocol minimizes the delay as well as the cost of the routing path. If the least-cost path does not satisfy the delay constraint, the protocol selects the least-delay path instead of the least-cost path. This helps to satisfy the QoS constraint by not forwarding the data packet on a path that will exceed the committed parameter constraint and violate the service agreement. In this protocol, each routing node uses cost as well as the delay vector table to decide on the next-hop path. The routing node first calculates the least-cost path and uses the same if it satisfies the delay constraint. If the delay constraint is not satisfied by the least-cost path, the routing node selects the least-delay path and forwards the packet to the next node by the least-delay path. An intermediate node can decide only between the following two choices of path – least-cost path or least-delay path. The routing node is not capable of selecting an optimum path by minimizing the cost as well as the delay. This has been done to retain the simplicity of the protocol and reduce the computational complexity. The protocol is helpful in communication between nodes in a distributed real-time application.

A special implementation of the protocol where all the nodes select the least-cost path satisfying the delay constraint will make this protocol similar to distance vector protocol. However, generally, all the paths may not be selected on the basis only of the least cost or only of the least delay. With either least cost or least delay, looping would not be present as in the case of distance vector routing. But as some of the paths are calculated on the basis of the least-cost vector and the remainder are calculated on the basis of the least-delay vector, looping may occur in the protocol. A loop is detected by a node if it receives back a packet that it has already forwarded for the source–destination pair. In such a case it informs the node from which it has immediately received the packet by sending it a 'remove loop' message. The 'remove loop' message propagates backward till the path recovers from the loop by following a least-cost path.

As DCUR cannot optimize delay and cost, it fails to select a link that may have a better overall delay as well as cost performance. Hence, the protocol may become very costly owing to the selection of the least-delay paths only in the case of a strong delay constraint. As the protocol depends on the cost as well as the delay vector routing tables, the accuracy of both tables should be maintained. It also increases the memory and computational requirements. Moreover, it has to keep a record of the packets forwarded along with its source–destination pair so as to initiate or participate in the loop avoidance process.

Delay-Constrained Least Cost (DCLC) Routing Protocol. DCLC routing protocol [30] is a source-based routing and overcomes the drawbacks of DCUR routing by optimizing the delay as well as the cost constraint during path selection. The two parameters are negotiated to reduce the overall delay. The protocol works on the assumption that there is a positive delay and a positive cost associated with each link in the network. The QoS puts a constraint on the maximum amount of delay (d) that can be tolerated in the network for a particular source–destination pair. The protocol finds a path with minimum cost, subject to satisfying the delay constraint of less than d between the source–destination pair. Each node keeps a cost and a delay vector for the best next-hop node for any destination in the network. The source node sends a control message for construction of a delay-constrained path. Any node can select either the least-cost path or the least-delay path. However, the least-cost path has priority over the least-delay path as long as the delay constraint is not violated. Thus, the protocol acts as a multiconstraint path problem that is bounded by delay constraint, and the path cost has to be optimized.

4.6 QoS Multicast Routing Protocols

QoS multicast routing protocols are required to provide a path from a source to many destinations under constrained parameters as defined in the SLA. Multicast QoS routing is required for applications that require group communication with an assured level of service in terms of certain QoS parameters. A few examples of such services are applying software upgrades in a time-bound manner to critical application, multiparty teleconferencing with assured maximum permissible delay in transmission, multilocation video conferencing with assured minimum bandwidth, jitter-free online diagnostics, and hard real-time image and video transmission for tele medicine and tele surgery. The nodes of the multicasting group can be confined within the LAN or can be spread across the WAN. In QoS multicast routing, the source node has to find the best feasible path from itself to all the nodes in the multicast group that satisfies the QoS constraints. The multicast path is in the form of a tree connecting the source to all the other multicast group nodes. In multicast QoS routing, the entire tree has to be optimized, unlike unicast QoS routing in which a single optimized path is calculated. Thus, the multicast routing protocol has to build a minimum cost distribution tree without violating the QoS constraints through which the data packets will be sent to the multicast group nodes.

The essential features [31] of a robust and scalable multicast routing protocol are as follows:

- The protocol should optimize between routing overhead and routing performance.
- Traffic should not be forwarded through the congested network areas.

- The protocol should attempt to minimize the state information required to be stored at each node.
- The routing protocol should not only adapt to the changing network load but also converge quickly in the case of changing load or topology.
- The protocol should support high scalability of the network.
- Decentralize the computation over several nodes for faster decision-making and prevent a single point of failure.
- It should enhance the probability of successfully locating a path from source to destination.
- The protocol should minimize the cost of the selected path as well as minimize the delay over the selected path.

Based on various properties, multicast groups have been differentiated [32] into the following:

- dense groups versus sparse groups, based on the number of multicast group members per link;
- open groups versus closed groups, based on the permission available to a sender to transmit a message to a multicast group even if it is not a member of the group or only if it is a member of the group;
- permanent versus transient groups, based on the duration of existence of the multicast group;
- static groups versus dynamic groups, based on the nature of membership of the nodes in the multicast group and their permission to join or leave a group and the adaptability of the group to accept them on joining and delete them on leaving the group.

A multicast routing protocol can be static or dynamic. In a static multicast routing protocol, all the members of the multicast group must be known to the protocol, and thereafter the multicast tree is constructed. Dynamic multicast routing protocol allows the nodes to join the multicast group while the tree is being formed. The tree is not entirely reconstructed for every node joining the multicast group, but an attempt is made to connect the newly joined member to an existing node of the multicast group. The choice of the routing protocol to be used is also dependent on the type of application using the multicast routing protocol, e.g. a dynamic protocol is preferred for an application such as online live radio, video streaming, and webcasting, where users keep joining and leaving the multicast group.

The following stepwise process [32] is followed for setting up a multicast group:

1) *Creation of a multicast group.* A unique address is provided to each group to avoid any clash of data between groups. An address assigned permanently is known as a static address, whereas an address assigned temporarily to the group is considered to be dynamic. Generally, static addresses are assigned to permanent multicast groups and dynamic addresses are assigned to transient multicast groups. This kind of matching of type of multicast group and type of address avoids insecure or unnecessary communication.

2) *Creation of a multicast QoS tree.* The next phase in multicast routing is the creation of a tree from the source covering all the receivers and reserving resources on the tree. The tree structure has been adopted as it allows parallel transmission down to the various nodes. The formulated tree should also satisfy the QoS requirements. A tree that satisfies the QoS requirements best is selected.

3) *Data transmission.* After creation of the multicast group and creation of the multicast QoS tree, the following tasks [32] are performed during data transmission:

- *Group dynamics.* The network should be able to detect each member joining or leaving the multicast group during the session's lifetime. Once the members have been detected, wasteful transmission of data can be stopped to members who have left the group and the data can be forwarded to newly joined group members.
- *Network dynamics.* The network should be able to detect node and link failures and restore the tree, bypassing faulty nodes with assured QoS.
- *Transmission problems.* Faulty packet transmissions are identified. Some control activities are necessary to overcome these transmission problems.
- *Competition among senders.* Resource contention occurs when a number of senders have to send their data along the same tree. A session control mechanism is required to prevent the service falling below QoS in such cases owing to unavailability of sufficient resources that can be reserved. Such a session control arbitrates transmission among senders.

Multicast routing protocols can be categorized into three main approaches: source-based protocols, center-based protocols, and hybrid protocols [32, 33]. The source-based approach creates a shortest path tree routed at the source, and each branch of the tree represents the shortest path from the source to each receiver. Owing to QoS constraints, the shortest path is the shortest delay path. However, source-based trees become too complex as the size of the network scales because a tree has to be created from each individual source spanning all the receivers. A few examples of source-based multicasting routing protocols are Distance Vector Multicast Routing Protocol (DVMRP), Protocol-Independent Multicasting Dense Mode (PIM-dense), Multicast Open Shortest Path First (MOSPF), and Explicitly Requested Single-Source Multicast (EXPRESS). The center-based protocol is also known as the shared-tree protocol, and it constructs a multicast tree spanning all the members and is rooted at a center node. The center-based protocol is highly scalable. When a node wants to send some data to all other nodes in the multicast group through the center-based protocol, it sends the data packets to the center. The data packet traversing from the source node to the center node passes through the other multicast group receiver nodes in the path from the source to the receiver. Once the data packet reaches the receiver, it is sent to those nodes along the tree that have not received the data packet earlier. Core-Based Tree (CBT) and PIM Sparse Mode (PIM-SM) are core-based multicast tree protocols. A hybrid protocol gives the receiver flexibility to switch between the center-based protocol and source-based protocols, and an example of this is Multicast Internet Protocol (MIP). The overall working of a source-based protocol and a center-based protocol is described to give an understanding of the procedure.

DVMRP. The protocol uses a reverse path multicast (RPM) algorithm [33] to build its source-based tree. In an RPM algorithm, when an intermediate routing node receives a packet, it forwards the packet to the other nodes connected to it only if it has received the packet from a link that is in the shortest path. This way the datagram is forwarded across the entire network from the source. A leaf node is one that does not have any multicast group nodes attached to it. When a leaf node receives the message, it sends a 'prune' message back along the path from which the packet was received. The 'prune' message traverses up to the source node and helps in constructing the forwarding table in all the intermediate routing nodes. As the 'prune' message times out after a certain periodicity, the packet is again forwarded throughout the network and the 'prune' message is received back from the leaf nodes. When a node joins a multicast group, the node to which it joins sends a 'graft' message upstream in the network to enable all the routing tables to update the forwarding table.

Core-based tree. The core-based tree [33] has a single shared tree spanning across all the nodes of the multicast group and is centered at the core. When new node joins a multicast group, the routing node to which it joins sends a 'join' message to the router above it in the shortest path towards the core and enters a transient join state for this new node in its forwarding entries. As the 'join' message reaches the core, it sends a 'join_ack' message back to the new node by the same path. When an intermediate routing node gets a 'join_ack' message for the new node, it forwards the same on the shortest path towards the router that initiated the 'join' message. The intermediate node also converts the transient entry for the new node into a permanent one, making the new node a part of the multicast spanning tree.

The way a new member is connected to the tree is used to classify the routing protocol into Single-Path Routing Protocols (SPRs) and Multiple-Path Routing Protocols (MPRs). SPR provides a single path for joining a new member to the tree, whereas MPR provides multiple candidate paths. Some of the recently proposed protocols [31, 32] based on MPR are as follows.

Spanning joins. A new member broadcasts 'join request' messages to all its neighbors, based on which a reply message is sent back to this new member by the on-tree nodes. The path of reply is treated as a candidate path. As the new member may receive multiple reply messages corresponding to multiple candidate paths, it selects the best path based on the QoS properties of the path received in the reply.

Quality-of Service-Multicasting (QoSMIC) Protocol. Two parallel procedures are followed for selecting the candidate paths – local search and tree search. A local search is the same as a spanning join and is used to search a small neighborhood. When an on-tree node is not found by the local search, a tree search is initiated. In the tree search, the new node contacts a designated core node which informs a few of the on-tree nodes to establish a path from them to the member. The new member then selects the best path out of these candidate paths.

Spanning joins and QoSMIC protocols are QoS-unaware protocols because QoS requirements are not considered for selection of the candidate path and the on-tree node is selected on the basis of minimum hop connectivity.

Quality-of-Service Multicast Routing Protocol (QMRP). QMRP is a QoS-aware protocol. QMRP consists of two searching modes for construction of the search tree single-path mode and multipath mode. The protocol uses single-path mode until it reaches a node that does not have sufficient resources for the assured QoS. At this stage, the protocol initiates the multipath mode for building the search tree further until it reaches the multicast tree. A QMRP tree is constructed from a new node that wants to join, and the construction of the tree moves towards the multicast tree. As QMRP considers only bandwidth constraints, the paths being built in the search tree satisfy the bandwidth requirement. A feasible path for connecting the new node is discovered when the growing search tree touches the pre-existing multicast tree.

These protocols either do not satisfy QoS constraints or satisfy them to a very little extent. A protocol that satisfies the properties of a good multicast routing protocol is RIMQoS.

Receiver-Initiated QoS Multicast Protocol (RIMQoS). This protocol constructs a multicast tree in a distributed fashion [34] and has the following characteristics:

- For a new node to join, no prior information about the existing nodes is required.
- Minimum resource reservation is needed.
- The existing tree is adjusted minimally.
- The 'join request' causes minimum overhead.
- The overall cost of the tree is to be minimized.
- RIMQoS works on top of any existing QoS unicast routing protocols that compute the QoS paths.

RIMQoS protocol is a single-source multicast application that satisfies QoS requirements. A router keeps forwarding information about incoming interface and outgoing interface set to all nodes in the network. When a new member wants to join, it sends a 'join request' to the source node. Additionally, in a forwarding entry, each outgoing interface is labeled as 'live' or 'on hold'. Multicast traffic is only forwarded to outgoing interface marked as live. Two types of notification are required: live notification and deny notification. When a new node wants to join in and the request is accepted, a live notification is sent to the new receiver. If the request is denied, a deny notification is sent to the receiver. The resources required for the assured QoS are reserved if the notification received is live.

4.7 QoS Best-Effort Routing

Best-effort routing attempts to provide the best feasible path from source to destination without guarantee of any assured level of service. Although a best-effort routing does not fulfill QoS requirements, it can be modified to satisfy minimum constraints to cater effectively for the assured level of service. On the Internet, which follows best-effort routing, a single best path towards the destination is computed by each intermediate routing node [35]. The selected path changes with change in the network topology or congestion in any part of the network. As the path is selected and the packets are delivered on a hop-by-hop basis, the packets may reach out of order of delivery and may even get dropped in any of the intermediate routers owing to load condition and unavailability of buffer space. Unlike the single-best-path selection strategy used on the Internet, best-effort QoS in a network requires multiple paths to a destination to support various levels of performance. The routing requires a path selection algorithm that efficiently computes a set of best-effort paths, and forwards the traffic over multiple paths to the destination. However, routing over multiple paths faces the challenge of loop detection and avoidance. Moreover, the best set of paths have to be recomputed if there are changes in topology, including node or link failures.

Load-sharing routing (LSR) [36] is one of the algorithms proposed for best-effort QoS routing. This algorithm tries to improve the QoS only in those regions of the network where it has deteriorated owing to congestion. The overhead of the routing algorithm is reduced by making only the neighboring nodes of a congested node implement the LSR algorithm. In this routing algorithm, each routing node maintains an active routing table and a passive routing table. The active routing table is created from the node's awareness of the network topology and gives the shortest path based on Dijkstra's algorithm from the node to any other node in the network. These paths are referred to as

OSPF paths in this protocol. The passive routing table keeps the shortest distance to all other nodes from each of its neighboring nodes as the source node. This is also calculated using Dikstra'a algorithm by keeping the neighbor as the source node. The entry in the routing table comprises the destination node, the next-hop node, the total cost, the hop count, the OSPF next hop, and the LSR flag. When the LSR flag is false, the node forwards the packet as per the OSPF algorithm and the next-hop node and the OSPF next-hop node are the same. When the LSR flag is true, the next-hop node is calculated on the basis of the LSR algorithm. The algorithm uses a number of different messages to calculate the path. The working of the algorithm can be well understood by referring to these messages.

When a node detects congestion beyond a threshold on any one of its links, it sends congestion notification about the node to all its neighbors. The neighbors that receive a congestion notification send a reroute request message to any one of the neighboring nodes to inform it that the router will reroute the packets meant for the node on the congested link to this router to which it is sending the reroute message. Once the congestion is cleared, a 'congestion over' message is sent to indicate normal behavior of the link. On receipt of the 'congestion over' message, the intermediate router sends a 'reroute over' message to its neighbor to which it had been rerouting the packet. When a router receives a reroute notification, it sets the LSR flag to true for the congested destination and starts rerouting the packets by the alternative path.

An alternative path routing algorithm [37, 38] that supports loop-free routing is LSR. An LSR coefficient, which is based on operating parameters, has to be decided for the LSR protocol, which then decides the number of multiple paths for a particular destination. The LSR parameter can be global across the network or local to a routing node. The aim of the approach is to maximize the number of alternative paths to enable at least one alternative path for each node. However, search for the LSR coefficient is exhaustive and time consuming.

A protocol that reduces the search for the LSR coefficient to only potential areas is E-LSR (efficient LSR) [39]. The cost of the links and the topology of the network are the parameters for deciding the alternative path. The algorithm uses the OSPF model to make the routing table. The entries in the routing table are ospfcost, ospfhopcount, and nexthop for a particular destination. Ospfcost is the cost of the ospf path to the destination. Ospfhopcount is the number of hops along the selected path. The nexthop is set to ospf nexthop to forward packets along the ospf path. There are two control messages used by the E-LSR algorithm. A 'congestion notification' message is sent by a node to all its neighbors (except the one connected to it over the congested link) when it detects congestion on that outgoing link. When a link that was congested earlier is no longer congested, a 'congestion over' message is sent out to all the neighbors. The working of the protocol is similar to that of the LSR algorithm.

References

1 L. Parziale. *TCP/IP Tutorial and Technical Overview*. IBM Redbooks. International Technical Support Organization, 8th edition, 2006.
2 R. Guerin, A. Orda, and D. Williams. QoS routing mechanisms and OSPF extensions. *2nd Global Internet Miniconference*, Phoenix, November 1997.

3 L. Peterson and B. Davie. *Computer Networks: A Systems Approach.* Morgan Kaufman Publishers, 2nd edition, 2000.

4 S. Misra. Quality-of-service routing, Chapter 509, pp. 3186–3190, in M. Khosrow-Pour (ed.). *Encyclopedia of Information Science & Technology*, IGI-Global, 2nd edition, 2008.

5 Q. Ma. *Quality-of-Service Routing in Integrated Services Networks.* PhD thesis, School of Computer Science, Carnegie Mellon University, January 1998.

6 R. Bellman. On a routing problem. *Quarterly of Applied Mathematics*, **16**(1):87–90, 1958.

7 A. Vasilakos, M.P. Saltouros, A.F. Atlassis, and W. Pedrycz. Optimizing QoS routing in hierarchical ATM networks using computational intelligence techniques. *IEEE Transactions on Systems, Man, and Cybernetics, Part C*, **33**(3):297–312, 2003.

8 P. V. Mieghem (ed.), F.A. Kuipers, T. Korkmaz, M. Krunz, M. Curado, E. Monteiro, X. M. Bruin, J. S. Pareta, and J. D. Pascal. Quality of service routing. *Quality of Future Internet Services*. Springer, 2003.

9 S. Chen and K. Nahrstedt. An overview of quality of service routing for next-generation high-speed networks: problems and solutions. *IEEE Network*, **12**(6): 64–79, 1998.

10 A. V. Kumar and S. G. Thorenoor. Analysis of IP network for different quality of service. *International Symposium on Computing, Communication, and Control (ISCCC 2009)*, CSIT, Singapore, 2009.

11 J. Saliba, A. Beresford, M. Ivanovich, and P. Fitzpatrick. User-perceived quality of service in wireless data networks. *Personal and Ubiquitous Computing*, **9**(6):413–422, 2005.

12 A. Jamalipour. *The Wireless Mobile Internet: Architectures, Protocols and Services.* John Wiley & Sons, Inc., 2003.

13 J. D. Chalmers and M. Sloman. A survey of quality of service in mobile computing environments. *IEEE Communications Surveys*, **2**(2):2–10, 1999.

14 Quality of service indicators. GSM mobile networks – quality of service survey, Portugal: Autoridade Nacional de Comunicações, October 2002.

15 W. Stallings. *Data and Computer Communications.* Prentice Hall, 8th edition, 2007.

16 S. Misra, Network applications of learning automata. PhD dissertation, School of CS, Carleton University, Ottawa, Canada, 2005.

17 Z. Wang and J. Crowcroft. QoS routing for supporting resource reservation. *IEEE Journal on Selected Areas in Communications*, **14**(7):1228–1234, 1996.

18 S. W. H. Wong. The online and offline properties of routing algorithms in MPLS. Master's thesis, Department of Computer Science, University of British Columbia, 2002.

19 F. Atlasis, M. P. Saltouros, and A. V. Vasilakos. On the use of a stochastic estimator learning algorithm to the ATM routing problem: a methodology. *IEEE GLOBECOM*, December 1998.

20 C. Segui, A. Zaballos, X. Cadenas, and J. M. Selga. Evolution of unicast routing protocols in data networks. *IEEE INFOCOM*, Barcelona, Spain, 2006.

21 S. Chen and K. Nahrstedt. Distributed quality-of-service routing in ad hoc networks. *IEEE Journal on Selected Areas in Communications*, **17**(8):1488–1505, 2006.

22 Z. Wang and J. Crowcroft. QoS routing for supporting resource reservation. *IEEE Journal on Selected Areas in Communications*, **14**(7):1228–1234, 1996.

23 Q. Ma and P. Steenkiste. QoS routing for traffic with performance guarantees. *IFIP International Workshop on Quality of Service*, pp. 115–126, Springer, 1997.

24 S. Chen and K. Nahrstedt. On finding multi-constrained paths. *IEEE International Conference on Communications ICC 98, Volume 2*, pp. 874—879, 1998.

25 B. Awerbuch, Y. Azar, S. Plotkin, and O. Waarts. Throughput-competitive online routing. *34th Annual Symposium on Foundations of Computer Science*, pp. 32–40, Palo Alto, CA, 1993.

26 R. Guerin and A. Orda. QoS-based routing in networks with inaccurate information: theory and algorithms. *IEEE INFOCOM '97. 16th Annual Joint Conference of the IEEE Computer and Communications Societies. Driving the Information* Revolution, Volume 1, pp. 75–83, Japan, 1997.

27 H. F. Salama, D. S. Reeves, and Y. Viniotis. A distributed algorithm for delay-constrained unicast routing. *IEEE INFOCOM '97. Sixteenth Annual Joint Conference of the IEEE Computer and Communications Societies. Driving the Information Revolution, Volume 1*, pp. 84–91, Japan, 1997.

28 I. Cidon, R. Rom, and Y. Shavitt. Multi-path routing combined with resource reservation. *IEEE INFOCOM '97. Sixteenth Annual Joint Conference of the IEEE Computer and Communications Societies. Driving the Information Revolution, Volume 1*, pp. 92–100, Japan, 1997.

29 K. G. Shin and C. C. Chou. A distributed route-selection scheme for establishing real-time channel. *6th IFIP International Conference on High Performance Networking (HPN'95)*, pp. 319–329, 1995.

30 S. Selvan and P. S. Prakash. Optimized delay constrained QoS routing. *Advances in Modeling – Computer Science and Statistics, Volume 14*, 2009.

31 S. Chen and Y. Shavitt. SoMR: A scalable distributed QoS multicast routing protocol. *Journal of Parallel Distributed Computing*, **68**(2):137–149, 2008.

32 A. Striegel and G. Manimaran. A survey of QoS multicasting issues. *IEEE Communications Magazine*, **40**(6):82–87, 2002.

33 B. Wang and J. C. Hou. Multicast routing and its QoS extension: problems, algorithms, and protocols. *IEEE Network*, **14**(1):22–36, 2000.

34 A. Fei and M. Gerla. Receiver-initiated multicasting with multiple QoS constraints. *IEEE INFOCOM 2000. 19th Annual Joint Conference of the IEEE Computer and Communications Societies, Volume 1*, pp. 62–70, 2000.

35 B. R. Smith and J. J. Garcia-Luna-Aceves. Best effort quality-of-service. *17th International Conference on Computer Communications and Networks, ICCCN '08, Volume 1*, pp. 3–7, 2008.

36 A. Sahoo. A load-sensitive QoS routing algorithm in best-effort environment. *IEEE MILCOM, Volume 2*, pp. 1206–1210, 2002.

37 A. Sahoo. An OSPF based load sensitive QoS routing algorithm using alternate paths. *Computer Communications and Networks*, pp. 236–241, 2002.

38 A. Tiwari and A. Sahoo, A local coefficient based load sensitive routing protocol for providing QoS. *12th International Conference on Parallel and Distributed Systems (ICPADS '06), Volume 1*, p. 8, Washington, DC, 2006.

39 A. Tiwari and A. Sahoo. Providing QoS support in OSPF based best effort network. *13th IEEE International Conference on Networks. Jointly held with the IEEE 7th Malaysia International Conference on Communication*, Malaysia, 2005.

Abbreviations/Terminologies

ATM	Asynchronous Transfer Mode
CAC	Connection Admission Control
CBT	Core-Based Tree
DCLC	Delay-Constrained Least Cost (Routing Protocol)
DCUR	Delay-Constrained Unicast Routing (Protocol)
DVMRP	Distance Vector Multicast Routing Protocol
E-LSR	Efficient Load-Sharing Routing
ETSI	European Telecommunications Standards Institute
EXPRESS	Explicitly Requested Single-Source Multicast
FIFO	First In First Out
FTP	File Transfer Protocol
HTTP	Hypertext Transfer Protocol
IETF	Internet Engineering Task Force
IPv4	Internet Protocol Version 4
ISP	Internet Service Provider
LA	Learning Automata
LAN	Local Area Network
LSR	Load-Sharing Routing
MIP	Multicast Internet Protocol
MOSPF	Multicast Open Shortest Path First
MPR	Multiple-Path Routing
NMS	Network Management System
NP	Network Performance
OSPF	Open Shortest Path First
PHB	Per Hop Behavior
PIM	Protocol-Independent Multicasting
PIM-dense	Protocol-Independent Multicasting Dense Mode
PIM SM	Protocol-Independent Multicasting Sparse Mode
PNNI	Private Network–Network Interface
QMRP	Quality-of-Service Multicast Routing Protocol
QoS	Quality of Service
QoSMIC	Quality-of-Service Multicasting
RIMQoS	Receiver-Initiated QoS Multicast Protocol
RPM	Reverse Path Multicast (algorithm)
RSVP	Resource Reservation Protocol
SELA	Stochastic Estimator Learning Automata (Routing)
SLA	Service Level Agreement
SMTP	Simple Mail Transfer Protocol
SPR	Single-Path Routing
SWP	Shortest Widest Path
TCP/IP	Transmission Control Protocol/Internet Protocol
ToS	Type of Service
UDP	User Datagram Protocol
VoIP	Voiceover Internet Protocol
WAN	Wide Area Network
WSP	Widest Shortest Path (Algorithm)

Questions

1 Why is QoS required and what are the advantages of QoS?

2 Mention various QoS parameters defined for various criteria and applications.

3 Describe the packet header used in differentiated services.

4 As defined by IETF, what are the two models for implementing QoS?

5 Explain the following five processes of traffic regulation – classification, metering, marking, shaping, and dropping.

6 Mention the advantages and disadvantages of differentiated services and integrated services.

7 What are the properties of the SELA routing algorithm?

8 Name a few unicast source routing algorithms.

9 What are the essential features of a robust and scalable multicast routing protocol?

10 Explain the process of creation of a multicast group.

11 Explain best-effort routing. Also, describe any best-effort routing protocol.

12 Differentiate between source-based and center-based multicast routing protocol, along with one example of each protocol.

13 Explain the following:
    i tree constraint for QoS in multicasting,
    ii traffic shaping,
    iii service level agreement,
    iv per hop behaviour,
    v Resource Reservation Protocol (RSVP),
    vi properties of the shortest widest path (SWP) algorithm,
    vii widest shortest path (WSP) algorithm,
    viii SELA routing algorithm,
    ix Delay-Constrained Unicast Routing Protocol,
    x Delay-Constrained Least-Cost Routing Protocol.

14 Differentiate between the following:
    i elastic traffic vs inelastic traffic,
    ii link constraint vs path constraint,
    iii interior router vs boundary router,
    iv differentiated services vs integrated services,
    v guaranteed service vs controlled load service in an integrated service,

    vi tentative vs confirmed list of nodes in the SWP algorithm,

    vii link optimisation routing vs link-constrained routing in QoS unicast routing,

    viii single-path vs multipath mode of construction of a search tree in QMRP.

15 State whether the following statements are true or false and give reasons for the answer:

    i The number of routing hops in a network can be one of the QoS metrics.

    ii The QoS routing algorithm uses information provided by QoS routing protocol to calculate the path by which the packet should be forwarded for delivery with assured quality.

    iii Some of the QoS parameters can be defined as excellent connectivity, low delay, assured bandwidth, and jitter-free service.

    iv The differentiated service code point uses eight bits that were previously used for type of service in the IPv4 packet header.

    v 'Flow' in an integrated service is the bidirectional data stream between two applications.

    vi The receiver can tear down the reservation of resources in the router of the path by sending a 'path teardown' message.

    vii Shortest widest path and widest shortest path are QoS-based path selection algorithms.

    viii Once a loop is detected by a node in Delay-Constrained Unicast Routing Protocol, a 'remove loop' message is used to recover from a loop.

    ix QoS best-effort routing guarantees an assured level of service.

    x Multicast routing protocols can be source based or destination based.

5

Routing and MPLS Traffic Engineering

5.1 MPLS Fundamentals

Traffic Engineering (TE) is a set of methods that work on the actual interconnected network and optimize the network resource utilization for efficient flow of traffic through the physical network. It is achieved by minimizing the overutilization of network capacity and distributing the traffic load on costly network resources such as links, routers, switches, and gateways. In the context of routing, TE is of very useful because traditional routing techniques are based on greedy shortest-path computation techniques that lead to overutilization of certain network resources, even when other resources remain underutilized.

Multiprotocol label switching (MPLS) is an IETF standard that merges the control protocols available at layer 3 and the speed of operations available at layer 2 of an IP network. MPLS creates an autonomous system wherein the data packet is tagged so as to ensure its faster processing in the path. These tags are called labels, and backbone switch is not required to see the IP header to calculate the path of the packet; rather the label is used for this, which increases the speed of packet forwarding by avoiding the need to read the packet header.

There are two types of router in an MPLS network. The routers at the periphery of the MPLS cloud are used to connect to the IP traffic coming into the MPLS cloud and going out of it. The MPLS routers at the periphery are called label edge routers (LERs), and

Network Routing: Fundamentals, Applications, and Emerging Technologies, First Edition.
Sudip Misra and Sumit Goswami.
© 2017 John Wiley & Sons Ltd. Published 2017 by John Wiley & Sons Ltd.
Companion website: www.wiley.com/go/misra2204

the routers inside the MPLS cloud create the MPLS backbone and are known as label switched routers (LSRs). A packet enters an MPLS network from an ingress LER and is tagged with a fixed-length label. It then moves through the LSRs until it moves out of the MPLS network through an egress LSR, which removes the label. The path traversed by a packet inside the MPLS network is termed the labeled switched path (LSP), and there is a Label Distribution Protocol (LDP) in the MPLS that monitors the exchange of labels and maps labels with destination. The LSPs are created by specifically adding one-by-one the LSRs in it, based on the assurance of service expected at each router to ensure the compliance of overall QoS from the ingress router to the egress router.

The traditional routing algorithms that forward packets based on the information about the destination address only are short sighted and are therefore constrained [1] by the following limitations:

- They do not take into consideration the current flows, or expected future flow demands in the network. These algorithms are greedy, in the sense that they would route a request through the default shortest path, the shortest-widest path, or the widest-shortest path, and would reject a demand when the precomputed routes using these algorithms are congested, even when other alternative routes are free to accept more requests.
- They do not take into consideration information about network infrastructure, network topologies, or traffic profiles to avoid loading bottleneck links in a network that might lead to rejection of future demands.
- The previous algorithms will perform negatively when they operate in an online routing situation, where the tunnel setup requests arrive one at a time and the future demand is unknown. These algorithms require knowledge of future demands to operate successfully.
- The previous algorithms are not adaptive to possible link failures. Therefore, in a situation where a link fails, these algorithms will not be able to route requests through alternative routes.

5.2   Traffic Engineering Routing Algorithms

Online routing using TE principles has drawn considerable attention. Of all the online TE algorithms [1–5], the one that has attracted most attention is the minimum interference routing algorithm (MIRA) designed by Kar *et al.* [2]. In addition to the MIRA algorithm, there are a few other TE routing algorithms that were proposed by other researchers, some of which are: (i) profile-based routing (PBR) [1, 6], (ii) the dynamic online routing algorithm (DORA) [7], (iii) Iliadis and Bauer's algorithm [5], (iv) Wang *et al.*'s algorithm [3], and (v) Subramanian and Muthukumar's algorithm [8].

Some of the properties that a 'practically useful' TE algorithm (based on MPLS) should have are as follows [2]:

1) The algorithm should be based on an online routing model, where LSP setup requests arrive one at a time (not all at once) and the future demand is unknown. On the other hand, in an offline model, all LSP setup requests are known *a priori*, and there are no demands for future LSP setup requests.
2) It should be able to use knowledge about the *physical locations* of the ingress–egress router pairs through which an LSP is set up.

3) The algorithm should be able to *adapt to possible link failures* in the network. In other words, a good TE algorithm should be able to reroute post-failure requests through alternative routes.

4) It should be able to route the requested bandwidth without splitting the demands as much as possible through multiple paths. This is necessary because it is often the case that the nature of the traffic does not allow a demand to be split. Splitting traffic is, however, a common practice in scenarios such as load balancing and network performance improvement.

5) The algorithm should, if possible, support a distributed implementation, where instead of performing the route computation in a centralized server the computations of each LSP's route request are distributed at the local ingress node.

6) It is quite desirable for such an algorithm to be capable of using different *policy constraints*. For example, service level agreements might impose a restriction that LSPs with less than a threshold value of flow guarantees should not be accepted.

7) Such an algorithm should operate under strict bounds of computational complexity. The algorithms should be very fast and execute within a fixed time constraint. The amount of computation involved with LSP setup requests should be minimized so that the algorithm can be implemented on a router or a route server.

5.3 Minimum Interference Routing Algorithm

The most influential online routing algorithm in TE is the MIRA algorithm [9]. MIRA [9–11] is online because it does not require *a priori* knowledge of the tunnel requests that have to be routed. The algorithm is targeted towards applications such as the setting up of LSPs in MPLS networks. The algorithm helps service providers set up bandwidth-guaranteed tunnels with ease in their transport networks. The term 'minimum interference' used in the nomenclature of the algorithm indicates that the tunnel setup request is to be routed through a path (or a path segment) that must not interfere too much with future tunnel setup requests. The algorithm aims to protect the critical links in a network from being overloaded, thereby reducing the chances of rejection of future requests. The critical links are identified by the algorithm to be those that, if congested because of heavy loading, might lead to rejection of requests. One characteristic that is particularly attributable to MIRA is that it is the first algorithm to use information about ingress–egress pairs. Unlike its predecessors, MIRA uses any available information about ingress–egress nodes for potential future demands. MIRA is based on the core concepts of max-flow and min-cut computations.

Max-flow. A *flow* [10–13] in a communication network can be defined as the rate at which information flows in the network. A *max-flow* can be defined as the maximum rate at which information can be transmitted over a link in the network by satisfying the capacity constraints. Intuitively, in a max-flow problem, we aim to design the network in such a way that the flow should not violate the capacity constraints, and the algorithm has to transmit the maximum possible flow from a source node to a sink node.

Critical links. If information has to pass through a minimum-hop route, there would be *interference* at the bottleneck link. MIRA attempts to minimize this interference due to routing of requests through the bottleneck links. The bottleneck links are called *critical links* [2, 11] and are computed using the min-cut concept described below. The set of critical edges between a pair of source and sink nodes is the minimum cut between that pair. MIRA defers the loading of critical links as much as possible.

Cut. The 'cut' in 'flow networks' indicates partitioning of the nodes in the network into two subsets [10]. If N is a set of nodes in a flow network, a cut involves partitioning [9] all the nodes of the network into two subsets, S and $S' = N - S$, such that $S \cup S' = N$.

s–t cut and min-cut. A cut is referred to as an *s-t* cut [10] if s belongs to S (i.e. one subset of the nodes) and t belongs to S' (i.e. the other subset of nodes N). The capacity of all the forward arcs in the cut is referred to as the *capacity* of the *s–t* cut. The minimum capacity among the capacities of all *s-t* cuts is referred to as the *min-cut.*

Max-flow min-cut theorem. This theorem states that: 'The maximum value of the flow from a source node s to a sink node t equals the minimum capacity among all *s–t* cuts' [10].

5.3.1 The Algorithm

Suppose we have a network represented by a set of N nodes, L links, and a set B representing the bandwidth between the links. Of all the N nodes, only a subset of them are ingress–egress nodes, of which, again, only a subset of all possible ingress–egress nodes will lead to the setting up of an LSP [2, 9]. A request for setting up an LSP is processed either directly by a route server or indirectly via an ingress node, both of which finally, through some interactions, know the explicit route from the ingress node to the egress node.

A request for setting up an LSP is denoted as the triple (o_i, t_i, b_i), where o_i represents the ingress router, t_i represents the egress router, and the amount of bandwidth requested for LSP is b_i. It is assumed that all the QoS requirements are converted into an effective bandwidth requirement, that all the requests for setting up an LSP come one at a time, and that the future demands are unknown. In such a model it is intended to devise an algorithm that will help to determine a path along which each demand for an LSP is routed along such a path so as to make optimum use of network resources.

Suppose there is a set of distinguished node pairs P, which can be perceived as potential ingress–egress router pairs. Suppose the pair $(s,d) \in P$. All LSP setup requests occur between these pairs. Now it is assumed that a request of D units of bandwidth arrives for the ingress–egress pair $(a,b) \in P$. Suppose now that the max-flow problem is solved between the ingress–egress pair (s,d) with the link capacities reflecting the current residual capacities of those links. The max-flow value determines the maximum amount of bandwidth that can be routed from s to d.

The *interference* between a path in (a,b) and an ingress–egress pair (s,d) can be defined as the reduction in the max-flow value in the ingress–egress pair owing to the routing of bandwidth on that path. As they do not make any assumptions about the knowledge of future demands, the demand of D units between a and b should be routed in such a way as maximally to utilize the smallest max-flow value for the ingress–egress pairs in $P \backslash (a,b)$.

To solve the problem, one approach is to compute the weights for the links in the network and route a demand along the weighted shortest path. The problem is to estimate the weights for all those links so that the current flow does not interfere too much with future demands. Without delving much into the procedure about how the link weights are estimated, the final equation that computes these weights for the links is as follows:

$$w(l) = \sum_{(s,d) \in P \backslash (a,b)} \alpha_{sd} \frac{\partial \theta_{sd}}{\partial R(l)} \tag{5.1}$$

where α_{sd} represents the weight for the ingress–egress pair (s,d), θ_{sd} represents the max-flow value for the ingress–egress pair, and $\partial\theta_{sd}/\partial R(l)$ represents the rate of change of the value of maximum flow between (s,d) with change in the residual link capacity $R(l)$.

Equation (5.1) can be further simplified by considering the notion of critical links. For an ingress–egress pair, a *critical* arc belongs to any min-cut for that ingress–egress pair. The min-cut is always computed with the present value of the residual capacities on those links. If C_{sd} represents the set of critical links for an ingress–egress pair (s,d), by the max-flow min-cut theorem it can be shown that

$$\frac{\partial \theta_{sd}}{\partial R(l)} = \begin{cases} 1 & \text{if } l \in C_{sd} \\ 0 & \text{otherwise} \end{cases} \tag{5.2}$$

which implies that

$$w(l) = \sum_{(s,d):l \in C_{sd}} \alpha_{sd} \frac{\partial \theta_{sd}}{\partial R(l)} \tag{5.3}$$

Equation (5.3) can be used to compute the set of critical arcs for all ingress–egress pairs, which is essentially the problem of determining the weights of the arcs as described by equation (5.1). The set of critical links for a predetermined set of ingress–egress pairs of nodes can be determined by a single execution of the max-flow algorithm between the ingress–egress pair.

For practical implementations [9], the value of α_{sd} can be chosen to signify the importance of an ingress–egress pair (s,d). The weights can be chosen to be inversely proportional to the max-flow values, i.e. $\alpha_{sd} = 1/\theta_{sd}$. In other words, critical arcs that have lower max-flow values are assigned more weights than those with higher max-flow values.

MIRA prevents loading of the critical links and routes requests along non-critical links, even if such a path is not a minimum-hop shortest distance path.

5.3.2 Limitations of MIRA

Although MIRA is quite influential to the TE community because it proposes an online algorithm for routing bandwidth-guaranteed tunnels using the idea of minimum interference, it is not 'robust'. The limitations of MIRA [1] are as follows:

1) MIRA fails to identify the effect of interference on a cluster of ingress–egress nodes. It just computes the effect of interference on single pairs of ingress–egress nodes.
2) MIRA does not keep track of the expected bandwidth requests between a pair of nodes, thereby rejecting requests even when there might be sufficient residual capacity available to route the request between the affected pair.
3) MIRA is computationally expensive in complex networks where thousands of max-flow/min-cut computations have to be performed for every request.

5.4 Profile-Based Routing Algorithm

The PBR algorithm was designed to work on a group of ingress–egress nodes. The PBR algorithm, however, uses the traffic profile of a network in a preprocessing step roughly to predict the future traffic distribution. PBR has two main phases of processing: (i) the offline preprocessing phase; (ii) the online routing phase.

Offline preprocessing phase. This phase is based on the solution to the multicommodity flow computation problem [10] on a set of traffic profiles. The traffic profiles can, typically, be obtained from a number of sources, e.g. the service level agreements (SLAs) between the customers and network service providers, historical trend statistics, or by suitably monitoring the network for a predetermined period of time.

Using the same notations that were used by the original authors, i.e. Suri *et al.* [1, 6], let us assume that there is a capacitated network, $G = (V, E)$, with the function cap(e) denoting the capacity of a link $e \in E$, and the set of links in the network with V vertices. It is also assumed that there is a table of traffic profiles obtained by using one of the methods mentioned in the last paragraph. Let us denote these traffic profiles by (*classID*, s_i, d_i, B_i), where *classID* denotes the traffic class to which an ingress–egress node belongs, s_i and d_i denote the ingress–egress pairs, and B_i is the aggregated bandwidth between the above ingress–egress pairs for that class. The term 'class' can typically be perceived as, for example, having requests between an ingress–egress pair to belong to one class, or as having all requests that have the same ingress or egress node to belong to one class [11]. In multicommodity flow preprocessing, each of these classes is classified as a distinct *commodity*. The requested bandwidth is treated as an aggregated bandwidth between a pair of ingress–egress nodes of the identified class as mentioned, and not as a single flow. Therefore, each profile (and *not* each flow) can be split, and consequently several flows belonging to a traffic profile can be routed on distinct paths.

Satisfying all bandwidth route requests may not always be possible. Thus, additional edges, called *excess edges*, are added to the network. This is done to avoid the situation where there is no feasible solution to the multicommodity flow problem. The excess edges are assigned an infinite capacity, so as to discourage routing of flows through these routes as much as possible. The rest of the network edges are assigned a cost of unity. Thus, the *cost* function [1] can be expressed as

$$\text{cost}(e) = \begin{cases} 1 & \text{if } e \text{ is a link in the network} \\ \text{cap}(e) = \infty & \text{if } e \text{ is an excess edge} \end{cases}$$

If there are k commodities, the multicommodity problem to be solved for the transformed network [1] from the last step can be expressed as

$$\text{minimize} \Sigma \left(\text{cost}(e) \sum_{i=1}^{k} x_i(e) \right)$$

subject to the following constraints:

1) All edges satisfy the capacity constraints. In other words, if e is not an excess edge, then the following inequality holds true: $\sum_{i=1}^{k} x_i(e) \leq \text{cap}(e)$.

2) All nodes conserve the flow for each node, except at the ingress–egress nodes corresponding to that flow.
3) B_i is the amount of commodity i reaching its destination d_i.

The $x_i(e)$ values that can be obtained from the multicommodity flow computation are used for preallocating the capacity of the edges of the network. The incoming requests in the online phase of the algorithm, will use these capacities as thresholds so as to route flows belonging to traffic class i.

Online routing phase. The online phase of the PBR algorithm is much simpler than the offline phase. The terminologies of the original authors [6] are used to explain the online phase of this algorithm. The reduced graph, with residual capacities from the offline preprocessing phase, is used as the input to this phase. Let this residual capacity of an edge e of class j be denoted by $r_j(e)$. The online routing phase takes this residual graph to process an online sequence of incoming LSP setup requests (id, s_i, d_i, b_i), where $id, s_i, d_i,$ and b_i are respectively the ID of the request to be routed, the ingress and egress node pairs between which the request is to be routed, and the bandwidth that is to be routed. While routing the requests, given a choice, the requests are routed through the minimum-hop shortest paths between the ingress–egress pairs of nodes.

PBR was identified to have the following limitations [11]:

- PBR is based on the assumption of the splitting of a group of flows, even if it is not a single flow. This will still remain problematic when an individual flow has a very high demand.
- The performance of the algorithm will be limited by the accuracy of information provided in the traffic profiles in the preprocessing phase of the algorithm.

5.5 Dynamic Online Routing Algorithm

DORA is another TE routing algorithm, proposed by Boutaba *et al.* [7], for the routing of bandwidth-guaranteed tunnels. DORA attempts to accommodate more future path setup requests by evenly distributing reserved bandwidth paths across all possible paths in the entire network, thereby balancing the load in an efficient manner. It does so by avoiding, as much as possible, routing over links that have a greater potential to coincide with the link segments of any other path, and those that have a low running estimate of the residual bandwidth.

DORA is designed to operate in two phases [7]. In the first phase, DORA computes, for every source–destination pair, the path potential value (PPV), which stores in an array the potential of a link to be more likely to be included in a path than other links. In other words, the algorithm tries to preprocess a set of links that are more likely to be included among the different paths that packets could travel between any pair of source–destination nodes. The PPV for any pair of nodes (S,D) is denoted as $PPV_{(S,D)}$. If a path can be constructed over a link L for any pair of source–destination nodes, $PPV_{(S,D)}$ is reduced by unity, whereas if different paths can be constructed over the same link L for different source–destination pairs, the value of $PPV_{(S,D)}$ is increased by unity.

In the second phase, DORA first preprocesses and removes those links that have a required bandwidth exceeding the residual bandwidth. It then provides weights to the different links in the network by taking into consideration the PPV arrays and the current residual bandwidths. As there are two parameters that may influence the weights that can be assigned to the links, namely the PPV and the current residual bandwidth, it introduces a new parameter called bandwidth proportion (BWP), which determines the proportion of influence either of these two parameters could have on the weight of the link.

5.6 Wang *et al.*'s Algorithm

Wang *et al.* [3] also proposed an algorithm for setting up bandwidth-guaranteed paths for MPLS TE, which is based on the concept of critical links, the degree of their importance, and their contribution in routing future path setup requests. Thus, like MIRA, the algorithm essentially belongs to the class of minimum interference algorithms. The algorithm takes into account the position of source–destination nodes in the network. Specifically, for any path setup request, the algorithm first considers the impact of routing this request on future path setup requests between the same pair of source–destination nodes. To do this, the algorithm assigns weights to the links that may be used by future demands between that pair of source–destination nodes. For any source-destination pair of nodes (s,d), it computes the maximum network flow, $\theta^{s,d}$, between s and d, and $f_l^{s,d}$, which represents the amount of flow passing through link l. Then, for all links in the paths between s and d, they compute the bandwidth contribution of each of those links, $f_l^{s,d}/\theta^{s,d}$, to the max-flow between s and d [3].

In addition to the link's bandwidth contribution, as it also considers the link's residual bandwidth $R(l)$, the algorithm proposes the parameter $f_l^{s,d}/\theta^{s,d}.R(l)$. It uses this normalized bandwidth contribution of any particular link l to compute the overall weight of a link l as

$$w(l) = \sum_{(s,d)\in L} \frac{f_l^{s,d}}{\theta^{s,d}.R(l)}$$

where $w(l)$ represents the total weight contribution of all the links between a pair of source–destination nodes. Subsequently, it assigns the weights to the different links using the above formula, and then, in the latter phases of the algorithm, eliminates those links that have requested bandwidth for any request greater than the residual bandwidth in that link. It then runs Dijkstra's algorithm on the reduced network (with the weights assigned using the formula above) to route a request through the shortest path between a pair of source–destination nodes in a network.

5.7 Random Races Algorithm

The RRATE algorithm [13] accepts certain packet transfer requests with an assured bandwidth and then efficiently uses the network to forward the packet at the committed bandwidth. The input to the algorithm comprises the routing bandwidth details, and the output of the algorithm is the path in the network to be followed as well as routing of the packets over that path. The algorithm is based on a reward function for each action. It also defines a value N that indicates the limit of reward an action can be granted. The algorithm is based on the theory of 'random races (RR)' [14].

The algorithm is divided into three phases: offline operations, online operations, and post-convergence operations. In the *offline operations*, the shortest paths between the routers are determined and the RR corresponding to each router pair is maintained. The maximum bandwidth utilization permissible on the link in terms of its percentage utilization is also specified. During the *online operations phase*, a decision is taken regarding accepting or rejecting a request for a path. The request for a path is accepted if the prerouting path utilization approximation is less than the threshold bandwidth utilization specified for the path. Further, path selection is done in a sequential manner by sorting the *k* paths in increasing order of available threshold bandwidth, starting the selection from the path with least available bandwidth. The path that has just a little more than the required bandwidth is selected. This reduces the amount of unutilized bandwidth as well as providing sufficient bandwidth to the requested service. If all the paths have been checked and none of them could be selected, the request is rejected. In the *post convergence phase*, the packet is routed through the first path providing the assured bandwidth.

References

1 S. Suri, M. Waldvogel, D. Bauer, and P. R. Warkhede. Profile-based routing and traffic engineering. *Computer Communications*, 26:351–365, 2003.

2 K. Kar, M. Kodialam, and T. V. Lakshman. Minimum interference routing of bandwidth guaranteed tunnels with MPLS traffic engineering applications. *IEEE Journal of Selected Areas in Communications*, 18(12):2566–2579, 2000.

3 B. Wang, X. Su, and P. Chen. Efficient bandwidth guaranteed routing algorithms. *Journal of Parallel and Distributed Systems and Networks*, 2002.

4 W. Szeto, R. Boutaba, and Y. Iraqi. *Dynamic Online Routing Algorithm for MPLS Traffic Engineering. Lecture Notes in Computer Science 2345*. Springer, 2002.

5 I. Iliadis and D. Bauer. A new class of online minimum-interference routing algorithms, pp. 959–971, in Enrico Gregori, Marco Conti, Andrew T. Campbell, Guy Omidyar, and Moshe Zukerman (eds). *NETWORKING 2002: Networking Technologies, Services, and Protocols; Performance of Computer and Communication Networks; Mobile and Wireless Communications. Lecture Notes in Computer Science 2345*. Springer, 2002.

6 S. Suri, M. Waldvogel, D. Bauer, and P. R. Warkhede. Profile based routing: a new framework for MPLS traffic engineering, pp. 138–157, in *Quality of Future Internet Services, Lecture Notes in Computer Science 2156*. Springer, 2001.

7 R. Boutaba, W. Szeto, and Y. Iraqi. DORA: efficient routing for MPLS traffic engineering. *Journal of Network and Systems Management*, 10(3):309–325, 2002.

8 S. Subramanian and V. Muthukumar. *Alternate Path Routing Algorithm for Traffic Engineering*. ICSENG, 15th edition, 2002.

9 M. Kodialam and T. V. Lakshman. Minimum interference routing with applications to MPLS traffic engineering. *IEEE, INFOCOM 2000. 19th Annual Joint Conference of the IEEE Computer and Communications Societies, Volume 2*, pp. 884–893, 2000.

10 R. K. Ahuja, T. L. Magnanti, and J. B. Orlin. *Network Flows: Theory Algorithms, and Applications*. Prentice Hall, 1993.

11 S. W. H. Wong. The online and offline properties of routing algorithms in MPLS. Master's thesis, Department of Computer Science, University of British Columbia, July 2002.

12 T. H. Cormen, C. E. Leiserson, and R. L Rivest. *Introduction to Algorithms*. MIT Press, 1990.

13 S. Misra, Network applications of learning automata. PhD dissertation, School of CS, Carleton University, Ottawa, Canada, 2005.

14 D. T. H. Ng, B. J. Oommen, and E. R. Hansen. Adaptive learning mechanisms for ordering actions using random races. *IEEE Transactions on Systems, Man, and Cybernetics*, 23(5):1450–1465, 1993.

Abbreviations/Terminologies

BWP   Bandwidth Proportion
DORA   Dynamic Online Routing Algorithm
IETF   Internet Engineering Task Force
IP   Internet Protocol
LDP   Label Distribution Protocol
LER   Label Edge Router
LSP   Labeled Switched Path
LSR   Label Switched Router
MIRA   Minimum Interference Routing Algorithm
MPLS   Multiprotocol Label Switching
PBR   Profile-Based Routing
PPV   Path Potential Value
QoS   Quality of Service
RR   Random Races
RRATE   Random-Races-Based Algorithm for Traffic Engineering
SLA   Service Level Agreement
TE   Traffic Engineering

Questions

1 What is traffic engineering and what is its usefulness?

2 Name five different traffic engineering routing algorithms.

3 Mention the properties that a traffic engineering algorithm should have.

4 Explain the following: max-flow, critical link, cut, and max-flow min-cut theorem.

5 Draw a hypothetical network and, using it, explain the working of MIRA.

6 Differentiate between the two main phases of processing in a PBR algorithm.

7 State the two phases of operation of DORA and explain both phases.

Exercises

1   Consider the network as indicated in the figure below. Calculate the min-cut edge for source = X and destination = Y. What is the capacity of the *s–t* cut and the value min-cut?

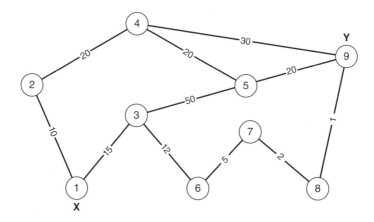

2   Design a network and mention the traffic flow over it to bring out the disadvantages of MIRA.

Part III

Routing on the Internet

6

Interior Gateway Protocols

6.1 Introduction

A network is termed an autonomous system (AS) [1-3] if the administrative decisions related to the network, such as the network topology, addressing scheme, assignment of addresses to the hosts, and routing decisions, can be taken by the administrator of the network. The scope of an administrative domain may not be confined to a single network but can be over a combination of networks connected to each other and sharing common routing information. Still, these networks remain under the administrative control of only one administrator for the purpose of sharing the common routing table in a centralized, shared, or distributive mode. An AS is characterized as a network in which the nodes are connected to each other in such a fashion that any node can communicate with any other node unless there is some link or node failure in the network. The routers in an AS share routing information among each other, using a common routing protocol, and the network is managed by a single organization.

Generally, two different ASs are connected across a geographical distance over a wide area network, and the Internet service provider (ISP) provides the connectivity to these two ASs. The technique used by an ISP for routing the traffic cannot be controlled by the administrators of the ASs. The ISP is responsible for routing the traffic between the ASs and does it in a way that is generally transparent to the ASs. The ASs can only enter into an agreement with an ISP to provide a minimum guarantee of the level of service being provided. These guarantees can be based on a number of parameters such as

Network Routing: Fundamentals, Applications, and Emerging Technologies, First Edition.
Sudip Misra and Sumit Goswami.

number of hop counts, delay, bandwidth, and jitter. The assurance is mutually agreed based on a service level agreement (SLA) assuring quality of service (QoS). The routing protocols that work within an AS are called Interior Gateway Protocols (IGPs). An IGP is also used to route the traffic within each separate network of an AS. The Exterior Gateway Protocol (EGP) handles the routing among the ASs.

The autonomous system is also known as a routing domain, and Interior Gateway Protocol supports the distribution of routing information among the routers of the routing domain and then helps in computing the best routing path from one node to another. The 'best' path can be in terms of one or more parameters as defined by the administrator or the protocol. In an IGP, two types of message are generally exchanged between the nodes – the neighbor discovery messages and the network reachability messages. The neighbor discovery messages are used to detect the neighbors of the router, their status, the routing mechanism used by the router, and the condition of the connectivity to the neighboring router. The network reachability messages help the router to get an idea about the topology of as big a portion of the network in which it is operating as possible. The protocol attempts to seek the neighbor and the reachability information as well as try to incorporate mechanisms to authenticate it and keep it updated so as to take care of link failure or node failure or any other change in the network topology.

The term 'gateway' in the Interior Gateway Protocol refers to the router. The 'gateway' was used historically to define the functions performed by the present-day router. The protocol guides how to reach from one router to another router within a network or a group of networks. There are two types of IGP: distance vector routing and link state routing. Distance Vector Routing Protocol gives each router in the network information about its neighbors and the cost of reaching any node through these neighbors. By exchanging information with the neighbors, a router can estimate the cost to reach a node and the neighbor to which it should forward the packet to enable it to construct the best path. In link state routing, each router has information about the topology of the network, by using which it calculates the best path to the destination and the outgoing (egress) link of the router to reach the destination using the shortest path.

The history of Interior Routing Protocol [4-6] can be traced back to the NSFNet of the National Science Foundation (NSF), USA. NSFNet was a part of ARPANET and was running a software called 'fuzzball' on five DEC LSI-11 computers across the network spread over five supercomputing centers at different academic institutes. The fuzzball software gave the hardware the capability to act as a modern-day router and it helped in connecting various networks to ARPANET. These fuzzball computers in the NSFNet acted as an AS, and thus used an Interior Routing Protocol.

The Interior Routing Protocol used by NSFNet was a variation of the present-day Distance Vector Protocol. The protocol used the network delay as the parameter for selecting the best path. The protocol was known as HELLO Protocol because the fuzzball routers on a regular basis sent hello packets to their neighbors and calculated the time required by the packet to reach the destination and then the acknowledgement to come back from the destination to them. This was done by time stamping the HELLO packets. A router knows that it can reach any other router that its neighbor can reach, but with an additional packet delay as calculated using the HELLO packet. However, using time delay as a route calculation metric led to spurious results in the protocol, as it was not directly related to the bandwidth owing to other parameters such as processing time and buffer. Even if there was no change in the network parameter, the link delay was not constant over a period of time, leading to flapping. HELLO Protocol was slowly replaced with RIP, and now HELLO Protocol is neither in use nor a part of TCP/IP.

6.2 Distance Vector Protocols

The Distance Vector Protocol gets its name from the two parameters that it uses – distance and vector. These two parameters are used to forward the packet from the source to the destination. The protocol has awareness about the distance between any two neighboring nodes as well as the distance from the source to the destination before it starts forwarding the packet. Vector indicates the direction in which the packet should move. Each intermediate routing node in the protocol is aware of the direction in which it should forward the packet so as to enable the packet to reach the destination using the shortest path.

The working of a distance vector routing protocol helps the routers to exchange their link information with immediate neighbors and update their routing table. When this routing table is shared recursively over time, each router gets to know the distance to every other router in the network and the direction towards each of the routers. In each level of recursion of sharing the routing table, information regarding one more level of router is added in terms of connectivity from the router separated by hops. At every step, as the levels of routers keep on increasing, the best path to the router is calculated and retained and the other paths to the router are ignored. Finally, when the recursion reaches the farthest level of router, the protocol stops and the final routing table is generated. However, there is frequent exchange of data between the neighboring nodes to incorporate topological changes. This exchange of data can be periodic or triggered by an event such as link failure or node failure [7, 8]. Some of the common terminologies used in distance vector routing are as follows.

Route convergence. A network is said to have achieved route convergence if all the routers in the AS contain in their routing tables the same information about the network. Route convergence continues if the routers do not have to change any information in their routing tables based on the regular exchange of routing information between the routers, i.e. no information of any topological changes is being received from their neighbors. If there are some topological changes or failure of a link or a node, changes in the routing table are triggered, disturbing the converged network. The neighboring node will transmit the information regarding the change in the form of a modified routing table. This will subsequently lead to changes in the routing table of all the routers, and the network again waits for all the routers to converge. The rate at which a network converges depends on two factors. Firstly, the rate at which the routing tables are exchanged between the routers, and secondly, the processing speed of the routing nodes to recalculate their routing tables based on the routing entries received from the neighbor. The faster a network converges, the faster it becomes operable.

Periodic updates. The routers send their routing tables to their neighbors at regular intervals of time for periodic update. A periodic update may be sent by a router to its neighbor even if there is no change in the routing table. However, this may lead to unnecessary transmission of data across the network and consumption of bandwidth by the data, which is not required. But at the same time, a periodic update from a neighbor informs the router that the neighboring router is active and the link to it is operational. The routing protocols may modify the form of the periodic updates and the router may just send 'no change' information to its neighbors instead of transmitting the entire routing table. Similarly, in the case of any change in the routing table, it may send only the changed parameters with proper sequencing to indicate the appropriate place where the changes are to be incorporated. This helps in reducing the amount of data required to be transmitted for a periodic update.

Triggered updates. Triggered updates are sent by a router to its neighboring nodes if any new activity or change is detected in the network, which the router should immediately report to its neighbors instead of waiting to transmit it along with the regular update. A triggered update may be sent by a router when it detects any one of the following:

- a node failure or a link failure making one of its neighbors unreachable,
- the introduction of a new link into the network, one end of which is connected to this router,
- a new node that has joined the network and is directly connected to the router,
- changes in the link parameter,
- a change in state of any interface of the router from active to inactive/failed, or vice versa.

Synchronized updates. These occur when all the routers in the network or a network segment exchange routing table updates at the same instance in time. A synchronized update causes an abrupt increase in bandwidth utilization and can cause collisions in the network.

Routing loops. A routing loop is created when a packet is forwarded between a set of routers without ever reaching the destination. In a routing loop, the packet continuously traverses through the same set of routers in a unidirectional manner. A routing loop is generally created if there is some flaw in the route configuration by the routing algorithm. It may occur in the case of user-defined routes using a static routing protocol in a huge network that is prone to configuration errors by the network administrator owing to its complexity. A routing loop is also created in the case of a slowly converging network, as the route forwarded to the routers may keep changing while the packet is in transit when the network has not converged. A packet in the routing loop may finally get lost. It unnecessarily consumes network bandwidth as well as processing capacity, processing time, and buffer space of the intermediate routers.

Count to infinity. In a count to infinity, packets are routed infinitely within the network. The packet may finally be delivered to the destination after a relatively huge amount of delay or it may continue being routed within the network. It is different from the routing loop as in this case the packet may not enter a loop and may be moving in a random path inside the network without any uniform pattern or routing loop. It creates the same set of problems in a network that is created by a packet in a routing loop.

Split horizon. Split horizon is a technique to prevent routing loops in a network. Split horizon states that when a router receives a route from its neighbor, the router should not propagate the route back to the neighbor from which it has received the route. Split horizon is achieved by a technique called route poisoning. In route poisoning the parameter value for the interface through which the update was received is set to an infinitely large or infinitely small value so as to make it the last choice in the best route selection or indicate the unavailability of the poisoned route.

6.2.1 Routing Information Protocol

Routing Information Protocol (RIP) [9] is a true distance vector protocol that exclusively works on metrics based on hop count. It is an intradomain routing protocol that, though not the best routing protocol, is the protocol with least overhead for a small to moderate-sized network. The router transmits its routing table along all its interfaces to the neighboring routers every 30 s. However, a router may be configured not to broadcast the routing information along one or more of its interfaces, but to continue to receive updates along the same interfaces. An RIP host is said to be in active mode when it receives routing updates from its neighbors as well as transmitting its routing table to its neighbors. A RIP host is defined as operative in passive mode if it receives routing information from its neighbors but does not transmit its routing table to its neighbors. The terminal routers are generally configured in passive mode.

The RIP works well in a small autonomous system. It is not an effective protocol for huge networks with plenty of routers or in networks with slow connectivity links. The protocol is ineffective in these operating scenarios owing to parameters such as a maximum hop count of 15, the transmission of routing table updates after every 30 s, and maintenance of timers such as route invalid timer, hold-down timer, and route flush timer. The default values and performance of these timers suit only a small network with optimum bandwidth links. A hop count of 15 is considered to be sufficient for RIP, as it is an Interior Gateway Protocol (IGP), and a network packet is not expected to cross the hop count in general operating conditions in a moderate-sized network.

Routing table. The routing table in RIP primarily has three columns: the destination network, the next node, and the metrics. The destination column has entries for the various destination network addresses available in the AS. The next node column contains information regarding the interface on which the traffic should be routed to reach the destination. Metrics are based on minimum hop counts required to reach the destination. Hop count is the number of links that have to be covered to reach the destination node. By default, the maximum reachable router in RIP requires a hop count of 15; a hop count of 16 is treated as unreachable. A packet is dropped as soon as it crosses a hop count of 15, and the router dropping the packet sends an ICMP message to the source of the packet indicating 'destination unreachable'. This is done to prevent infinite loops and endless travelling of a packet on the Internet owing to faulty routing. If a router is configured for a few more metrics, a network packet may be dropped even before reaching a hop count of 15.

Protocol timers. RIP involves regular transmission of route updates to its neighbors and rebuilding its routing table based on the route information received from its neighbors. It has to maintain a few timers to indicate the time between forwarding updates, awaiting updates, declaring a link unreachable, or flushing the entries from the routing table.

The route update timer is used by every router to determine the time when it should transmit its entire routing table to its neighbors. The default value of the route update timer is 30 s, i.e. the route update information is sent out by each router along its exit interfaces to the neighbor routers every 30 s.

The route invalid timer is maintained to determine the validity of an entry for a link in the routing table of a router. If a router does not receive information about a link from the route updates of any of the neighbors for six consecutive updates, i.e. 180 s, then it classifies the route as invalid and broadcasts this information to its neighbors during the next route update.

A route flush timer determines the time between declaration of a route as invalid and its removal from the routing table. The entry for this invalid route is not immediately removed from the routing table of the router that has declared it invalid as the router may receive an update about the link from any of its neighbors after it has declared the link as invalid. The router also takes some time to propagate the information about the invalid link to its neighbors. This type of implementation generally happens in the case of unstable links.

The hold-down timer defines the period for hold-down. Hold-down is used to prevent any change in the routing table owing to routine update messages, which may wrongly reinstate a link that has gone down. As a link goes down, it is detected by the routers connected to it. The routers in turn recalculate the routing table and forward it to their neighbors in the form of triggered updates. The triggered updates are further transmitted by the neighbors that have just received them, and this initiates a wave of triggered updating. However, a router that has recently modified its routing table from a triggered update may receive a routine routing table update from a router that is not aware of the particular link failure. This routing update may lead to reinstating the metrics for the unavailable link. Hold-down tells the routers to hold down any further changes until the hold-down period is over.

Types of Packet

In RIP, a router may send a routing table update to its neighbor in response to a message received from it or it may send a routing table update that does not correspond to any routing update request. Therefore, RIP packets are of the following two types:

Request packets. The request packet is sent to a neighbor router requesting its complete or partial routing table.

Response packets. The response packet is sent by a router in response to a request packet received from any of the neighboring routers. The response packet may contain the entire routing table or a partial routing table. In addition to transmission of the response packet on receipt of the request packet, the response packet is also sent every 30 s to all the neighbors to share its routing table.

RIP Versions

RIP version 1, defined in RFC 1058, supports only classful routing, while RIP version 2, defined in RFC 2453, supports classless routing [2, 10]. Thus, in RIP version 1, the subnet information is not transmitted during routing table updates and the default subnet has to be used as per the defined class. The first three bits of the IP address are used to determine the class of the network address, and thereafter the subnet mask corresponding to that class is used. RIP version 1 uses classful routing because it was introduced before the concept of subnet was introduced or before classless interdomain routing (CIDR) was implemented. RIP version 2 is also known as prefix routing as it sends the subnet information during route updates. RIP version 1 and RIP version 2 are fully interoperable, with forward compatibility as well as backward compatibility.

RIP uses UDP for exchange of update information among routers. The format of the RIP message is different in version 1 and version 2. One RIP datagram can carry

Bytes 0   8   12   512

Figure 6.1 UDP datagram for RIP message.

Bytes 8   9   10   12

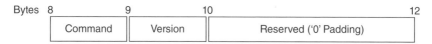

Figure 6.2 RIP header.

information of up to 25 entries of the routing table. As shown in Figure 6.1, the size of a UDP datagram is 512 bytes, eight bytes of which are used by the UDP header and the remaining 504 bytes can be used by the RIP. Followed by the UDP header is a 1 byte 'Command' field and a 1 byte 'Version' field followed by two reserved bytes generally padded with '0' as indicated in Figure 6.2. These four bytes are known as the RIP header, which is followed by the RIP message. The RIP header is common in version 1 and version 2. The RIP message starts with two bytes for the 'Address Family Identifier'. This field is common in version 1 as well as version 2. The 'Command' field, which is of 1 byte, can have values ranging from 1 to 5. The details of the values are:

1- indicates that it is a request message asking the recipient to send partial or full routing table to the sender of the message.
2- indicates that it is a response message to a request and contains the entire or partial routing table of the sender in response to the request received from the receiver. A value of 2 in the command field is also transmitted when the sender sends an update message without any specific request message.
3- indicates Trace On. This is obsolete and not in use.
4- indicates Trace Off. This is obsolete and not in use.
5- is reserved for use by Sun Microsystems for its use.

The values of 6, 7, and 8 were defined in RFC 1582 and are used to indicate triggered request, triggered response, and triggered acknowledgement respectively. The values of 9, 10, and 11 were defined in RFC 2091 and are used to indicate update request, update response, and update acknowledgement respectively.

The 'Version' field indicates the RIP version and generally has a value of 1 or 2 to indicate the version of the RIP. The 4 bytes 'Address Family Identifier' stores the information regarding the type of network address in use. As the IP address scheme is used in general, the 'Address Family Identifier' field should have a value of 2 as defined in RFC.

In RIP version 1, the 'Address Family Identifier' is followed by two reserved bytes generally with '0' padding, and thereafter it has four bytes for the IP address of the host or network, eight bytes for '0' padded reserved fields, and four bytes of 'Metric'. As shown in Figure 6.3, this entire RIP message comprising the 'Address Family Identifier', 'IP Address' and 'Metric' can be repeated 25 times with a common RIP header in an RIP datagram.

The format of the RIP header remains the same in RIP version 2. In RIP version 2, the RIP message comprises two bytes of 'Address Family Identifier' followed by two bytes of 'Route Tag' followed by four bytes each for 'IP Address', 'Subnet Mask', 'Next Hop', and

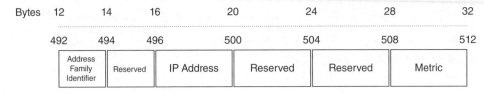

Figure 6.3 RIP version 1 message format.

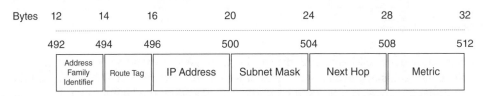

Figure 6.4 RIP version 2 message format.

'Metric'. The RIP version 2 message format is shown in Figure 6.4. The 'Route Tag' field is used to support multiple routing protocols and to distinguish between the RIP-based routes and other protocol-based routes. The 'Subnet Mask' introduces the classless addressing.

Limitations of RIP. A few limitations of RIP [3] are as follows.

The protocol is not dependent on the bandwidth of the link in deciding a route to destination. As shown in Figure 6.5, if router A is directly connected to router B on an 8 kbps public switched telephone network (PSTN) link and router A is connected to router B through router C with a 155 Mbps link across both hops, RIP would route the packet on the direct link of 8 kbps between router A and router B rather than on a much faster link of 155 mbps between router A and router B through router C. This directly affects the quality of transmission in the case of real-time traffic carrying voice and video packets, which calls for QoS parameters. However, a network administrator can deceive the RIP by manipulating the hop count entry in the routing table and increasing the value of the hop count for the slower link to a value greater that the hop count for the faster link.

The protocol supports only equal-cost load balancing and does not provide for a mechanism to balance the traffic load across links with different costs. The protocol sends all the packets through the cheapest path in terms of hop count. In the case of a lower-bandwidth link, this may lead to congestion, but that does not change the routing strategy for RIP by selecting a less congested path as it will keep trying to send all the packets through the least hop count path only. The traffic is also not distributed across alternative routes to reach the destination as the routing table does not contain any information to support the load balancing. In the case of Figure 6.5, the entire traffic will be forwarded directly from router A to router B and will never be shared across both routers, i.e. directly from router A to router B as well as from router A to router B via router C.

As the routers broadcast their routing tables to all their neighbors every 30 s, this leads to network congestion and consumption of bandwidth. Owing to the delay of 30 s in transmitting the routing updates, it takes time for the network to converge. If the network has a few unstable links, convergence becomes very difficult. Moreover, it also

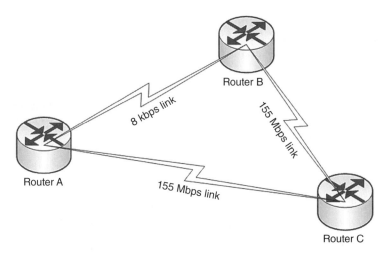

Figure 6.5 A sample network of routers.

leads to the forwarding of packets to unavailable links, information about which has not yet been received by the router on account of the slow convergence.

The protocol has a mechanism for loop detection by using hop count and dropping a packet completing 15 link hops, but it does not have any mechanism for avoidance of routing loops.

6.2.2 Interior Gateway Routing Protocol

Interior Gateway Routing Protocol (IGRP) is a distance vector routing protocol that was invented by Cisco Systems Inc. in the mid-1980s and is proprietary to it. IGRP can be used in a network that comprises all Cisco routers. The protocol was introduced to overcome the drawbacks of RIP and introduce a protocol that overcomes the limitations of RIP to an optimum extent. The protocol supports a maximum hop count of 255 with a default value of 100 hops. This makes the routing protocol highly scalable, and it can be implemented in huge networks.

The metrics to determine the most suitable path are not simply based on a single parameter, but can be calculated based on a number of parameters, commonly referred to as 'composite metrics'. The default parameters used for calculation of metrics are based on the bandwidth of the path and the cumulative interface delay, which is the delay in the network. The bandwidth of the path is based on the minimum bandwidth link on the path. In addition to these default parameters, a few more parameters such as channel occupancy of the link (load), reliability of the link based on current error rate, hop count, and maximum transmission unit (MTU) can be used for metric calculation. MTU indicates the maximum packet size that can be transmitted through the path without fragmentation. These parameters have different permissible ranges of values. The value of bandwidth can range between 1200 bps and 10 gbps; the value of delay can range between 1 and 2^{24}, and the values of load and reliability can range between 1 and 255. Cisco has predefined links and associated bandwidth for token ring, ethernet, and T1. A reliability scale of 255 indicates a 100% reliable link or 100% load. In addition to these metrics, IGRP permits the configuration of a few user-defined constants to add to the computational algorithm for composite metrics. All these parameters are combined by an algorithm [2] that is tuned by assigning specific weights to each parameter finally to arrive at the resultant metrics. The composite metrics are calculated as follows:

$$\text{Metrics} = \left[\left\{ \left(\text{constant 1 / unloaded path bandwidth} \right) \times \left(1 / \left(1 - \text{channel occupancy} \right) \right) \right\} + \left(\text{constant 2} \times \text{composite delay} \right) \right] \text{reliability}$$

where

$$\text{Composite delay} = \text{switching delay} + \text{circuit delay}$$
$$+ \text{transmission delay} \left(\text{on no-load condition} \right)$$

Based on the above calculation, the smaller the value of the metrics, the better the path.

IGRP is a classful network protocol and so does not transmit any subnet-related information as it deals with the default subnet based on network class. RIP as well as IGRP, being classful, waste a lot of IP address space, and this is a major disadvantage of these protocols. All the routers in an autonomous system use a common AS number to identify and exchange routing information among them. The value of the AS number ranges between 1 and 65 535. Identification of the network by an AS supports convergence of various networks and flexibility to run different routings in the same network.

The protocol can load balance [11] among six unequal links. The bandwidth as well as the other parameters can be used to decide on the load balancing across variable metric links ranging from the best link to the worst link. However, the traffic will be routed through the worst link only if it is above the predefined range of metrics acceptable for transmission along the link. The range of metrics by which a load can be balanced across unequal cost paths can be defined by the network administrator and is known as variance. The amount of traffic routed across these links with different metrics is in the same ratio as the metrics of the link, i.e. the better the metrics, the more traffic is routed through it as compared with a link with poorer metrics. Although RIP can also support load balancing, it is based on a single parameter, which is hop count, and can share data across links with the same number of hops to the destination. IGRP also supports sharing of a single stream of traffic across two equi-bandwidth links in a round robin order. If one of these eqi-bandwidth links fails, the entire traffic is automatically switched over to the other link. IGRP also uses a few techniques such as hold-downs and split horizons to enhance the stability of the protocol.

Protocol timers. IGRP uses the same set of protocol timers as defined in RIP, but with a difference in their values. The use of these timers, however, remains the same as that in RIP. The route update timer has a default value of 90 s, as against a default value of 30 s in RIP. This leads to lesser exchange of routing tables between the routers, leading to more effective utilization of bandwidth for actual data than for exchange of routing tables between the routers. This leads to reduced flooding of the network by the routers themselves.

The value of the route invalid timer is 270 s, which is calculated as 3 times the route update timer. The value of the flush timer is 630 s, which is calculated as 7 times the value of the update timer. The value of the hold-down timer is 3 times the route update timer with an additional buffer of 10 s.

Key improvements over RIP. IGRP was developed as an attempt at improvement over RIP. The key advantages of IGRP over RIP are as follows:

- It can be used for larger networks.
- The routing table is updated every 90 s to reduce flooding, link congestion and bandwidth utilization.
- It uses bandwidth and cumulative interface delay as minimum composite metrics.
- It can load balance between six different links.
- It uses the AS number as a unique identifier to identify all the routers in an autonomous system.
- It shares a single stream of traffic across two equi-bandwidth links in a round robin fashion, with switchover to one of the links in the case of failure of the other.

6.3   Link State Protocols

In a link state protocol, the router exchanges its network topology information in terms of its links and interfaces with its neighbors. The neighboring nodes further transmit this information to its other neighboring nodes, and the process continues to enable the topological information about each node to reach across the network to all the routing nodes. Each router in the autonomous system has complete information about the topology of the network, which it uses to determine the best path to any other node in the network. In link state routing, the router decides the next hop router or the egress interface link from it on the basis of the best end-to-end path to the destination.

As the routers in the link state protocol have an entire topological view of the network, it is the protocol most suited to traffic engineering and implementation of QoS. As the router is aware of the end-to-end links, various parametric constraints can be imposed on the links to cater for the assured level of service agreed in the SLA. But knowledge of the entire network requires memory and processing capability at each routing node. As the size of the network increases, the requirement of memory for storage of the topology and processing capacity to calculate the entire network topology increases. The size of topological data exchanged between the neighbors also increases, leading to traffic congestion. Considering these requirements, the link state protocols are not highly scalable and are confined within the AS for intradomain routing.

Each router participating in link state routing creates a packet called a link state PDU (LSP). An LSP contains information about the neighbors of the router, the type of link, and the distance or cost of reaching the neighbor. The router discovers its neighbor after booting by using a reachability protocol that uses a 'hello' packet. LSPs contain a serial number to help in sequencing and timeout by unique identification because LSPs are frequently generated by each router. An LSP is generated when the router boots up and whenever changes are detected in its interfaces or in its links. LSP packets are flooded in the entire network. The flooding of the packet is through the neighbors. Each routing node sends its LSPs to its neighbors. Each neighbor checks the sequence number of the LSP and, if it detects it to be a new LSP that the node has not transmitted earlier, it transmits it to all its neighbors except the one from which the LSP was received. In this way, the neighbors generate a wave of LSP transmission across the network, covering all their neighbors in one wave, these neighbors cover all their other neighbors, and the process continues until the LSP reaches the entire network. Therefore, each router receives the PDUs of all other routers in the network. These LSPs are stored in the link state database of each router.

All the routers compute the entire network topology using these PDUs stored in the link state database and creates the shortest path tree and the routing table. Thus, the link state database should be the same across all the routers, as it is built by collecting the LSPs of all the routing nodes. However, the routing table generated by each router would be different as the routing table will contain entries related to the shortest path from the routing node to every other node in the network by forming the shortest path tree in which the routing node is at the root.

The modern link state routing protocols resolve the scalability issue of the link state protocol by dividing the AS into smaller areas and running the link state routing protocols within smaller areas with the provision of communication among these areas. Splitting the AS into smaller areas also reduces the memory and processing requirement of the routing protocol.

The major advantage of link state routing over distance vector routing are as follows:

- All the routers in the network have an entire topological view of the network.
- Each router in the network builds its own complete network map based on the LSPs received from all the routing nodes.
- As the LSPs are flooded, they immediately spread across the network and thus help in quick convergence of the network owing to their faster receipt at the routing nodes.
- The LSPs are not sent at regular intervals, but only on detection of change in the link or interface of the router, and once when the router boots up.
- Scalability issues of implementing link state routing in a huge AS can be resolved by using the hierarchical or the layered approach of separating the network into smaller areas.

6.3.1 Open Shortest Path First Protocol

Open Shortest Path First (OSPF) [2, 9] routing protocol version 2 is defined in RFC 2328. OSPF is an open standard routing protocol that supports only IP routing and converges fast even in a large network. The protocol supports classless IP addressing, VLSM, and multicasting. It can select multiple equal-cost paths between the source and the destination and hence can perform load balancing by distributing the traffic across the equal-cost paths. The metrics of the route refer to the cost of the path and can be based on a single parameter or a combination of parameters such as delay and throughput. A router can have multiple routing tables in it, each based on a different metric. The default metric used by the protocol is bandwidth. The metric formula is

$$Cost = 100\,000\,kbps\,/\,link\,speed$$

where the numerator value is configurable.

The protocol separates a bigger network into smaller networks called areas, the routing is dealt with separately within each area, and all the areas are interconnected by a backbone. As OSPF is a hierarchical protocol, it is fast, scalable, and stable. OSPF prevents the corruption of routing tables as it can authenticate the node transmitting the route advertisement.

Areas. In OSPF, an area [2] is the grouping of routers, links, and the associated OSPF network. Each area is uniquely identified by its 32 bit area ID. An area ID '0' represents the backbone area. The backbone area must be contiguous as there cannot be separated backbones in the AS with the same area ID. As a router can be a member of one or more areas, the area ID is associated with the particular interface of the router that belongs to the area. As the routers lie on the border of two separate areas, any link in OSPF will belong only to one of the areas. The non-backbone area is also known as the secondary area, and the backbone area is known as the primary area. The non-backbone area must be directly connected to the backbone area through one of its routers that is common to the non-backbone area as well as the backbone area. The router inside an area should maintain the topological information of only the area to which it belongs. The information is stored in the form of a topological database. The topological database of all the routers within an area should be the same.

Routers. For unique identification of a router, it is given a 32 bit router ID (RID), which is different from its IP address. However, as the IP addresses of the routers are also unique, many implementations of OSPF use the IP address of one of the interfaces of the router, e.g. the lowest- or the highest-number IP address, to generate the RID. Depending on the connectivity of the router in the area, which is also depicted in Figure 6.6, the following kinds of router have been defined in OSPF:

- *Internal router.* All the interfaces of this router are in the same area. Internal routers may be connected to each other or to the area border router. They maintain the topology database only of their area.
- *Backbone router.* This is a router with an interface in the backbone area, i.e area ID 0.

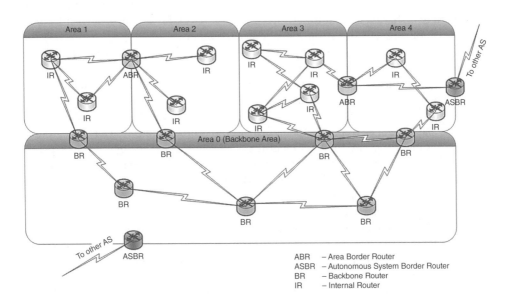

Figure 6.6 OSPF routers and their areas.

- *Area border router (ABR).* This router connects two or more areas. It has at least two different interfaces in two different areas. There is flooding of information in an area. The ABR gets the range of addresses in the area from the flooding and passes it on to the other areas. ABRs can also be backbone routers. They maintain the topology database of each of area to which they are attached and run a shortest path first algorithm for each area separately.
- *Autonomous system border router (ASBR).* These routers connect the AS to which they belong to some other AS. One of the interfaces of such routers is generally located in the backbone, i.e. area '0', or these are also the ABR routers.

A multiaccess network has more than two routers attached to it. The network may allow broadcasts using a broadcast address recognized by all the routers, or the network may not possess the broadcast capabilities and each packet is addressed to only one destination router. In the case of a multiaccess router, a designated router (DR) and a backup designated router (BDR) are deployed to reduce the flooding of the network with exchange of link state advertisements (LSAs) between the routers [10]. The interface of each router in the network has a priority attached to it, depicting its ability to become a DR or BDR. The router with the highest priority is selected as a DR. Generally, the default priority attached to the router is 1, and a 0 priority attached to the router indicates its inability to serve as a DR.

- *Designated router.* In a multiaccess network, adjacencies are formed between the router and the DR to prevent bandwidth consumption of the network by the formation of adjacencies between every pair of routers. The DR has adjacencies with all the routers in the network and is the only source to forward LSAs.
- *Backup designated router.* The BDR assumes the role of a DR when the DR fails. As it is in active standby mode, it has adjacencies with all the routers in the network, as in the case of a DR.

Neighbor routers. Two or more routers can be neighbor routers [10] if they are in the same area and share a common network segment. As OSPF supports authentication, they may define a security passphrase among each other.

The *neighbor discovery* is done by regular exchange of 'hello' packets on each of the interfaces. The packet contains the RID of the routers whose 'hello' packet has been received by the router. When a router receives a hello packet with its RID in it, the two routers enter into neighbor relationship.

The multiaccess network should *elect a DR* and a BDR. In the case of such a network, the 'hello' packet will also contain the information required for electing a DR, i.e. router priority, DR identifier, and the BDR identifier. The router with the highest router priority in the segment becomes the DR. The tie in the priority, if any, due to the same router priority for DR is resolved by electing as the DR the router that has the higher RID. The same election process is thereafter followed for electing the BDR, but in this case the DR is not eligible to participate in the election of the BDR.

The neighbor routers form an *adjacency relationship* when they synchronize their topology database with each other. Synchronization is obtained by exchanging link state information, and it is done to ensure the same contents of the topology database in the neighbor routers. In the case of a multiaccess network, the router does not establish a direct adjacency relationship with all its neighbors, but it is done through DR and BDR as shown in Figure 6.7.

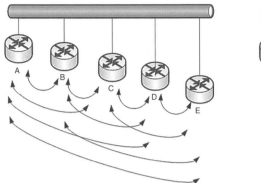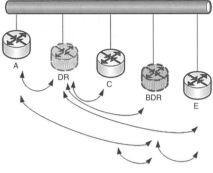

Figure 6.7 Neighbor relation in point-to-point network versus multiaccess network.

The process of topology *database synchronization* is done by exchange of database information as well as the exchange of database entries between the routers. The neighbors that desire to enter an adjacency relationship exchange database description packets with each other. The database description packet contains the listing of the LSAs available with the router. The router that receives the database description packets checks its topology databases to detect whether the neighbor has any more recent LSA or any LSA that is missing in the receiver's topology database. The router then requests the updated information using a link state request. On receiving the link state request, the recipient of the request sends those specific LSAs to the router that has requested the LSAs. On receiving an LSA, the router sends an acknowledgement to the sender.

Links. A connection in a network is referred to as a link. It is the network or router interface assigned to a network through which a packet may be received by the network or forwarded by the network. Every interface of a router is associated with a link. However, a single link may have one or multiple IP addresses. A link is generally in one of two states: up or down. A link can sometimes be in a state that is neither up nor down, but has a varying degree of congestion. Four different types of link [9] are used in OSPF:

- *Point-to-point link.* This directly connects two routers without any other router, host, or network in-between.
- *Transient link.* This is a network with multiple routers connected to it.
- *Stub link.* This is a network connected to a single router.
- *Virtual link.* This is a link created by the administrator between two routers when the actual link between the two routers goes down. The virtual link is created by connecting the two routers with a number of intermediate routers in-between, thus forming a continuous, but longer path.

Operation. The protocol exchanges link state information every 30 min, and the changes occurring in the network during this time are communicated to other routers using link state advertisements (LSAs). The LSAs are exchanged only within their area. The routers maintain the topological database, which is known as the link state database (LSDB). The LSDB stores the LSAs received from all the other routers in the area. The LSDB of all the routers in the area should have the same contents. The RID is used to tag an LSA in the LSDB to recognize the router from which it was received. Owing to the hierarchical design of OSPF, the routers are not required to maintain the path for all the other routers in the AS.

The border router advertises the range of addresses available in the area and not the individual address of each of the routers in the area. Thus, all the border routers maintain a database of the range of addresses corresponding to every border router and the shortest route to the border router in their area. This helps to prevent the processing of the entire address by the border router to detect the area to which the packet should be forwarded, as the same can be done by processing a portion of the address only. OSPF uses Dijkstra's algorithm to create the shortest path tree for each router. Although the LSDB is the same in all the routers in the area, the shortest path tree will be different in all the routers, as it has the corresponding router in the root node of the tree and thereafter the tree is generated connecting all the other routers. The routing table is built from this tree by detecting the shortest path from the root node, i.e. the router itself, to every other router in the network [1]. This routing table is the final lookup table used by the router for forwarding the packets by selecting one of its interfaces based on the routing table entries.

6.3.2   Intermediate System to Intermediate System Protocol

Intermediate System to Intermediate System Protocol (IS-IS) defines an intermediate system as 'a device used to connect two networks and permit communication between end systems attached to a different network'. Intermediate system is ISO terminology for a router. IS-IS is a link state protocol that is highly scalable and has fast convergence, and the complete topological view supports implementation of traffic engineering, which makes it a favorite protocol for the service providers. IS-IS Protocol supports two different network layer protocols – IP and OSI connectionless network service (CLNS) [2, 3]. The IS-IS Protocol can also be used in a dual-network environment comprising IP as well as OSI. By supporting both IP and OSI, the protocol can interconnect dual routing domains with dual or pure (IP or OSI) routing domains delivering the packets to IP hosts, OSI end system, or a dual-end system.

IS-IS was designed in 1987 as a dynamic routing protocol for CLNP under the ISO 10589 standard and later adapted for IP as per RFC 1195 in 1990. In the year 2008, IPv6 support was added to IS-IS by RFC 5308, and RFC 5120 permitted the multitopology (IPv4 as well as IPv6) concept. IS-IS supports classless interdomain routing with variable subnet length masking.

Two-level protocol. IS-IS is a two-level hierarchical routing protocol. The entire network is split into small areas, and the routing within an area is taken care of by level 1 routing and the routing outside the area is taken care of by level 2 routing [12]. Thus, the level 2 routers form the backbone of the network, and each area should have at least one level 2 router to connect to the backbone. Each node belongs to one area only, and there is no overlapping of the areas. However, the same router can be in level 1 as well as in level 2, or it may belong to any one of the levels only. The border between the areas is based on the links that connect routers belonging to separate areas. The formation of areas and the levels of the routers can be visualized from Figure 6.8. Level 1 routing is responsible for routing within the area. Level 2 routing is responsible for delivery of the packet to the area that has the destination host. When a packet is sent from a source to a destination, level 1 routing forwards the packet from the source that is contained in its area to the nearest level 2 intermediate system. Now the level 2 IS forwards the packet

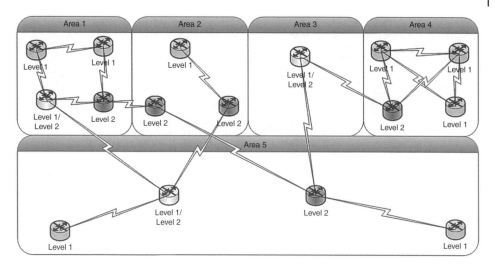

Figure 6.8 Areas in IS-IS depicting interconnections between level 1 and level 2 routers.

to the destination area where it is sent from the level 2 IS to the level 1 IS containing the destination. Then the level 1 IS routes the packet to the destination within the area.

In link state routing, the routers are aware of the entire network topology. But as the network is divided into areas in IS-IS, the level 1 router is aware only of the topology of its area, which comprises level 1 and level 2 routers in the area. As the level 1 router is responsible only for the intra-area routing, the link state database of the level 1 router comprises information related only to the routers in its area. However, this information is not used to find the optimal path if a packet has to be sent outside the area because the level 1 router sends the packet to its nearest level 2 router if the packet has to be sent outside the network. Forwarding the packet to the nearest level 2 router and thereafter routing at level 2 may not lead to the shortest path. This can be well understood from the example of a sample network depicted in Figure 6.9, on the assumption that all link costs are the same. For example, node B has to send a packet to node H. Node B being a level 1 router will forward the packet to its nearest level 2 router, i.e. node D, and thereafter the route followed by the packet will be node D → node E → node F → node I → node C → node J → node H. Alternatively, if node B is not to forward the packet to the nearest level 2 router for interarea forwarding, the packet would follow the path node B → node A → node C → node J → node H, which is optimal by comparison with the previous route. IS-IS specification mentions about four different types of metric: cost, delay, expense (monitory cost involved in using the link), and error (measured in terms of the probability of residual error associated with the link).

A router at level 2 is connected to other level 2 routers that may be in the same area or in a different area. As the level 2 router is responsible for interarea routing, the link state database of the level 2 router contains information about all other level 2 routers in the network. If a router is only in level 2, it does not contain any topological information about its area, as intra-area routing is beyond its scope. However, a level 1/level 2 router should have topological information about its area as well as information about all other level 2 routers (backbone). The level 1/level 2 router can have its neighbors in the same area (level 1 routers, level 2 routers in the same area) as well as in other areas

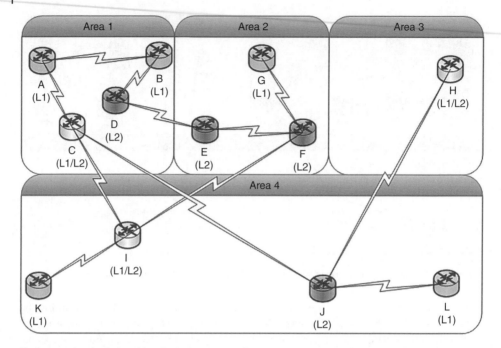

Figure 6.9 Level 1 (L1) and level 2 (L2) routers in the areas in IS-IS routing to depict suboptimal routing from level 1 to level 2.

(level 2 routers in other areas). It has to maintain a link state database for intra-area routing and another link state database for interarea routing, and run two separate shortest path first algorithms separately on each of these databases.

Packets. There are four different types of IS-IS packet used in the protocol:

- *IS-IS hello (IIH) packet.* This is used to discover neighbors and maintain adjacencies. The IIH packet is sent every 10 s. It is different on P2P links and LANs. In order to be of full MTU size, the IIH packets are padded.
- *Link state packet (LSP).* An LSP has a fixed header and LSP contents. The LSP header comprises the LSP ID, the sequence number, the remaining lifetime, the checksum, the type (level 1, level 2), the attached bit, and the overload bit. The LSP contents comprise all the information pertaining to a router, such as adjacencies, connected IP prefix, area address, and/or OSI end systems. There is not only one LSP per router, but there is only one LSP per LAN network. There is only one LSP per LAN because a virtual node is assumed for an LAN, and this virtual node is called the 'pseudo node', which imitates a router and generates an extra LSP.
- *Partial sequence number packet (PSNP).* This is used for requesting LSPs and confirming the receipt of the link state information and acts as acknowledgements on P2P links.
- *Complete sequence number packet (CSNP).* This is used while distributing the complete link state information over LANs. The CSNP contains all the LSPs from the link state database. As all the routers should have the same LSPs in their link state database, CSNP helps to synchronize the link state database of those routers that have outdated or missing LSPs.

IS-IS operation. Being a link state protocol, IS-IS follows the outlined procedure of a link state routing for packet forwarding. An IS on booting or joining a network in IS-IS discovers its neighbors on all its interfaces by sending 'hello' packets. The two ISs across a data link become neighbors only in the case of matching authentication and IS level. Each IS generates its LSPs based on the connectivity at its interfaces, and these LSPs are flooded in the network. All the ISs in the network construct the link state database from the flooded LSPs. The link state database is identical in all the level 1 routers of an area or in all the level 2 routers because it depicts the topology of the network or the area and it should be the same all across. The LSP database is used to construct the shortest path tree, which helps to build the routing table of each IS. Multiple equal-cost shortest paths to the destination can also be computed by the algorithm. The LSP in a link state database has a lifetime of 1200 s, and the generator of the LSP should usually refresh the LSP every 900 s to prevent it from expiry and deletion from the link state database.

The protocol recalculates the shortest path in the case of any topological change in the network. The two-level routing in IS-IS provides an additional layer of security. The packets are confined within the area for level 1 routing and the packets are confined only at the backbone for level 2 routing. The topological information of one area is not known to the other area or the backbone. Moreover, as the broadcast is confined to the level and the area, the amount of traffic generated by the protocol is reduced and confined to its broadcast domain (area) only.

Flooding. Flooding of an LSP on a P2P link is different from the flooding of an LSP on an LAN. In the case of a P2P, when link adjacency is established, the two neighboring ISs exchange CSNP. Thereafter, missing LSPs are sent if they are not present in the CSNP after a request has been received for the same using PSNP. An LSP is active for its 20 s lifetime, before which it is refreshed in 15 s. For flooding in an LAN, a designated router (DR) is elected for each LAN based on priority. DR creates a pseudo node representing the LAN, which is shown in Figure 6.10, and the pseudo node has its LSP. DR also performs flooding in the LAN and multicasts the CSNP every 10 s. The ISs in the LAN check their link state database with reference to the CSNP and request specific LSPs by requesting the DR using PSNP.

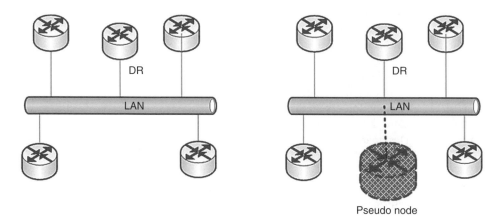

Figure 6.10 Creation of a pseudo node by DR.

References

1 S. Mueller. *Upgrading and Repairing Networks*. Techmedia Publishers, 4th edition, 2004.
2 T. Lammle. *Cisco Certified Network Associate Study Guide*. BPB Publishers, 4th edition, 2003.
3 W. Stallings. *Data and Computer Communications*. Prentice Hall of India Publication, 8th edition, 2007.
4 C. M. Kozierok. *The TCP/IP Guide: A Comprehensive, Illustrated Internet Protocols Reference*. No Starch Press, 2005.
5 National Science Foundation, the launch of NSFNET. http://www.nsf.gov/about/history/nsf0050/internet/launch.htm.
6 National Science Foundation, Fuzzball: the innovative router. http://www.nsf.gov/about/history/nsf0050/internet/fuzzball.htm.
7 R. Graziani and A. Johnson. *Routing Protocols and Concepts: CCNA Exploration Companion Guide*. Cisco Press, 1st edition, 2012.
8 A. Johnson. *Routing Protocols and Concepts: CCNA Exploration Labs and Study Guide*. Cisco Press, 2007.
9 B. A. Forouzan. *Data Communications and Networking*. Tata McGraw-Hill Publication, 4th edition, 2006.
10 L. Parziale, D. T. Britt, C. Davis, J. Forrester, W. Liu, C. Matthews, and N. Rosselot. *TCP/IP Tutorial and Technical Overview*. IBM Redbook, IBM Corporation, International Technical Support Organization, 1998.
11 Cisco system, interior gateway routing protocol, document ID-26825. http://www.cisco.com/en/US/tech/tk365/technologies_white_paper09186a00800c8ae1.shtml.
12 Cisco system, intermediate system-to-intermediate system protocol. http://www.cisco.com/en/US/tech/tk365/technologies_white_paper09186a00800a3e6f.shtml.

Abbreviations/Terminologies

ABR	Area Border Router
ARPANET	Advanced Research Projects Agency Network
AS	Autonomous System
ASBR	Autonomous System Border Router
BDR	Backup Designated Router
CIDR	Classless Interdomain Routing
CLNP	Connectionless Network Protocol
CLNS	Connectionless Network Service
CSNP	Complete Sequence Number Packet
DR	Designated Router
EGP	Exterior Gateway Protocol
ICMP	Internet Control Message Protocol
IGP	Interior Gateway Protocol
IGRP	Interior Gateway Routing Protocol
IIH	IS-IS Hello (Packet)

IS-IS	Intermediate System to Intermediate System
ISO	International Organization for Standardization
ISP	Internet Service Provider
LSA	Link State Advertisement
LSDB	Link State Database
LSP	Link State Packet
MTU	Maximum Transmission Unit
NSF	National Science Foundation
OSI	Open Systems Interconnection
OSPF	Open Shortest Path First
P2P	Point to Point
PDU	Protocol Data Unit
PSNP	Partial Sequence Number Packet
PSTN	Public Switched Telephone Network
QoS	Quality of Service
RFC	Request for Comments
RID	Router ID
RIP	Routing Information Protocol
UDP	User Datagram Protocol
VLSM	Variable-Length Subnet Mask

Questions

1 Describe an autonomous system. Why is routing within an AS different from the routing between ASs?

2 Differentiate between a neighbor discovery message and a neighbor reachability message.

3 Explain the working of distance vector protocols.

4 State the difference between the routing loop and count to infinity.

5 Mention at least five factors that can trigger an update in a distance vector routing.

6 How is the active mode of operation of an RIP node different from a passive mode of operation?

7 Describe the format of a UDP datagram for an RIP version 2 message. It should include details about the UDP header, RIP header, and RIP message.

8 State the advantages and limitations of RIP.

9 Mention the improvements in IGRP over RIP.

10 Explain the working of a link state protocol.

11 Describe the various kinds of router in an OSPF along with their functionality.

12 What are the different types of IS-IS packet used in the protocol?

13 What are the advantages of a two-level protocol over a single-level protocol?

14 State whether the following statements are true or false and give reasons for the answer:
A Interior Gateway Protocol supports routing between ASs on the Internet.
B Synchronized updates lead to flooding.
C Split horizon is a technique used to prevent flapping.
D The frequency of exchange of routing tables between neighbors in RIP is 30 s.
E The edge/terminal routers in an RIP should necessarily be active routers.
F A route flush timer in RIP directly helps in routing loop avoidance.
G IGRP is a proprietary protocol.
H IGRP is a classless protocol.
I In IS-IS, a level 1 router in one area can communicate with a level 1 router in another area only if there is a direct communication link between the two.
J IGRP and IS-IS are distance vector protocols.

15 For the following, mark all options that are true:

A Routing within an AS can be termed:
 • intra-AS routing,
 • intradomain routing,
 • neighbor discovery,
 • route convergence.

B The maximum hop count implemented in RIP is:
 • 8,
 • 15,
 • 100,
 • 255.

C The following are Interior Gateway Protocols:
 • OSPF,
 • BGP,
 • IS-IS,
 • EGP.

D A routing loop can be created by:
 • flooding,
 • faulty route configuration,
 • split horizon,
 • slow converging network.

E The following are types of link in OSPF:
- stub link,
- active link,
- virtual link,
- point-to-point link.

Exercises

1 An RIP host is in active mode when it receives routing updates from its neighbors as well as transmitting its routing table to its neighbors. An RIP host is in passive mode if it receives routing information from its neighbors but does not transmit its routing table to its neighbors. The terminal routers are always in passive mode. Draw a state diagram for the RIP hosts.

2 Assume that a network as indicated in Figure 6.6 (ignore the areas and consider only the routers) performs routing based on RIP with route poisoning implemented over it. In the network diagram, along with each of the routers, write down the routing tables exchanged between the routers in the first two rounds after booting. Necessary assumptions regarding IP addresses may be made.

3 Assume that a network as indicated in Figure 6.9 (ignore the areas and consider only the routers) performs routing based on RIP. Select any three nodes and show the values of all the protocol timers every 10 s for the first 2 min. The network with all the nodes boot up and become operational at time $t = 0$. At time $t = 15$ s, node A fails; at time $t = 30$ s, node A becomes operational and node B fails; at time $t = 45$ s, node B becomes operational and node C fails; at time $t = 60$ s, node C becomes operational and node D fails; the sequence of failure of nodes and their subsequent coming up, involving node E, node F, node G, and node H, continues in the same manner for the first 2 min. Any other value, if required, may be assumed. What will be the values of the protocol timers in the same time period if the network is using IGRP?

4 Assume the existence of a network running IGRP with node failure as explained in exercise 3. What will be the values of the protocol timers after every 10 s during the first 2 min? Now assume that the nodes fail and come up every 2 min in the same sequence as explained in exercise 3. Write down the values of the protocol times after every 1 min in a 15 min period for any three nodes.

5 Consider the network given below, which uses RIP with load balancing. 100 MB data has to be transferred from node A to node B. What will be the amount of data that will flow over each of the links in the network? If the metrics are based linearly only on hop count, what will be the amount of data flowing over each link if the network has IGRP implemented with load balancing?

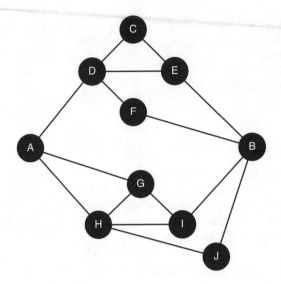

6   Consider the network depicted in the figure below. It is planned to implement OSPF in the network. Divide the network into appropriate areas, including the backbone area, as depicted in Figure 6.6, and then identify the nodes as – internal routers, backbone routers, area border routers, and ASBR. Also, identify the point-to-point link, transient link, stub link, and virtual links (assumptions may be made to identify virtual links).

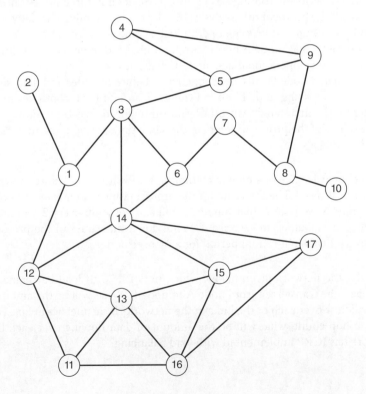

7 In the solution network designed for exercise 6, identify the point-to-point link, transient link, stub link, and virtual links (assumptions may be made to identify virtual links).

8 The network depicted in exercise 6 above has to implement IS-IS routing, and so divide the network into suitable areas and identify the level 1 and level 2 routers.

9 Consider the network given below. What would be the contents of the RIP message from each of the routers after bootup?

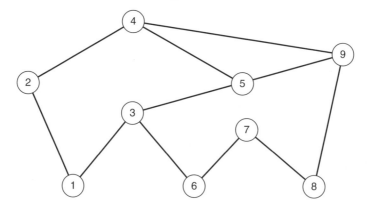

10 Consider the network given in exercise 9. Assume that the network has stabilized and every node has its RIP routing table. Draw the RIP routing table of each node. Now the link between node 3 and node 5 goes down and the link between link 4 and link 9 also goes down. What would be the RIP message from node 3, node 4, node 5, and node 9 to its neighbor? What would be the contents of the RIP routing table of each node after the new topology stabilizes?

7

Exterior Gateway Protocol

7.1 Introduction

The Advanced Research Project Agency (ARPA) was convinced about the importance of networking in the early 1960s. ARPA collaborated with Massachusetts Institute of Technology (MIT) for the realization of a reliable network to connect computers in academic institutions with those in the Department of Defense (DoD) of the United States and its research agencies. It was theoretically established that such a network could be based on packet switching over low-speed dial-up telephone links. The concept was practically tested in 1965 by connecting a computer in Massachusetts with another in California. By 1968, ARPA had finalized the design and specification of the planned network, which was called ARPANET [1]. ARPA is an organization that, on behalf of the US DoD, primarily grants funding to academic institutes and R&D divisions of private industry for next-generation research, design, and development. BBN Technologies (originally Bolt, Beranek, and Newman), which is an American company started by two professors of MIT and their students, bagged the contract from ARPA to

Network Routing: Fundamentals, Applications, and Emerging Technologies, First Edition.
Sudip Misra and Sumit Goswami.
© 2017 John Wiley & Sons Ltd. Published 2017 by John Wiley & Sons Ltd.
Companion website: www.wiley.com/go/misra2204

establish ARPANET. By the end of 1969, a node each in the following organizations was connected in ARPANET:

- University of California, Los Angeles (UCLA) – Network Measurement Center;
- Stanford Research Institute (SRI) – project on 'Augmentation of Human Intellect';
- University of California, Santa Barbara (UC Santa Barbara) – application visualization projects investigating methods for display of mathematical functions using storage displays to deal with the problem of refresh over the net.
- University of Utah – project on investigating methods of 3D representations over the net.

From the time of conceptualization of the ARPANET to the present-day Internet, the name of the initial coordinating and funding agency, i.e. ARPA, has changed a number of times. ARPA was renamed the Defense Advanced Research Projects Agency (DARPA) in 1971. DARPA reverted to the name ARPA in 1993, and was again named DARPA in 1996.

Host-to-host protocol was developed in 1970 by the Network Working Group (NWG) for implementation over ARPANET. In 1972, email communication started over ARPANET. Until the 1980s, ARPANET formed the core backbone network for the Internet to which various networks were connected [2-4]. There were two categories of routers in ARPANET – the core routers and the non-core routers.

Core routers. The backbone was formed by the core routers, which communicated with each other using Gateway-to-Gateway Protocol (GGP). All the core routers were monitored, controlled, and maintained from a centralized facility known as the Internet Network Operations Center (INOC), which was established by BBN.

Non-core routers. The non-core routers connected the networks in their respective autonomous system to any of the core routers of the ARPANET backbone. These non-core routers used Exterior Gateway Protocol (EGP) to communicate with the core router with which they were connected. The non-core routers were maintained by the respective organization that controlled the autonomous system.

With the architecture of ARPANET based on core and non-core routers, each non-core router was to be connected to a core router to connect a network in the autonomous system to the Internet backbone, which was singled out as ARPANET. Hence, communication between two different networks on separate autonomous systems had to pass from their respective non-core routers as well as through a few core routers of the backbone. A non-core router used to make an EGP connection only with a core router. Two EGP non-core routers, even though in proximity and directly connected, leading to a single-hop connectivity, would not communicate with each other as two non-core routers were not capable of achieving an EGP connection. This led to the compulsory addition of a core router in all connectivity that otherwise could have been a direct connectivity. This additional hop over a core router in the backbone was referred to as the 'extra hop problem'. To eliminate the problem of extra hop, BBN established three EGP core servers, and the external networks were asked to connect to at least two EGP core servers instead of each connecting randomly to the backbone network through a non-core router communicating with any core router over EGP. If each network was connected to at least two core EGP servers, any destination network information would be available in at least one of the EGP servers to which the source was connected, and hence the destination could be reached directly based on the information of the correct source–destination route, eliminating the essential non-core–core–non-core router hop.

In 1983, ARPANET was divided into two separate networks – MILNET and ARPANET. MILNET encompassed the network of computers connected to the defense establishments, while ARPANET was a scaled-down version of the erstwhile ARPANET with the military sites removed from it and only the academic and research institute nodes remaining connected to this ARPANET. However, MILNET and ARPANET remained connected, and nodes in one network could communicate with nodes in the other network as Internetwork traffic routing was enabled using Internet Protocol.

7.1.1 Hosts vs Gateways

The present day router, which was earlier commonly known as the gateway, has multiple interfaces and can forward packets across interfaces. A host is differentiated from a router by its inability to send packets across its interfaces. The present-day hosts can be configured to act as routers as well by simultaneously taking up the processing work of the host and the packet-forwarding activity of the router. Even though a host may have multiple interfaces, it should not be configured by default to act as a router in parallel. A network should preferably separate the functionality of host and router for the following reasons:

- The host computers have application software running on them with an interface for the user and provide various services to the user. The router has networking services running in it with the aim of interconnecting networks.
- The number of hosts in a network is too huge as compared with the number of routers. The routers are involved in exchange of routing information among themselves. If all the hosts simultaneously act as routers, the amount of routing information exchanged will be too huge and will lead to bandwidth, memory, and processing constraints.
- The host computers generally have application software running on them, while the routers have a networking operating system and the routing protocols installed in it. Upgrade of the routing protocol would be easier if the version upgrade or patches had to be applied only on the routers and not on all the hosts.
- Routers are meant for 24×7 operations, while the hosts are used on demand. Participation of the hosts in routing and their frequent switching on and off will lead to regular change in the topology, making the network initiate the route update process. It will be very difficult for the network to converge.
- There are certain protocols and processes that have a separate policy for communication between host, communication between the routers, and communication between a router and a host. An example of router-to-host communication is the 'redirect mechanism', which is explained in the following section to indicate the need to separate host from router.

Redirect Mechanism

The redirect mechanism [2, 5] is the process by which a host is notified to select an alternative route if it has selected a wrong router to forward the traffic. This occurs when a host has a choice of two or more paths to send the data packet towards the destination. A host now leaves aside path P2 and selects a path P1 that has the immediate routers as R1 and then R2. Now assume that the very next router in path P2 is R2. The originating host should have sent the data packet directly to R2 if it had selected path P2. The Internet architecture includes a redirect mechanism that will be used by R1 to inform the host to update its routing table so that it can forward the packet to the destination directly through R2. As every host maintains the route information in its cache, it should keep building, updating, and maintaining routes based on received redirect messages. A redirect message can be sent only by a router to a host, and not vice versa.

7.1.2 Gateway-to-Gateway Protocol

The Gateway-to-Gateway Protocol was developed in 1982 by BBN Inc. for DARPA and set out in RFC 823 [4, 6]. Although the protocol is not in use any more, it was a stepping stone in the development of the present-day distance vector protocols. GGP was used for communication of reachability and routing information only between the gateways, i.e. the core routers, and hence was given the name Gateway-to-Gateway Protocol.

In the early Internet architecture there were core routers connected with each other, which acted as the backbone, and non-core routers were connected to the core routers to provide Internet connectivity to the local network [7]. The routing protocol that was used among the core routers was GGP. The system of core routers for the Internet backbone as well as the GGP, which was the routing protocol used in it, is now obsolete. But the working of GGP gives an idea of the evolution of routing. GGP can be taken as an example of a distance vector protocol with hop count as the metric.

When a GGP router boots up, it tries to discover its neighbors by sending a GGP echo message to all its neighbors at a regular predefined frequency, which is generally 15 s. The GGP neighbors receiving the GGP echo respond by sending a GGP echo reply message. When the GGP router that has sent the GGP echo message receives K-out-of-N GGP echo [8] replies from a neighbor, where K is the number of GGP echo replies sent by the destination neighbor router and N is the number of GGP echo messages sent by the originating router (generally, $K = \frac{1}{2}N$ or $K = \frac{3}{4}N$), the neighbor is considered to be connected and the link is treated as ready for communication.

The routing update message that is shared by the GGP routers contains the following two pieces of information – N and D, i.e. the IP address of the networks that are connected through the router (N) and the distance (D) to that network in terms of number of hops required to reach that router. This information is sent to the neighbors, sorted in groups based on hop counts, i.e. the group of neighbors at a hop distance of 1 from the GGP router, then the group of neighbors at a hop distance of 2 from the router, and so on. The GGP router that receives this update information compares this information individually for each of the network entries already available in it and updates the routing table if a low hop count path has been received through the neighbor or information about any new connected network has been received. The update message received from a neighbor carries a sequence number to avoid processing of outdated or duplicate information. On receiving an update message that has a new and higher sequence number than the one already processed by the receiving GGP router from the originating BGP neighbor, the GGP router sends a positive acknowledgement message. This acknowledgement message sent back to the originating neighbor GGP router also carries the sequence number of the received update message. If an update with a lower sequence number is received, the message is ignored and a negative acknowledgement is sent back to the originating neighbor. The negative acknowledgement message also carries with it the sequence number of the last correctly received message and not the sequence number of the last received update message.

Thus, GGP was a simple routing protocol with only four types of message – echo request, echo reply, acknowledgement, and routing update. Whenever there was a change in topology, the network had to exchange update messages and allow time for the propagation of topology change information across the entire network, and thereafter the network would converge. GGP being a protocol used over slow-speed links with lesser reliability, a slow convergence process was a major drawback of the protocol.

7.1.3 Autonomous System

An autonomous system (AS) [9] is formed by a collection of networks that are under the administrative control of a single organization. As it comprises a collection of routers forming the network, the AS is also referred to as the routing domain. The routers within an AS generally follow the same routing policy, and they may implement the same or different routing protocols for internal routing.

The Interior Gateway Protocol (IGP) is used for routing within an AS, while the Exterior Gateway Protocol (EGP) is used for routing between ASs [10] on the Internet backbone.

EGP is used for routing in a wide area network supported by a number of Internet service providers (ISPs). EGP supports routing among different service provider networks. EGP supports redundancies in the network paths by implementing routing policies supporting transfer of data through different ISPs simultaneously. It can also implement load balancing of traffic by distributing data over two or more different paths. The EGP can scale to extremely huge networks such as the Internet.

EGP connects two gateway hosts each having its own border router by enabling exchange of information in a network of autonomous systems. The neighboring EGP routers, each belonging to a separate AS, exchange routing tables for inter-AS routing. Inter-AS routing is also referred to as interdomain routing. The routing tables exchanged between the routers have entries giving the list of reachable routers, their addresses, and the path cost associated with reaching a router.

AS identifier. The EGP should be able to identify and distinguish between ASs, and so there should be unique AS identifiers. Each AS that is connected to the public domain through the Internet is assigned a 16 bit identifier by the Internet Assigned Numbers Authority (IANA). The AS identifier allocated by IANA has a value ranging between 1 and 65 535. RFC 4893 and RFC 5398 propose an increase in the AS identifier pool range by converting it from 16 bit to 32 bit size. The AS connecting to a single ISP may use an AS identifier from a private pool as the identification has to be mutually agreed only between the ISP and the organization. But an organization connecting to multiple ISPs over the Internet requires the AS identifier from the IANA-assigned pool. IANA does not allocate an AS identifier directly to the organization's AS, but through the five Regional Internet Registries (RIRs) [11], each of which is responsible for the assignment of the AS number in its geographical region. Each RIR has a huge geographic region under its coverage so as to cover the entire world. Presently the five RIRs are:

- African Network Information Center (AFRINIC)
  [Africa, portions of the Indian Ocean]
- Asia-Pacific Network Information Center (APNIC)
  [portions of Asia and Oceania]

- American Registry for Internet Numbers (ARIN)
  [United States, Canada, and many Caribbean and North Atlantic islands]
- Latin America and Caribbean Network Information Center (LACNIC)
  [Latin America, portions of the Caribbean]
- Réseaux IP Européens Network Coordination Center (RIPE NCC)
  [Europe, Middle East and Central Asia]

Unlike Interior Routing Protocol, the path selection in EGP is not based on minimizing the path cost between two ASs. The routing policy between ASs is based on a number of other parameters [12-16]. As each AS is under the administrative control of a separate ISP or organization, there may be varying relationships between the organizations. Moreover, there might also be an assured level of service agreement between the organizations. A certain level of secrecy regarding the internal network of the organization has also to be kept, and the internal topology of the network should not be shared with another organization. Taking all this into consideration, the path metrics are based on a set of policies that differ for each AS. The administrator of the AS defines the path metrics between the AS and its neighbors so that they satisfy the assured level of parameters. The policy may even impose restrictions on disclosing certain routes to its neighboring ASs. An AS may have a different routing policy for each of the neighboring ASs connected to it. Thus, the routing policy may not necessarily have a technical foundation but be based on political, organizational, security, maintenance, or management reasons. To enable EGP to detect the organization, it carries the customer prefixes or the Internet prefixes during routing. The router sends this routing table to its neighbors in response to a request received from them. Generally, a router requests its neighbors every 2–8 min to send the routing table entries.

The router participating in EGP routing is also known as the Internet facing router and requires the following information [11] to perform the routing:

- A list of neighbor routers for exchange of routing information. These neighbor routers will belong to different ASs.
- A list of networks to advertise as directly connected. This may be different from the list of ASs to which it is actually connected owing to organizational policy and security.
- The identifier of the AS to which it belongs.

EGP can be described as a policy-based routing, as the AS administrator defines the policies on whose basis a packet received from a particular source may be forwarded to the required destination. In the policy-based routing used in EGP, multiple packets received at an AS for the same destination may be forwarded through different routes or even multiple routes with a varying number of replicas for each packet, based on the predefined routing policy. The policy may be based on the address of the source from which the packet has been received or any intermediate AS through which the packet has travelled. The intermediate ASs play an important role in policy-based routing, which is very particular about the security aspect as the packet may have passed through a non-friendly AS and might be bugged or prepared for an attack or sniffing. The policies are also based on the size of the payload, the status of flags in the packet header, the protocol of the packet, hops, and timers.

7.1.4 Characteristics of EGP

The three significant characteristics [2] of the EGP are as follows:

Independence of operation. The EGP provides independence of operation not only to all autonomous systems with the routers running in it but also to the gateway routers. An autonomous system is provided freedom from the other autonomous systems on the Internet or the routers therein. The routers within an AS are not dependent on EGP or on the routers located in other ASs for data transmission within the AS or even beyond that. The gateway router takes care of data transmission beyond the AS. The EGP also provides freedom of operation to the gateway router as the frequency of receiving updates is not dictated by the protocol or any other AS or gateway router. The gateway router itself negotiates the parameter specifying the frequency of receiving updates from neighboring ASs. The gateway router cannot be interrupted with updates from other gateway routers other than those already agreed to at the predefined frequency.

Association of metrics with each independent path. EGP is based on distance as well as vector, i.e. direction from the source towards the destination. The metric is associated with each independent route. In a normal distance–vector routing, the cost of the destination from the source is maintained without entry of any intermediate routers, whereas in EGP the reachability from source to destination is associated with the sequence of intermediate gateway routers in the path and the associated metric. EGP maintains the information about various paths from source to destination, and hence it should know the cost associated with each path. EGP can send different packets through different paths or even the same packet from different routes, and hence the protocol should be able to select the route. Therefore the protocol maintains a metric specifying that the destination can be reached from the source via an intermediate gateway 'X' with a certain metric and can reach the same destination from the same source through another gateway 'Y' with another metric, and so on, to enable information about all available route metrics.

Domination of reachability over routing. In an exterior gateway, all destinations with a metric of non-infinity are termed reachable. EGP does not specify a common technique to compare one route with another based on metrics. Although an update received by a gateway has a metric in it, this is rarely used to compare one route with another. Moreover, the path for forwarding a packet from source to destination is not distinctively selected on the basis of the metric. Neither is the path changed, even when a burst of packets are being sent from source to destination and during the transmission an update is received indicating an alternative path with much better metrics. The routing architecture of the Internet does not have common criteria for metrics interpretation.

Border Gateway Protocol (BGP) is the most common EGP protocol in use across the Internet by all ISPs. A protocol also exists by the name of EGP, which was specified in RFC 827 and RFC 904 and was used to connect the autonomous systems over the Internet in the 1980s, much earlier than BGP. Thus, EGP in general refers to the class of protocols for interdomain routing, and BGP and EGP are the two most familiar routing protocols of this class. BGP version 4 is the common protocol being run by the Internet facing routers forming the backbone for the Internet, and this has replaced EGP version 3, which was used to interconnect ASs in the initial years of the Internet.

7.2   Exterior Gateway Protocol

The initial architecture of the Internet comprised a few core routers in the backbone, which communicated with each other using GGP. The non-core routers, which connected to these core routers, were at the periphery connecting the hosts to the network. These non-core routers communicated with the core routers using EGP. Slowly, the centralized architecture of the core routers in the backbone became decentralized, with independent autonomous systems connected to each other. The EGP class of protocols with modifications and newer versions continued to be used for the communication among ASs. It was sheer chance that the first protocol developed for the Exterior Gateway Protocol class of routing was also named the Exterior Gateway Protocol (EGP), which became obsolete with time and was replaced with another protocol named the Border Gateway Protocol (BGP). BGP was an advanced and improved kind of EGP. It can be said that there are two major Exterior Gateway Protocols – EGP and BGP. Routers were commonly referred to as 'gateways' in the past, and hence all these routing protocols [17-19] have been known as 'gateway protocols'.

7.2.1   Evolution of EGP Standards

The EGP was first documented in RFC 827, published in October 1982, with the aim of creating specifications for communication between gateways. Some of these gateways may not even have been trusted or known by the other communicating gateways. The RFC had introduced the requirements and limitations of EGP, and it was known as EGP version 1. The RFC described the process for neighbor acquisition, a technique for detecting neighbor connectivity using 'hello' and 'I heard you' messages, and called this process neighbor reachability. Specifications for exchange of neighbor reachability messages (polling) and the process to send neighbor reachability messages were also mentioned. The RFC also defined the term 'indirect neighbor' and indicated its difference from a direct neighbor. Although RFC 827 defined the 'stub gateway' and its communication with core gateways, RFC 888 was released in January 1984 exclusively to explain in detail the neighbor acquisition, neighbor reachability, message formats, and polling for EGP-based connectivity of a stub gateway to a core router. RFC 888 marked the beginning of version 2 of EGP. RFC 904 published in April 1984 was an improvement of RFC 827 and RFC 888, and this was the final and formal set of specifications of Exterior Gateway Protocol (version 2) that was in use before the BGP replaced it.

When the Internet was initially designed, it was planned that the connectivity would form a tree structure. The core routers were designated to form the root of the tree, and the non-core routers would be connected in a hierarchical architecture with the root of the tree. However, the present inter-AS linking with mesh structure disrupted the planned tree structure, and hence deviated from the basic topology over which the EGP was designed. EGP, being designed for a tree structure, had no provision for loop avoidance in any arbitrary topology, which was one of the major reasons for BGP taking over from EGP.

7.2.2   EGP Terminology and Topology

EGP routers that are neighbors of a particular EGP router are known at its peers. The peers can be internal or external, i.e. peers within the same AS or in different ASs. The EGP does not have a defined mechanism for neighbor discovery by itself. The IP address of each router has to be manually configured and the connectivity established between the neighbors to make the network ready for routing of packets between them. However, in EGP, two neighboring EGP routers may not necessarily share a common link, i.e. two EGP routers may not be directly connected with each other, but through one or more in-between non-EGP routers. Based on the number of external peers in different ASs to which a gateway is connected, it is classified as a stub gateway or a core gateway [19, 20]. A stub gateway is connected to only one AS other than the AS to which it belongs. There may be only one link to the other AS or multiple links from the stub gateway to the same or different gateways in that external AS. The core gateways connect to more than one external AS. In EGP, the stub gateway can forward updates related only to its network, while the core gateway can forward updates related to its network as well as the information learnt from other gateways. However, both core gateways and stub gateways can receive updates from external gateways.

7.2.3 EGP Operation Model

An EGP router will always be in one of the following states – idle, acquisition, down, up, or cease. There are ten different messages that are exchanged between the routers in various states, and, based on the message received or sent, the router changes its state or continues to remain in its present state. These messages are:

- request,
- confirm,
- refuse,
- cease,
- cease-ack,
- hello,
- I-H-U,
- poll,
- update,
- error.

The various messages that a state can receive and the messages or responses that an EGP router sends out when it is in a particular state are depicted in Table 7.1.

A gateway is initially in the idle state when it does not receive any message except the cease message, which it may optionally receive and respond with a cease-ack. The router is brought from the idle state to the acquisition state by a trigger, which can either be system generated or manual in the form of a request command or start event. When moving to the acquisition state from the idle state, the router sends a request command to its neighbors and thereafter periodically keeps sending the request command to its neighbors until it receives a confirm response. On receiving a confirm response, it moves to the down state, and on receiving a refuse response from the neighbor it moves back to the idle state. The router can also move directly from the idle state to the down state on receiving a request command.

When the router is in the down state, it accepts request messages, cease messages, and hello messages and responds to them. It also receives confirm messages, which it receives in response to the request messages sent by it. The router may also periodically send hello messages in this state, as well as receive unsolicited updates.

Table 7.1 States of EGP gateway along with messages related to the respective states.

Initial state	Messages received	Response/messages sent	Changed state
Idle	Request	Confirm	Down
	Request	Refuse	
	Cease	Cease-ack	
		Request	Acquisition
Acquisition		Request	
	Confirm		Down
	Refuse		Idle
Down	Request	Confirm, Refuse, error	
	Cease	Cease-ack, error	
	Hello	I-H-U, error	
		Hello	
	Update		
Up	(All messages)	(Respond all)	
		Hello	
		Poll	
Cease		Cease	
	Cease-ack		Idle

A neighbor reachability protocol working in EGP declares the router to be in the up state. The router in the up state can receive and respond to all types of EGP message.

The router can come to the cease state from the acquisition state, the up state, or the down state by sending a cease command to the router in that state. If the router in the cease state receives a cease-ack message or another cease message, it moves to the idle state.

When two gateways enter into a neighbor relationship, one of them becomes the *active neighbor* and the other becomes the *passive neighbor*. The active neighbor can only send various request messages such as an acquisition request and hello message, while the passive neighbor can only respond to these messages, and it cannot send these messages to the active neighbor. However, both neighbors can send a cease message or respond to it by sending a cease-ack message.

EGP does not describe the process of neighbor discovery and hence it is generally configured manually. Once an EGP router establishes connection with its neighbor, it has to keep checking regularly whether the neighbor is still connected and reachable. This is done by sending a hello message to the neighbor and receiving back an I-H-U reply message from the neighbor. If a poll message or update message is received in-between, it can also be treated as a reachability confirmation message between the neighbors and used in lieu of hello and I-H-U for that reachability confirmation cycle.

To update its routing table, the router sends a poll message to its reachable neighbor. The neighbor responds by sending an update message, which contains the details of the networks reachable from it. This is used by the router that sent the poll message to update its routing table entries.

The router sends an error message in response to any message received by it that is not as per the defined format for the header or data field. An error message is also sent if it receives 'poll' or 'hello' messages above a maximum rate. Any of the neighbors can decide to cease the connection with or without giving a reason for disconnection.

7.3 Border Gateway Protocol

There are a number of choices with an autonomous system to select an interior routing protocol from RIP, IGRP, OSPF, IS-IS, and EIGRP. However, for routing among autonomous systems, either compatible inter-AS routing protocols should be used or a common protocol should be used interconnecting the ASs. The latter option has gained popularity over other exterior routing protocols owing to the use of Border Gateway Protocol (BGP) almost throughout the Internet. BGP [7] was designed for decentralized routing and came as a replacement for EGP to support transition from ARPANET to NSFNet. RFC 1105 documented BGP in the year 1989. As BGP was the main protocol behind the Internet, its operation was regularly refined, complexities removed, features added, errors rectified, and adaptations made to any changes in TCP/IP. This led to regular work in this protocol, leading to the release of versions of BGP. BGP version 2 (RFC 1163), version 3 (RFC 1267) and version 4 (RFC 1654, RFC 1771) were released in the years 1990, 1991, and 1994 respectively. RFC 1654 documents the initial version of BGP 4, but thereafter classless interdomain routing was introduced in BGP version 4 by RFC 1771, which was published in March 1995. After the publication of RFC 1771, three more RFCs were published that were not a standard but were meant to support the protocol. These were RFC 1772, which defines 'Application of the Border Gateway Protocol in the Internet', RFC 1773, which concerns 'Experience with the BGP-4 protocol', and RFC 1774, which mentions the key features of BGP and analyses the protocol in terms of performance, scalability, link bandwidth, CPU utilization, and memory requirement. The applicability of BGP was introduced in RFC 1771 and documented in detail in RFC 1772. Furthermore, RFC 1774 also dedicated its last section to 'Applicability of BGP'.

7.3.1 Router Connectivity and Terminology

To participate in BGP routing, each AS should have at least one router running BGP. An AS may also have more than one router running BGP, and these routers may even communicate within the AS using BGP. A router running BGP is also referred to as a BGP speaker. However, the BGP speaker may not necessarily be a router, but it can be a host terminal running some other protocol, e.g. a static routing protocol to pass on BGP information received by it to some other BGP router. A boundary router in an AS is a router that connects an AS to another AS. Not all routers running BGP in an AS may be boundary routers, as only a selected few of the BGP running routers in an AS may be connected to some other AS. There may also be a few BGP running routers in the AS that are connected only to other BGP routers in the same AS but not in any other AS. In contrast to the boundary routers, the routers in an AS that connect only to routers within the same AS but not to any router in other ASs are known as internal routers. The other BGP routers to which a BGP router is directly connected are known as its neighbors. The neighbors of a BGP router can thus be in the same AS or in different ASs. The neighbors that are in the same AS are known as internal peers, and BGP communication between these internal peers is referred to as internal BGP. Thus, the routers participating in internal BGP will have the same AS number. Analogous to this, the neighbors of a BGP running router in another AS are referred to as external peers, and BGP communication between the external peers is referred to as external BGP. An AS that is connected to only one other AS through a single BGP link is known as a stub AS, while an AS that is connected to two or more other ASs is known as a dual-homed or multihomed AS respectively. Most of these BGP terminologies are depicted in Figure 7.1 and Figure 7.2.

Topology. A BGP router may be an internal router or it may be a stub or multihomed external router. This conveys that a BGP router may be connected to any number of other BGP routers. Generally, the designers of the network ensure that a router is connected to at least two other routers to ensure path availability even in the case of link failure. The support by BGP to this open-ended connectivity leads to variation in connectivity architecture between the ASs. As shown in Table 7.2, the boundary routers may be connected in complete mesh, partial mesh, tree, ring, bus/chain, or any other random architecture to enable them to communicate with each other. Adaptability to variation in topology leads to high scalability of the BGP network. Furthermore, BGP also supports dynamic change in topology, i.e. if the link between any two ASs goes down and two other ASs are connected, BGP will discover these changes and rediscover efficient paths. The protocol also ensures that the routing is free from loops.

Traffic type and policy. A data packet that originates within an AS or is meant for delivery in the AS is called local traffic, while packets that enter the AS just to pass through

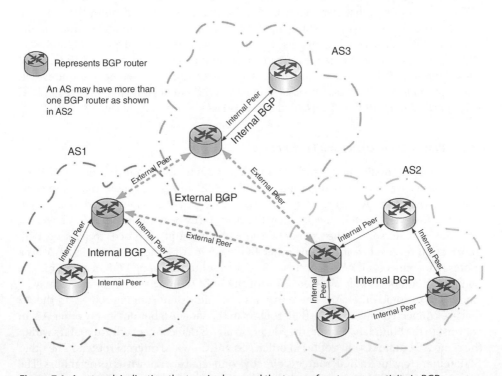

Figure 7.1 A network indicating the terminology and the types of router connectivity in BGP.

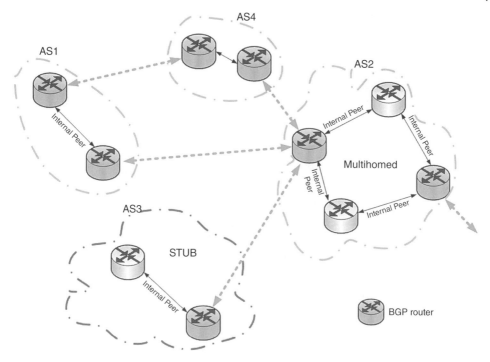

Figure 7.2 A network indicating stub and multihome routers in BGP.

it are known as transit traffic. Transit traffic is generated in some other AS and reaches this particular AS as this intermediate AS falls in the path from source AS to destination AS. BGP allows an AS to have variation in routing policy for different types of traffic. An AS may have a policy of not allowing any transit traffic or it may allow transit traffic only from selected ASs while restricting transit traffic from any other AS. The traffic policy may not necessarily be based only on the AS, but may be criteria based. These criteria may be single parameters or a weighted combination of a number of parameters such as the amount of data transferred, the number of hops traversed, the internal load on the routers, the time of the day, the number of available internal paths, and the number of external connections with boundary routers. An AS can also specify how the packets originating in this AS should be handled by the intermediate ASs. It is necessary to allow every AS to have its own policy on transit traffic, because if it is not allowed, there will be certain dual-homed or multihomed ASs as depicted in Figure 7.3 where the internal network might become choked by transit traffic. But as this policy allows the AS to have a different policy for different transit packets, to save on resources, no AS will like to allow any transit traffic to pass through it. At the same time it will want all other ASs to permit traffic generated by this AS or meant for delivery to this AS to pass through them, and that too on priority. This selfish behavior might lead to deadlocks and blocking of inter-AS routing. To avoid such situations, necessary arrangements have been made. An AS has mutual agreements with other ASs for transit traffic. In addition to this, there are certain backbone ASs that have the capability to carry a huge volume of traffic at a greater speed, and the ASs connect to these backbone ASs for transit traffic.

Table 7.2 Topologies for AS connectivity.

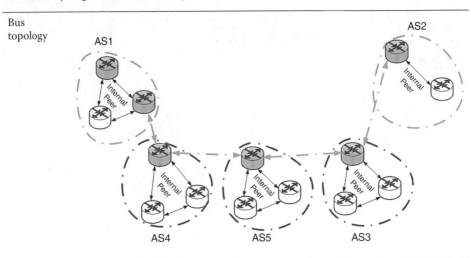

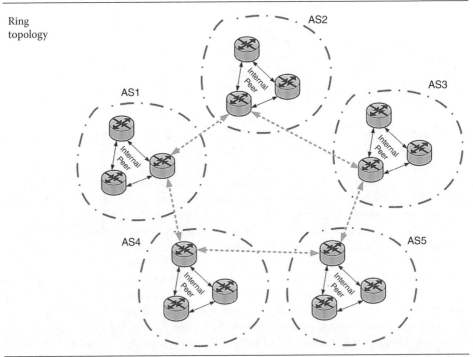

(Continued)

Routing. The BGP router maintains routing information to reach other ASs along with the information about the routers in the path. This information is stored in the form of a routing information base (RIB). Routing information is exchanged by each router throughout the network by forwarding it through its neighboring routers. All the BGP routers use this information to find single or multiple paths to other ASs, and also

Table 7.2 (Continued)

Star
topology

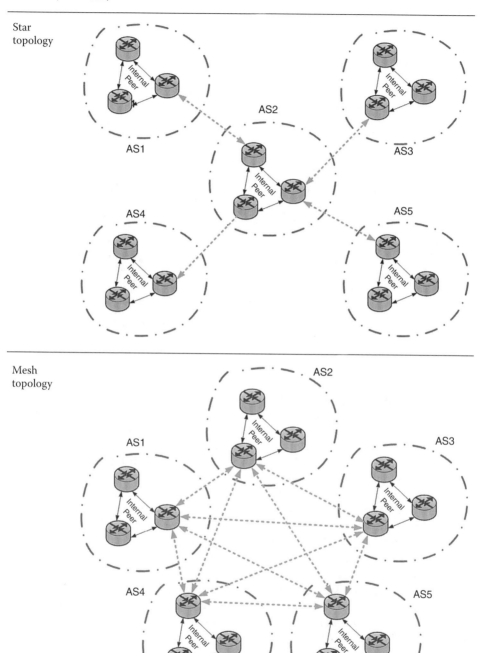

Mesh
topology

Table 7.2 (Continued)

Partial
Mesh
Topology

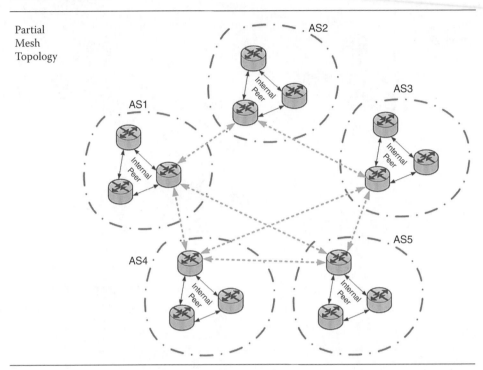

regularly to update this path information. As the packet has to pass through a number of different ASs, similar to path vector routing, BGP has to take into account the information about the intermediate ASs owing to policy and security issues of the originating AS, the forwarding AS, and other ASs in the upcoming path. Thus, the protocol is aware of the entire path and the intermediate ASs before forwarding a packet, and it is not concerned only with the next hop as in the case of distance vector routing. The selection of path by BGP is not simply based on the cost metrics, but is a mix of efficient path selection along with compliance with routing policies across ASs. These policies are based on a number of parameters, each of which may be affected by one or more of the intermediate ASs. Still, BGP cannot guarantee efficient delivery of packets through any particular route as the routing within an AS is not known to the BGP. Every AS has its policy to allow or disallow traffic received from other designated ASs. The internal route traveled by any packet inside an AS may also depend on the originating AS or any particular intermediate AS where the packet might have landed during its traversal.

There are BGP routers connected within an AS as well as BGP routers connected to BGP routers of other ASs. Hence, BGP routing can be categorized [21, 22] into three different scenarios:

i) **Intra-autonomous routing.** In an intra-autonomous routing, the BGP routing is confined within the same AS. The peer routers in intra-autonomous routing use BGP to have a view of the topology within the AS for effective packet delivery or forwarding the packet inside the AS. The intra-autonomous routing decides the

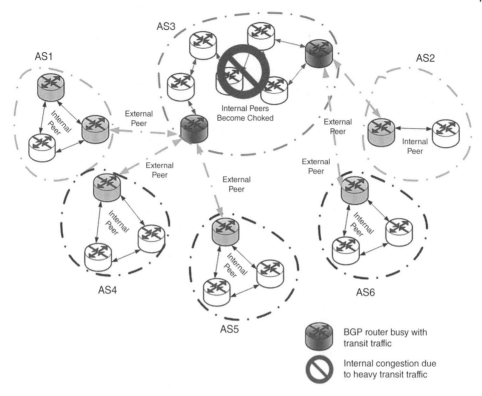

AS3

AS1

AS2

External
Peer

Internal Peers
Become Choked

External
Peer

External
Peer

Internal
Peer

External
Peer

External
Peer

External
Peer

AS4

AS6

AS5

BGP router busy with
transit traffic

Internal congestion due
to heavy transit traffic

Figure 7.3 Congestion in AS due to transit traffic.

routers that will act as boundary routers connecting with external peers. This routing can also be used for effective routing within an organization controlling an AS.

ii) **Interautonomous routing.** The interautonomous routing deals with routing between different routers located in separate ASs. The peer routers in this type of routing use BGP to have a view of the Internetwork without having any details about the internal network of any AS. The Internet is the most common and the biggest example of interautonomous routing.

iii) **Pass-through routing.** When a packet neither has originated in the AS nor is to be delivered within the AS, but rather is just passing through the AS en route from source to destination, it is known as pass-through routing. In pass-through routing the intermediate routers generally ensure that the packet is forwarded in such a way that it reaches nearer to its destination. The AS participating in pass-through routing may not necessarily have BGP running in its routers, including its boundary routers.

Confederation of AS. AS confederation is a collection of ASs that are represented by the same AS number. Although there may be two or more ASs in an AS confederation, the ASs outside the confederation visualize this collection of ASs as a single AS. An AS confederation helps in the splitting or merging of ASs. An AS may be split into several ASs for a number of reasons. A huge AS under the administrative control of a single organization may want to split into a number of separate ASs each under the administrative control of the organization's local region for effective management of the ASs as well as for maintenance of a separate region-based policy for each AS. An AS may also become too huge to be managed by Interior Gateway Protocol owing to scalability-related issues leading to memory, processing power, and processing time constraints. Splitting an AS into smaller ASs would resolve the problems arising out of these situations. On the other hand, it might be required to merge two ASs for administrative or technical reasons. An organization might take over a number of other smaller organizations, or a few Internet service providers might merge. In such a scenario, the organization might want the external BGP routers to view all the ASs under the administrative control of this organization as a single AS. The requirement of AS confederation was also felt to be an alternative to the initial defined structure of BGP where it mandates complete mesh connection between BGP routers within the same AS. If an AS is split into a number of smaller ASs, this reduces the number of interconnections inside the ASs and the associated overhead of intra-AS routing. Also, if an AS is split into a few ASs and the associated AS confederation is not formed, this would increase the AS path information.

AS confederation [23] was first documented in RFC 1965 in June 1996, and it was revised in February 2001 as RFC 3065. Finally, RFC 5065 was introduced in August 2007 and made RFC 3065 obsolete. Figure 7.4 depicts an example of AS confederation where the confederation of two ASs appears to be a single AS to ASs that are not members of the AS confederation. A router within an AS confederation possesses the *AS confederation identifier* of the AS confederation of which it is a member, as well as its *member AS number* within the confederation. The AS confederation identifier is the identifier of the AS confederation that is known to non-member ASs and is used by them in the path information. The member AS number is unique to each member AS within the AS confederation and is used only within the AS confederation.

Route aggregation. The process and requirement of route aggregation [24] in BGP is explained by the example shown in Figure 7.5. Let there be a scenario where

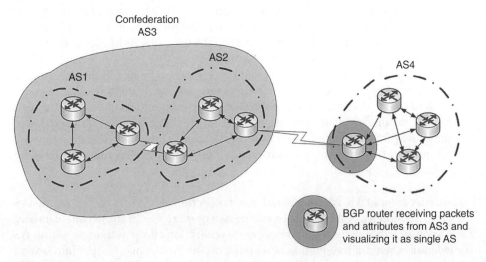

Figure 7.4 AS confederation.

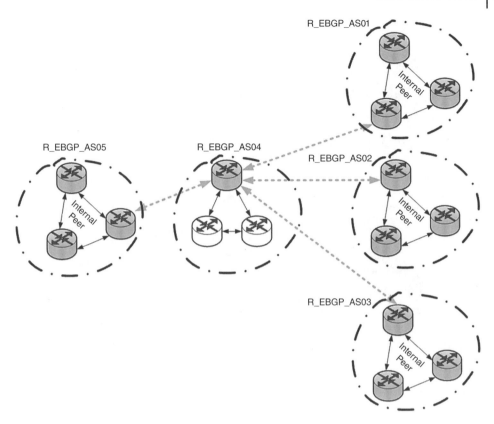

Figure 7.5 Route aggregation in BGP.

three external BGP routers R_EBGP_AS01, R_EBGP_AS02, and R_EBGP_AS03, each belonging to a separate AS, are connected to an external BGP router R_EBGP_AS04 in another AS. R_EBGP_AS04 is further connected to R_EBGP_AS05, which is another external BGP router in another AS. R_EBGP_AS04 aggregates the routes received from R_EBGP_AS01, R_EBGP_AS02, and R_EBGP_AS03 and forwards them to R_EBGP_AS05. Now, when R_EBGP_AS04 forwards the route to R_EBGP_AS05, it cannot be mentioned as (AS4, AS3, AS2, AS1) as this will indicate that the path has passed through all three ASs i.e. AS1, AS2, and AS3. Neither can it be indicated as (AS4, AS3) or (AS4, AS2) or (AS4, AS1), as this will give incomplete path information and may also lead to routing loops. Such a network is indicated and advertised using a set representation of ASs as (AS4, {AS1, AS2, AS3}), without any preference for order within the set.

Classification of path attributes. In BGP, the path to destination is not decided upon solely on the basis of the cost metrics of bandwidth, hop, and selection of the next-hop router. The BGP router is aware of the entire path and the characteristics of the ASs in the path. In order to maintain the path information along with the characteristics of the intermediate ASs, the routers should also share this additional set of attributes with each other during advertisements and path updates. Based on these attributes, each router decides on the path selection as well as on sending update information to peers [24–27]. As BGP is an inter-AS routing protocol, there may be a selected group of

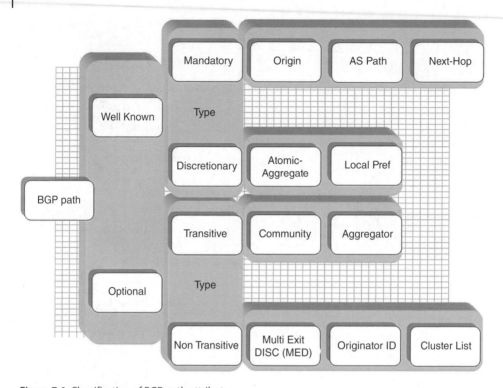

Figure 7.6 Classification of BGP path attributes.

ASs that use a proprietary attribute to share some private information, but there will be certain attributes that must be recognized globally by all BGP routers to enable exchange of mandatory information. Furthermore, a BGP router itself decides whether to process an update received from its peer. In a similar manner, it also decides which attributes to share with its peers. Hence, it should be indicated to the router which attributes are essential and which are not. Similarly, the BGP router should also be informed of those attributes that must not be dropped but rather passed on to its peers during updates. Based on these background factors, as shown in Figure 7.6, the path attributes are divided into two major categories – well-known attributes and optional attributes. The well-known attributes must be recognized by all BGP implementations, while the optional attributes may be recognized only by selected BGP implementations and not all, as these may be proprietary.

The well-known attributes are further divided into two subcategories – mandatory and discretionary. The *well-known mandatory* attributes, as shown in Figure 7.7, should be processed by the BGP implementation as well as advertised to the peers in the update message. If an update message is received without a well-known mandatory attribute, an error message is generated. The well-known mandatory attributes are the origin of path information (ORIGIN), the list of the AS sequence that describes the path taken (AS_PATH), and the IP address of the next-hop router (NEXT_HOP). The ORIGIN attribute can have three values – internal, external, or incomplete, and indicates how the BGP router learnt about the route, i.e. from an interior routing protocol or from an

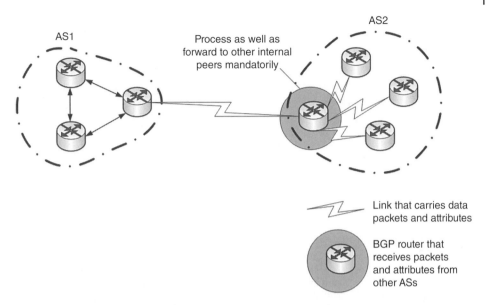

Figure 7.7 Processing of well-known mandatory attributes by BGP router.

external BGP or by some other known or unknown means other than EGP. The value of ORIGIN is introduced by the BGP router initiating the routing and should not be changed in the route by any BGP routers. As AS_PATH contains information about the BGP routers traversed in the route, it helps to prevent routing loops. An update received by a peer with its AS number in the AS_PATH indicates that the information has already been received and it is dropped. Before forwarding a packet to its external peer, a BGP boundary router adds in the beginning its AS number to the AS_PATH. NEXT_HOP is an attribute that is treated differently by external BGP and internal BGP. In the case of an external BGP, the boundary router puts the IP address of its egress interface to reach the external peer. In the case of internal BGP, the internal BGP routers generally do not change the NEXT_HOP and carry it without any change within the AS unless specifically configured to remove the previous entry and add its local IP address [27].

The *well-known discretionary* attributes, though recognized by the BGP implementation, may or may not be included in the update message to the peers, at the discretion of the implementation. The well-known discretionary attributes are local preference (LOCAL_PREF) and atomic aggregate (ATOMIC_AGGREGATE). The LOCAL_PREF is used within an AS and hence forwarded only in internal BGP to identify the most preferred internal path if there exist multiple paths to the destination. The LOCAL_PREF value is not sent to the external BGP peers. The values that LOCAL_PREF can take range from 0 to 4 294 967 295, indicating the degree of preference. A higher value indicates greater preference. The default value of LOCAL_PREF is 100. The ATOMIC_AGGREGATE is used to intimate that specific route information has been lost during route aggregation. When the ATOMIC_AGGREGATE flag is set, it indicates that routers receiving this update should not remove this attribute.

The optional attributes are divided into two subcategories – transitive and non-transitive. The *optional transitive attributes*, as shown in Figure 7.8, have to be passed

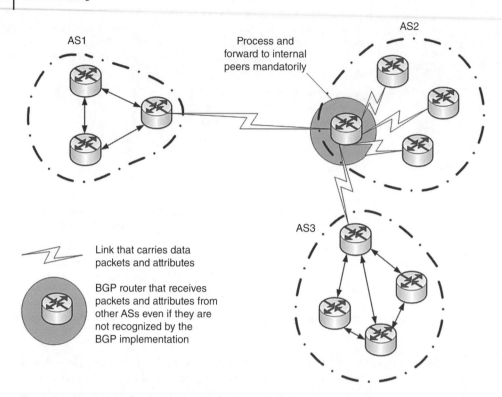

Figure 7.8 Processing of optional transitive attributes by BGP router.

on to the peers, whether they are recognized or not recognized by the BGP implementation. The only advantage of them being recognized is that the implementation can understand the meaning of an attribute that it is passing on to the peer, failing which the attribute is marked as partial by the BGP implementation. The optional transitive attributes are AGGREGATOR and COMMUNITY (No-Export, No-Advertise, Internet, NO_EXPORT-SUBCONFED). The AGGREGATOR attribute records the AS number and the BGP router within the AS that performed route aggregation. The information provided by this attribute indicates the source of origination of the less specific route. COMMUNITY is a 32 bit attribute accommodating two values – the local AS number and the community value in the first 16 bits and in the last 16 bits respectively. COMMUNITY defines a group of routes with similar policies. The default community in a BGP route is the Internet. If the community of the route has to be changed to any other value, it has to be done locally. There are a few other well-defined communities such as NO_EXPORT, NO_ADVERTISE, and NO_EXPORT-SUBCONFED, which restrict advertising the attribute outside the AS or confederation advertisements to peer routers and advertisements to any external ASs, even including ASs that are within the same confederation.

The *optional non-transitive attributes* [7], explained in Figure 7.9, are passed on to the peers if they are recognized by the implementation. They are dropped and not forwarded to the peers if the BGP implementation does not recognize them. The optional non-transitive attributes are Multi-Exit Discriminator (MED), Originator ID, Cluster

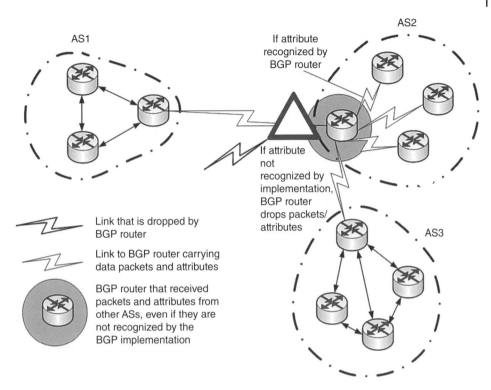

Figure 7.9 Processing of optional non-transitive attributes by BGP router.

List, Multiprotocol Reachable NLRI (network layer reachability information), and Multiprotocol Unreachable NLRI. The MED attribute helps adjoining ASs with multiple links between them to mark the most preferred link. The value of MED can range between 0 and 4 294 967 295, with higher preference to lower values. The default MED value is 0. The MED value is applicable only between two external BGP neighbors.

7.3.2 Routing Information Base

The prime objective of BGP is to determine routes across ASs. Searching for routes as well as the exchange and updating of routing information between the BGP routers is assisted by the routing information base (RIB). The routing information base has three sections – Adj-RIB-In, Loc-RIB, and Adj-RIB-out, the role of each of which is set out below.

Adj-RIB-In. This holds the RIBs received (inwards) from the peers (adjoining routers), and hence the name Adj-RIB-In. The RIB received from each peer is stored separately. Adj-RIB-In is used to update the routing information stored within the BGP router in Loc-RIB.

Adj-RIB-Out. This holds the routing information that the BGP router plans to share (outwards) with its peers (adjoining routers). The Adj-RIB-Out sent by the BGP router to its peers helps in 'route advertisement' and 'route updating' in the peers.

Loc-RIB. This is the main database, which is used locally by the BGP router in 'Route Selection'. The Loc-RIB is regularly updated from the selected information received through Adj-RIB-In.

The structure and the operations performed by the RIB are complex and may be stored in a single database or in separate, but related databases, depending on the internal implementation of the protocol.

7.3.3 BGP Operation

Before two BGP routers can establish connectivity between themselves and start operating, they should be linked to each other through a BGP-enabled network. Even though two BGP routers may be linked with each other, these BGP peers may or may not be connected or may be in the process of establishing connectivity or closing a connection. Hence, as indicated in Figure 7.10, the BGP session is in one of the following states [28] based on the status of connectivity:

- idle state,
- active state,
- connect state,
- open sent state,
- open confirm state,
- established state.

The process of session establishment, session retention, and session closure explains all these states very clearly and is shown in Figure 7.11. When a BGP peer boots up, it is in the *idle state*. The BGP router in the idle state cannot accept any BGP connections. When a start event is triggered manually or automatically, the BGP router initiates a TCP connection on port 179 with its peer and enters into the *connect state*. This TCP connection remains intact throughout the time messages are exchanged between the peers.

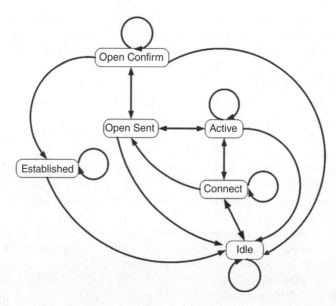

Figure 7.10 State diagram of BGP operation.

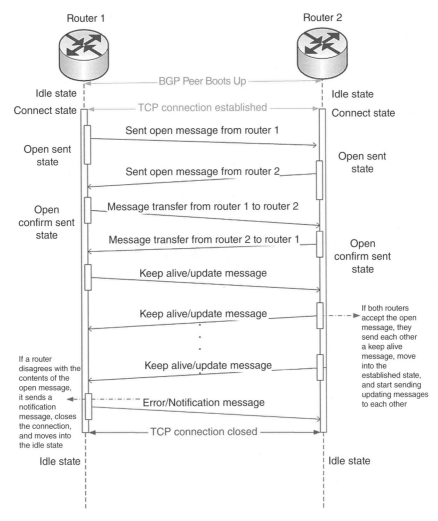

Figure 7.11 BGP session indicating various states.

While in the connect state, i.e. when an attempt has been initiated to make a TCP connection to a peer, if the connection fails, the BGP router tries again to establish the connection and enters into the *active state*, i.e. although the connection has not been established, it is actively trying to establish the connection.

When a BGP connection is established with a peer, the router exchanges BGP open messages and enters into the *open sent state*, i.e. the router has sent a BGP open message and is awaiting a response. The contents of the BGP open message includes the identifier of the router, its AS number, and the parameters that the BGP router desires to use for the session connectivity.

When a router receives an open message, it checks the contents of the message, and if it disagrees with the contents of the open message, it sends a notification to its peer, closes the connection, and goes back to the *idle state*.

If both peers agree on the contents of the open message, each of them sends a keep alive message back to its peer and enters into the *open confirm state*, i.e. the router has replied to the open message in confirmation and is awaiting further action.

When the first keep alive message is received from the peer, it enters into an *established state*, i.e. the connection has been established between the peers and the BGP session is said to be open between the peers. Once a router has entered into an established state, it cannot continue to remain in the state for ever without regular confirmation of its being alive. The frequency of update messages is much lower. Update messages may be exchanged between two routers once in a few seconds. Augmentation of regular communication between the peers is done by sending keep alive messages at a predefined frequency. The frequency varies between 1 and 21 845 s, but the most commonly used frequency of keep alive message transmission is 30 or 60 s.

If a BGP router does not receive any message, i.e. an update message or a keep alive message, for a specified amount of time known as the 'hold time', it closes the connection to that peer and deletes the routes through that peer in its RIB.

When a router establishes TCP connection with its peer for the first time, it exchanges its complete routing information with the peer, and thereafter only updates are exchanged between them to save bandwidth requirement and processing.

7.3.4 Decision Process

In BGP, the decision process is responsible for deciding which among the routes received by the BGP router in its Adj-RIB-In should be selected for storing in Local-RIB. It also decides on the data that should be put in the Adj-RIB-Out to share with its peers. An UPDATE message is exchanged between the BGP routers to share routes among themselves. The UPDATE message is also used to initiate entry for a new route, as well as to advertise routes that can no longer be reached and hence the paths through them should be withdrawn. The decision process helps to update and store the route table, select the preferred route, and advertise routes to peers.

Among the routes available in the Adj-RIB-In of the router, the various routes to each of the networks received from different external peers are analyzed and a preference level assigned to each route. Thereafter, the most preferred route to each of the networks is selected and forwarded to the internal peers. The best route to each network as available from the incoming route information available from Adj-RIB-In and compliant with its input routing policy is used to update the Local-RIB. Routes are selected from Local-RIB based on output policies of the AS for advertising to other BGP routers in adjoining ASs. The set of rules used to define the preferred path for forwarding to the internal peers may be different from the set of rules used to define the preferred route for external peers. This leads to differences in the preferred path to any particular network being sent to the internal peers and that to external peers.

The preference of routes may be based on a set of criteria such as number of in-between ASs, policies of the AS, source AS initiating the path, and reliability of intermediate ASs traversed by the packet. Even though BGP selects the most suitable path to each of the distant networks connected to it through multiple in-between ASs, the protocol cannot guarantee it to be the best path in terms of time, cost, or congestion because BGP is an inter-AS routing protocol and is not aware of the internal routers, links, and policies within any of the ASs. Selecting the best path based only on the

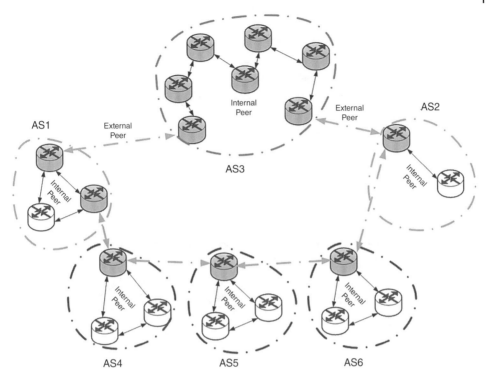

Figure 7.12 A sample network of BGP with similar link connectivity between the routers.

minimum number of in-between ASs may lead to scenarios of doing away with paths with a much larger number of in-between ASs but with a better overall performance. For example, in Figure 7.12, the preferred path between AS1 and AS2 will be through AS3 and not through (AS4, AS5, AS6), but the latter path would have better performance than the former one assuming a similar link connectivity between the routers. Another example is shown in Figure 7.13, where the path from AS1 to AS2 will be selected through AS3 and not (AS4, AS5, AS6). However, the links between AS1–AS4–AS5–AS6–AS2 are much faster, and hence this path would have better performance.

7.3.5 Route Selection Process

The route selection process helps BGP to decide which path should be preferred over other paths when multiple paths exist for the same destination. The BGP router will receive multiple paths from its peers through update messages sent by them and store them in the Adj-RIB-In. The router cannot store all routes towards any network in its Local_RIB, so the most preferred path has to be selected for storing. The stages for screening of paths and selecting the preferred path are depicted in Figure 7.14 and also listed below. Some of the stages might be optional or specific to certain settings, configurations, or protocols:

1) The reachable next-hop IP address of the path.
2) The local AS number is not in the AS_Path (loop prevention).
3) Links with higher weight.

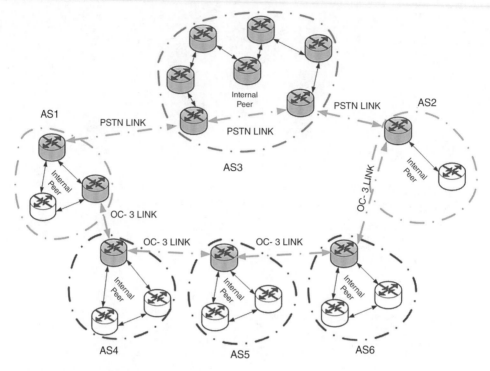

Figure 7.13 A sample network of BGP with variation in link connectivity between the routers.

4) Route with highest local preference.
5) Locally originated route preferred over externally generated route.
6) Shortest AS path.
7) Routers generated through BGP preferred over routes generated through EGP and the last preference for origin from other protocols.
8) Route with lowest MED value.
9) External BGP preferred over internal BGP.
10) Minimum cost to the next hop.
11) Oldest route.
12) For routes learned through EBGP – select route with lowest router ID.
13) For routes learned through IBGP – select route with lowest router ID.
14) Lowest interface IP address.
15) Shortest cluster list.
16) Lowest BGP neighbor IP address.

The major features and advantages of BGP can be summarized as follows:

- The protocol runs over TCP/IP.
- It is highly scalable.
- It avoids routing loops.
- It avoids count-to-infinity problem
- It supports multihomed networks.

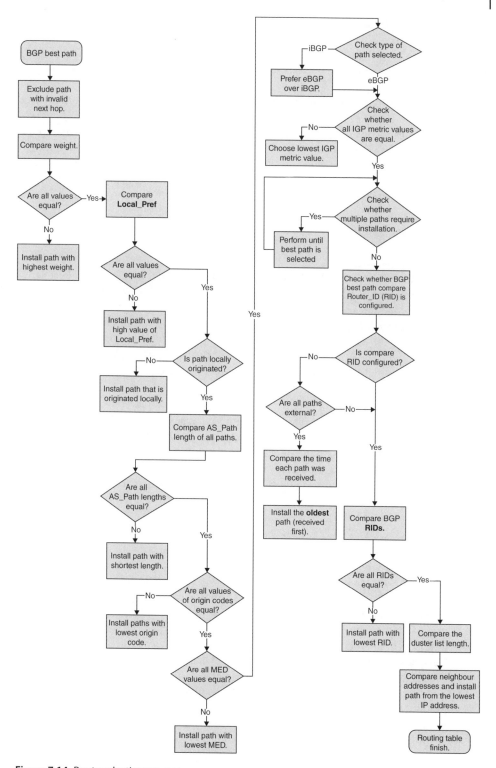

Figure 7.14 Route selection process.

- The protocol can detect a node or link failure and redirect paths by withdrawing from failed routes.
- It gives the shortest path in terms of hop count.
- A number of attributes can be associated with the route to help in the process of path selection.
- BGP is not entirely a shortest path routing algorithm but also has policy-based routing.
- It allows multiple options for route selection, optimization, avoidance, or manipulation using routing policies.
- BGP can work between ASs (external BGP) as well as within the AS (internal BGP).
- The neighboring routers are manually configured to ensure security.
- The present day Internet works on BGP.

References

1 B. M. Leiner, V. G. Cerf, D. D. Clark, R. E. Kahn, L. Kleinrock, D. C. Lynch, J. Postel, L. G. Roberts, and S. Wolff. A brief history of the internet. *ACM SIGCOMM Computer Communication Review*, 39(5):22–31, 2009.

2 T. Narten. Internet routing. *Symposium on Communications Architectures & Protocols, ACM SIGCOMM '89*, pp. 271–282, New York, NY, USA, 1989.

3 Y. Rekhter and T. Li. A Border Gateway Protocol 4 (BGP 4). *RFC 1771*, March 1995.

4 M. Murhammer and E. Murphy. *TCP/IP Tutorial & Technical Overview*. Prentice Hall, 6th edition, 1998.

5 R. Braden. Requirements for Internet hosts – communication layers. *RFC 1122*, 1989.

6 R. M. Hinden and A. Sheltzer. The DARPA Internet gateway. *RFC 823*, 1982.

7 C. Kozierok. *The TCP/IP Guide: A Comprehensive, Illustrated Internet Protocols Reference*. No Starch Press, San Francisco, CA, USA, 2005.

8 R. Braden and J. Postel. Requirements for Internet gateways. *RFC 1009*, 1987.

9 C. Hedrick. Routing Information Protocol: Internal Gateway Protocol. *RFC 1058*, 1988.

10 P. Baran. *On Distributed Communication: External Gateway Protocol, Volume 1*. Rand Corporation, August 1964.

11 A. Balchunas. *CCNP ROUTE: Implementing IP Routing*. Cisco Press, Cisco Networking Academy, 2012.

12 African Network Information Center (AFRINIC). www.afrinic.net.

13 Asia Pacific Network Information Center (APNIC). www.apnic.net.

14 American Registry for Internet Numbers (ARIN). www.arin.net.

15 Latin America and Caribbean Internet Addresses Registry (LACNIC). www.lacnic.net.

16 Réseaux IP Européens (RIPE). www.ripe.net.

17 E. C. Rosen, B. Beranek, and N. Inc. Exterior Gateway Protocol. *RFC 827*, 1982.

18 L. J. Seamonson and E. C. Rosen. 'Stub' Exterior Gateway Protocol. *RFC 888*, 1984.

19 D. Mills. Exterior Gateway Protocol formal specification. *RFC 904*, 1984.

20 J. Doyle and J. D. Carroll. *CCIE Professional Development: Routing TCP/IP*. Cisco Press, 2001.

21 S. Nivedita, R. Padmini, and R. Shanmugam. *Using TCP/IP*. Que, 2nd edition, 2002.

22 J. F. Kurose and K. W. Ross. *Computer Networking: A Top-Down Approach*. Pearson Education, 6th edition, 2004.

23 P. Traina, D. McPherson, and J. Scudder. Autonomous system confederations for BGP. *RFC 5065*, 2007.

24 R. White, D. McPherson, and S. Sangli. *Practical BGP*. Addison Wesley Longman Publishing Company, CA, USA, 2004.

25 W. Odom, D. Hucaby, and K. Wallace. *CCNP Routing and Switching Official Certification Library*. Cisco Press, 1st edition, 2010.

26 BGP attributes. http://training.apnic.net/docs/eROU04_BGP_Attributes.pdf.

27 K. Solie and L. Lynch. *CCIE Practical Studies, Volume II*. Cisco Press, 2003.

28 Y. Rekhter, T. Li, and S. Hares. A Border Gateway Protocol 4 (BGP 4). *RFC 4271*, 2006.

Abbreviations/Terminologies

ARPA	Advanced Research Project Agency
ARPANET	Advanced Research Projects Agency Network
AS	Autonomous System
BBN	BBN (Bolt, Beranek, and Newman) Technologies [a company]
BGP	Border Gateway Protocol
DARPA	Defense Advanced Research Projects Agency
DoD	Department of Defense (of USA)
EBGP	External Border Gateway Protocol
EGP	Exterior Gateway Protocol
EIGRP	Enhanced Interior Gateway Routing Protocol
GGP	Gateway-to-Gateway Protocol
IANA	Internet Assigned Numbers Authority
IBGP	Internal Broader Gateway Protocol
IGP	Interior Gateway protocol
IGRP	Interior Gateway Routing Protocol
I-H-U	I Hear You
INOC	Internet Network Operations Center
IS-IS	Intermediate System to Intermediate System (a routing protocol)
ISP	Internet Service Provider
MED	Multi-Exit Discriminator
MILNET	Military Network
MIT	Massachusetts Institute of Technology
NLRI	Network Layer Reachability Information
NWG	Network Working Group
NSFNet	National Science Foundation Network
OSPF	Open Shortest Path First
RFC	Request For Comments
RIB	Routing Information Base
RIP	Routing Information Protocol
RIR	Regional Internet Registry
SRI	Stanford Research Institute
TCP/IP	Transmission Control Protocol/Internet Protocol
UCLA	University of California, Los Angeles

Questions

1. Explain the connectivity of core and non-core routers in ARPANET during the 1980s.

2. Give reasons to justify separation of a host computer from running gateway applications.

3. GGP was built when the links had low bandwidth and were unreliable. Justify the statement by giving an example of any mechanism adopted in GGP to overcome this problem.

4. Explain the 'extra hop problem' of ARPANET.

5. Name the five Regional Internet Registries. Identify the RIR of your area.

6. What does the statement 'EGP is a type of EGP' mean?

7. State an advantage of being a stub router or a stub AS. Think in terms of types of traffic.

8. Why does an AS allow transit traffic?

9. **Differentiate between the following:**
   A external BGP and internal BGP,
   B stub router and multihome router,
   C intra-autonomous routing and pass-through routing,
   D route aggregation and AS confederation,
   E active state and connect state (in BGP).

10. Classify the path attributes of BGP and explain each of them.

11. State whether the following statements are true or false and give reasons for the answer:
    A All the stages are mandatory for screening of paths and selecting the preferred path in BGP.
    B The Internet works on BGP.
    C BGP avoids routing loops and the count-to-infinity problem.
    D While in the connect state, if the connection fails, the BGP router enters into the idle state.
    E The routing information base has three sections – Adj-RIB-In, Loc-RIB, and Adj-RIB-out. All three of these should be stored in a single database.
    F MED is an optional path attribute in BGP.
    G A stub AS does not have a BGP boundary router.

Exercises

1 Assume a network of core routers as shown in the figure below and running GGP. The entire network boots up at time $t = 0$. Show the exchange of messages across all the links in the first 20 s.

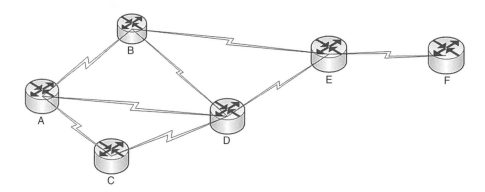

2 Consider the network mentioned in exercise 1 running GGP. Indicate the series of routing update messages, including the sequence numbers exchanged between the core routers, until the topology information propagates across the entire network and the network converges. Now assume that links between core router B and core router E and between core router B and core router D fail. What will be the sequence of messages exchanged among the routers until the new topology propagates across the entire network? Necessary assumptions may be made regarding the network addresses.

3 Consider the network below. Path cost is based only on the number of hops. The address of the router may be assumed. Write down the routing tables exchanged between the routers for inter-AS routing.

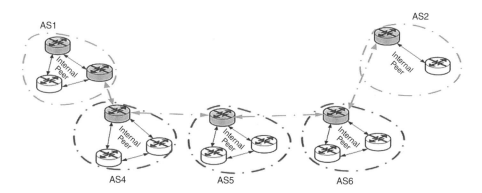

4 Consider the network given in this exercise. The routing between ASs is based on a policy that the linkages of the AS with any router in the AS with an even number will not be informed to the neighboring AS. The path metrics are based on hop count. What

is the routing table shared by each Internet facing router to its neighbor in response to a request received from it. Addresses may be assumed.

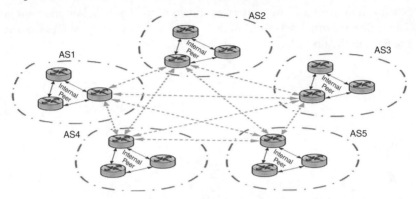

5 Draw a timeline to indicate the contribution of each RFC in the Exterior Gateway Protocol (all associated protocols such as GGP, EGP, BGP).

6 Draw the state diagram of an EGP gateway.

7 EGP makes no mention of the process of neighbor discovery, and hence it is generally configured manually. Assign the IP address to each port of the routers in the network depicted in exercise 1 so that connectivity can be established between the neighbors to make the network ready for routing of packets between them.

8 There are a number of ASs interconnected with each other. Each AS has only one border router running BGP, and thus the interconnection of the ASs has been depicted as the interconnection of the boundary routers in the figure below. The AS number has been depicted as the router ID. Identify the external peers of all the routers. For each AS, identify whether it is a stub AS, a dual-homed AS, or multihomed.

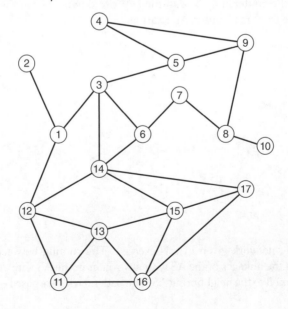

9   In the network explained in exercise 8, use proper representation to indicate route aggregation in any three different routers.

10 Consider the network along with the policy given in exercise 4. The network runs BGP. Write the RIB as a single database for the network.

Part IV

Other Routing Contexts

8

Routing in ATM Networks

8.1 Introduction

Asynchronous transfer mode (ATM) is a connection-oriented switching and network transmission technique introduced in the late 1980s. ATM was introduced in the days when public switched telephone networks (PSTNs) were in common use supporting only voice or data. ATM brought in the technology to support voice, video, and data over a single network. This is a full duplex transmission technology that supports a variety of physical media such as UTP cable and fiber optic cable. ATM can be used in a small LAN as well as between two or more LANs spread across wide distances and connected over a WAN.

As ATM is connection oriented, it ensures delivery of cells in order and with high accuracy. Most of the ATM switching function is implemented in the hardware, and the least number of operations are performed through software. This enhances the speed of the system. Operability over fiber optic cable makes it free from noise interference during transmission and can help ATM reach gigabit transmission rates.

An ATM packet size [1] is fixed at 48 bytes for payload and 5 bytes for the header, leading to a total size of 53 bytes. The fixed packet size leads to removal of information

Network Routing: Fundamentals, Applications, and Emerging Technologies, First Edition.
Sudip Misra and Sumit Goswami.
© 2017 John Wiley & Sons Ltd. Published 2017 by John Wiley & Sons Ltd.
Companion website: www.wiley.com/go/misra2204

regarding size of the payload in the packet header. The fixed-size packet also supports the design of ATM-specific switching hardware to work at higher speed, as the packet size is fixed and known *a priori*. Even the operating software running on the ATM switches is much simpler and hence faster, as it requires neither any procedure to detect the size of the packet nor any procedure to determine where one packet ends and the other begins. The fixed-size, 53 byte packet of an ATM is referred to as a 'cell'. Generally, a small packet size is good for transmission of voice and video over the network, while a big packet size better supports data traffic. A large packet size enables a huge amount of data to be carried with comparatively less overhead of carrying its header. However, a fixed-size large packet may lead to wastage of data carrying space in the packet owing to non-availability of sufficient data to fill the complete data space of each packet. The cell size of 53 bytes was decided so as to make the size optimum to support voice as well as data.

The mechanism for handling variable packet size in a network is complex and hence cannot be completely implemented in the hardware. The operating software running on network equipment, such as switches or routers, has to handle a variable packet size and read each bit passing through it to determine the start of the packet, read the information about the size of the packet from the header, detect the end of the header and the start of data payload, distinguish between the actual data and the padding in the data payload, and finally detect the end of the packet in the stream [2].

The packets from various network links are multiplexed over a single line for long-distance transmission. A sample output of multiplexing packets from three different networks over a single network is depicted in Figure 8.1. The largest-size packet 'G' gets preference over any other packet as it was the first to reach the multiplexer, while the smallest-size packet 'F' has to wait until packets 'G' and 'B' are transmitted. This leads to a huge waiting time before transmission of the small packets in the example. Fixing priority to the lines would also not help in the scenario because, even if the second line could have been assigned the highest priority, 'F' would still be transmitted after 'G' as it arrived after 'G' and the multiplexer would not keep waiting for 'F' to arrive when 'G' had already done so, and the time and probability of arrival of 'F' are not known to the multiplexer. The only difference that priority-based multiplexing with highest priority to the second line can make here is that 'F' will be transmitted after 'G' but before 'B'. As data packets are generally larger in size than voice and video packets, this example clearly shows the disadvantage that voice and video traffic can have over this type of network.

Using the same size packets in the network helps in avoiding delays due to the larger packets making the smaller packets wait for channel availability. The larger-sized packets can be broken into smaller fixed-size packets, and if the entire network uses the same size packets as depicted in Figure 8.2, the problem depicted in Figure 8.1 is avoided. This

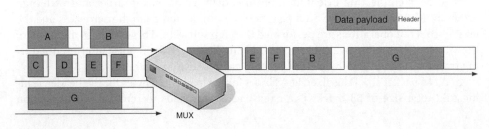

Figure 8.1 Multiplexing packets of variable size from three networks.

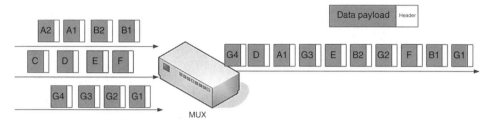

Figure 8.2 Multiplexing packets of fixed size from three networks.

also enables picking a packet from each of the channels one after the other, leading to a small waiting time and a small delay between the transmission and reception of two consecutive packets from the same network. The working of ATM is similar to the example depicted in Figure 8.2 and is known as asynchronous time division multiplexing, wherein a number of input channels are multiplexed over a single channel. It is called asynchronous because it picks the data packet, i.e. the 'cell', from any of the input channels that has a cell awaiting transmission. The output channel at a particular time slot is empty only if all the input channels do not have any cell to transmit at that time slot.

8.1.1 ATM Frames

The ATM network has two different interfaces, one for connection between the ATM switches and the other for connection between the ATM switches and the endpoints. The interface between ATM switches is called the network-to-network interface (NNI). The endpoints in an ATM network can be computers, workstations, and servers. The interface between ATM switches and endpoints is known as the user-to-network interface (UNI). The UNI and NNI are depicted in a sample network shown in Figure 8.3.

   Although the ATM cell header is 5 bytes, there are two different types of cell header in ATM, depending on the endpoints between which the ATM cell travels. When an ATM cell travels between an ATM switch and a network endpoint, the cell uses a UNI header, which is depicted in Figure 8.4. When the ATM cell travels between two ATM switches, it uses an NNI header as depicted in Figure 8.5. As can be seen from the figures of the UNI and NNI headers, the first four bits of the UNI header are for 'generic flow control', while even these four bits are used for the virtual path identifier in the NNI header. This is the only disparity between the two headers. Apart from this, the name and the size of all other fields in the headers are the same in both UNI and NNI. The fields in the headers [1] are described in Table 8.1.

8.1.2 ATM Connection

ATM is a connection-oriented switching network. Hence, a connection has to be established between the endpoints before transmission of the cell stream. Two different types of connection can be established in an ATM – permanent virtual circuit connection (PVC) and switched virtual circuit connection (SVC). PVC between the endpoints can be established only through the network service provider, while SVC is created by the ATM with support from any network layer protocol every time the endpoints want to communicate with each other.

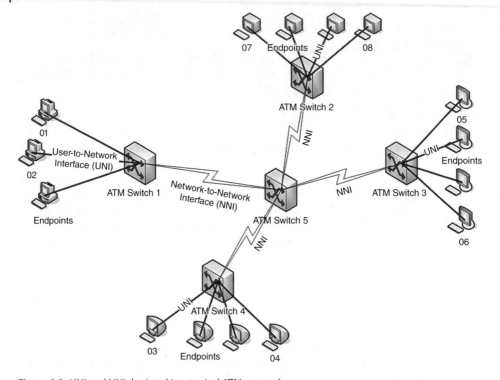

Figure 8.3 UNI and NNI depicted in a typical ATM network.

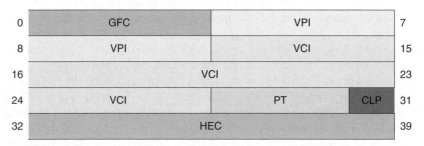

Figure 8.4 ATM cell header for the user network interface (ATM switch–endpoint).

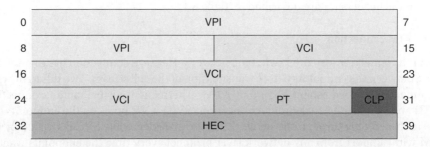

Figure 8.5 ATM cell header for the network node interface (ATM switch–ATM switch).

Table 8.1 The fields in the ATM cell header.

Field	Interface	Bits	Description
Generic flow control (GFC)	UNI	4	This is used to identify the individual computer at the endpoint. The default value is '0000' and the field is rarely used these days.
Virtual path identifier (VPI)	UNI, NNI	8 UNI 12 NNI	Uniquely identifies the virtual path through the network for the cell. In the case of a control cell for call setup or termination, the value of the field is 0.
Virtual circuit identifier (VCI)	UNI, NNI	16	This field along with the VPI uniquely identifies the path of the cell through the switched network. Field values of 0 to 15 are reserved for use by the International Telecommunication Union (ITU), and field values of 16 to 32 are for the use of the ATM forum for signaling and control operations.
Payload type (PT)	UNI, NNI	3	All three bits have separate respective indications. A value of '0' in the first bit indicates that the cell has a data payload, and '1' indicates a control payload. For a data payload, the second bit, which has an initial value of '0' from source, reports network congestion by being changed to '1' by the switch facing congestion. For data payloads, the third bit indicates the last cell of the series in the AAL5 frame. For control payloads, the bit is used for control purposes.
Cell loss priority (CLP)	UNI, NNI	1	This indicates priority for dropping the cell during network congestion. A field value of '0' indicates a preference for not being dropped, while cells with a field value of '1' are selected for dropping.
Header error control (HEC)	UNI, NNI	8	This stores the CRC value of the first four bytes of the header.

For creating the connection channel, logically three different hierarchies of connection exist in ATM. These are the virtual circuits (VCs), virtual path (VP), and transmission path (TP). The VC between two endpoints carries all the cells of a message and is transmitted from the source to the destination in order through the VC. Virtual circuits are uniquely identified by a 16 bit virtual circuit identifier (VCI). The set of virtual circuits over the same path between switches are bundled together to form the VP. The creation of a virtual patch for the bundled virtual circuits helps to perform common control and management functions together for all the VCs in the VP. The VP is uniquely identified by the virtual path identifier (VPI). All the virtual circuits that are bundled in the same VP have the same VPI. There can be different virtual circuits with the same VCI in two or more different virtual paths. These VCs with the same VCI can be identified uniquely with the help of the VPI. TP is the connection between switches or between the switch and its endpoints. A logical cross-section of the virtual connection depicting VC, VP, and TP is represented in Figure 8.6. Breaking up the connection into VP and VC helps to create a hierarchy to support better routing strategies. The ATM switches, when communicating with each other, have to use

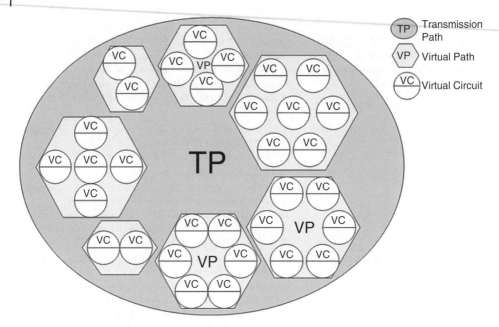

Figure 8.6 Cross-section of a virtual connection indicating VC, VP, and TP.

Input			Output		
Interface No.	VPI	VCI	Interface No.	VPI	VCI

Figure 8.7 Structure of an ATM switching table.

only the VPI, while only the boundary switches that connect to the endpoint have to use VPI and VCI.

In Figure 8.3, if endpoints 1 and 2 want to communicate with endpoint 3, two virtual circuits are created, one from endpoint 1 to endpoint 3 and the other from endpoint 2 to endpoint 3. A single virtual path will bundle both these virtual circuits, and this virtual path will be from ATM switch 1 to ATM switch 4 through ATM switch 5. However, if endpoint 4 wants to communicate with endpoint 5 and endpoint 6 wants to communicate with endpoint 7, two virtual circuits and two virtual paths will the formed. The first virtual path will be from ATM switch 4 to ATM switch 3 through ATM switch 5, and the second virtual path will be from ATM switch 3 to ATM switch 2 through ATM switch 5. There may be one or more links between the ATM switches. Assuming that there is only a single link between the switches, as depicted in Figure 8.3, there is a transmission path each from switch 1, switch 2, switch 3, and switch 4 to switch 5.

In an ATM network, the ATM switches perform cell routing with the help of a switching table maintained in it. The boundary switches that are connected to the endpoints at one interface and some other switch or endpoint at the other interface have to use the virtual circuit identifier as well as the virtual path identifier for cell switching. As shown in Figure 8.7, the switching table has six columns of information per row, three each for input and output respectively. These columns are the interface number, VPI, and VCI for both the input and the output. When a cell arrives at the interface of a switch, the VPI and the VCI are known through the header of the cell, and the arrival interface number is determined from the interface on which the packet has arrived. The switch checks the routing table entry corresponding to this triad of < Interface No. (input), VPI (input), VCI (input) > and, on locating the entry corresponding to this information, gets the switching information in the form of < Interface No. (output), VPI (output), VCI (output)>. Based on this output information, the switch forwards the cell through the interface number mentioned in the switching table with the changed VPI and VCI as obtained from the switching table [2]. Thus, it is observed that the VCI and VPI are of local significance, restricted to a single link and changing with each hop. When a cell is forwarded from one ATM switch to another ATM switch, the switching table is used to change the value of VCI and VPI of the cells passing through the ATM switch.

8.1.3 ATM Architecture

ATM is a three-layer protocol that is represented in Figure 8.8, and these layers from top to bottom are:

- ATM adaptation layer,
- ATM layer,
- physical layer.

ATM adaptation layer. The ATM adaptation layer can be divided into two sublayers – the convergence sublayer (CS) and the segmentation and reassembly sublayer (SAR). The CS sublayer receives the data frame from the upper layer and encapsulates it in a format for supporting reassembly at the other end. The SAR sublayer divides the data into 48 byte segments and passes it on to the ATM layer, where a 53 byte cell is created with this 48 byte payload. As the layer divides the data into 48 byte segments at the transmitting end and reassembles the 48 byte data into the complete message at the receiving end, it has been named the segmentation and reassembly sublayer.

Several versions of AAL have been defined. These are AAL0, AAL1, AAL2, AAL3/4, and AAL5. The most commonly used among these are AAL1, which is used for time-dependent applications at constant bit rates such as voice and video, and AAL5, which also has the error control mechanism for applications with variable bit rates. AAL2 was designed for variable-data-rate bit streams for applications such as compressed voice and video, and the layer was later modified for low-bit-rate, short-frame traffic. AAL3/4 was designed for a variable-bit-rate connection-oriented (AAL3) and variable-bit-rate connectionless (AAL4) service for applications such as data transfer over LAN.

ATM layer. The ATM layer takes a 48 byte segment from the AAL, adds the 5 byte header to it, and makes a 53 byte ATM cell. The ATM layer is responsible for multiplexing-demultiplexing, switching, routing, flow control, and traffic management. The ATM layer is also responsible for monitoring the connection for QoS. However, unlike equivalent layers in the other protocols, the ATM layer is not responsible for error correction.

Layer	Sublayers
ATM Adaptation Layer	Convergence (CS)
AAL 1 | AAL 2 | AAL3/4 | AAL5	Segmentation and Reassembly (SAR)
ATM Layer	
Physical Layer	Transmission Convergence (TC)
	Physical Medium Dependent (PMD)

Figure 8.8 Three-layered ATM protocol stack.

Physical layer. The physical layer is responsible for actual transmission of the cells in the network. ATM cells can be carried by any of the commonly used physical layer transmission mediums such as metallic wire or fiber optic cable. Although ATM was initially designed to work over SONET at the physical layer, now it is not limited to any transmission medium, and even wireless transmission can be used at the physical layer for ATM. The physical layer can be subdivided into two sublayers – the physical-medium-dependent sublayer (PMD) and the transmission convergence sublayer (TC). PMD is responsible for interfacing with the actual transmission medium and performing the signal encoding for the same. The design of the physical layer has enabled ATM to transmit over different types of physical network by creating and defining a variety of PMDs. TC is the interface between the ATM layer and the PMD. The TC layer takes the cell from the ATM layer and maps it to the specific frame for the PMD.

The ATM switches do not require the AAL while communicating with each other, and they use only the two lower layers, i.e. the physical layer and the ATM layer. All three layers, including the AAL, are required for communication between endpoints.

8.1.4 Service Categories

Initially ATM reserved a specified and fixed amount of bandwidth for a connection to ensure the service quality. However, all the applications do not require a fixed amount of bandwidth throughout the period of connection as every application has got its different traffic pattern, bandwidth requirement and bandwidth consumption pattern. The requirement can be for a real time support or non real time support for a variety of traffic patterns such as traffic at constant rate, bursty traffic or it can even manage with any available bandwidth. Though all these traffic arrive in the form of 53 bytes cell streams, each has a different requirement for traffic flow based on the application

Table 8.2 ATM service categories.

Service Category	Application	Bandwidth Requirement	Service Level
CBR	Real-time applications – voice, video, videoconference, telephone call, video on demand, radio, TV.	The maximum bandwidth is required whenever the application is in use. The rate of flow of information received at destination is equal to the rate of transmission at source.	Guaranteed constant bandwidth.
rt-VBR	Compressed voice or video, teleconferencing.	Traffic varies with time, bursty in nature, but time sensitive.	Transfer delay and delay variation are tightly controlled. Network resources allocated at minimum sustainable cell rate.
nrt-VBR	Reservation system, process monitoring, store/compress and forward video.	Bursty traffic that can tolerate delays. The peak cell rate, average cell rate, and expected frequency of burst are specified.	Delay variation is not controlled, cell loss is controlled. Network resource is allocated to provide low delay and minimal cell loss.
ABR	Critical file transfer, financial transactions, fax.	The bandwidth requirement of the application changes with the traffic condition. It may specify a minimum required bandwidth, but may use more if available.	Best-effort service with congestion control.
UBR	File transfer, remote terminal, mailing, network monitoring.	No specific bandwidth or QoS required. Use the available bandwidth.	Best-effort connection.

and hence ATM has specified the following service categories to handle different types of traffic:

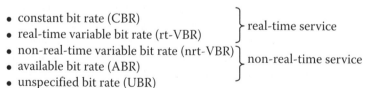

- constant bit rate (CBR)
- real-time variable bit rate (rt-VBR) } real-time service
- non-real-time variable bit rate (nrt-VBR)
- available bit rate (ABR) } non-real-time service
- unspecified bit rate (UBR)

The application scenario, bandwidth requirement, and service levels for the five service categories specified in ATM are explained in Table 8.2.

At any point in time, out of the available link bandwidth, a certain amount of bandwidth is reserved for carrying CBR traffic. Thereafter, from the remaining bandwidth, first the rt-VBR and then the nrt-VBR are assigned bandwidth. From the remaining small amount of bandwidth left in the link, ABR is assigned the bandwidth as it has

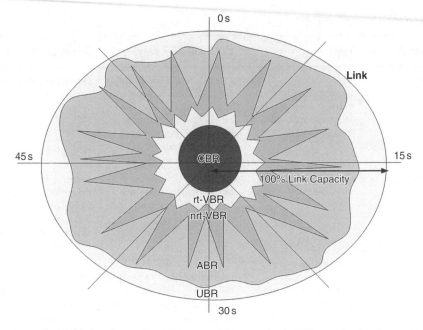

Figure 8.9 Bandwidth distribution for different service categories in ATM over circular time-variant display.

specified the bare minimum bandwidth it requires which is generally very small and it may use more bandwidth if available. If some bandwidth is still available owing to non-utilization by the remaining four services or underutilization by these four services, that bandwidth is assigned to the UBR, which has the least priority. The concept can be seen in Figure 8.9, which is a logical representation of time-varying bandwidth allocation to the different service categories in ATM.

8.2   PNNI Routing

Routing in ATM networks is more complex than routing in IP networks because in IP networks only the routes for packet forwarding have to be discovered, while in ATM routing, not only does the route have to be determined, but also the QoS has to be guaranteed. Before selecting a route in the ATM network, it has to be ensured that sufficient network resources are available and reserved for the data to reach the destination. There are a number of QoS parameters in ATM routing, which further complicates the routing process. ATM is a source-based routing and not a hop-by-hop routing. As this source-based routing should also ensure QoS, the QoS information about all the intermediate switches in all the possible paths should be known to the source before forwarding the cell [3]. There can also be variation in the network load and the network condition. Thus, it requires a regular exchange of resource availability information between the switches.

ATM is a connection-oriented protocol and hence a virtual connection is established between the source and the destination before sending the data. The entire data is sent to the destination through this path, and hence the individual cells are not required to carry the destination address in the header and all the cells follow the same path. As all the cells will follow the same virtual circuit between source and destination, the path selection algorithm should also attempt to select the optimum path from among the multiple available paths.

In ATM the routing is in terms of signaling messages. The signaling messages help to establish the switched virtual connection among the ATM switches. The routing can be between two ATM switches or between two ATM networks. To support routing between two ATM switches, the private network–node interface (PNNI) is used. The private network-to-network interface (PNNI) is used for routing between two ATM networks. PNNI helps in the interconnection of switches from various vendors over different networks. PNNI was introduced after Interim Interface Signaling Protocol (IISP), which is a static routing protocol for the ATM network.

PNNI comprises a routing protocol and a signaling protocol. The routing protocol defines the mechanism for creation of a hierarchical topology, neighbor discovery, sharing of topology information among the nodes, creation of peer groups, selection of peer group leaders, link aggregation, node aggregation, and finally path selection. The signaling protocol is responsible for establishing the connection between nodes, indicating any resource starvation at nodes or links before establishment of the path through them, and thereafter setting up alternative routing paths.

8.2.1 PNNI Interface

An ATM network comprises ATM switches, physical links, and end systems. An end system can be a computer or a server. End systems are connected to a switch, and the switches are connected to each other to extend the network. The connectivity between the switches or between the end system and the switch can be point to point or point to multipoint through the physical links. These physical links can be over fiber, copper cables, or even wireless. The point of connectivity of a link with the switch is termed the 'port'. Thus, for clear identification of the connectivity between two switches, the switch identifier and the port identifier have to be mentioned. The links are bidirectional and duplex in nature. The traffic flowing in each direction is generally different in terms of the amount of data being carried, the time of link utilization, and the amount of data to be transferred. This leads to differences in the characteristics of the link in the two directions. Further to variation in the traffic load through the link in the two directions, the link characteristics may themselves be different in the two directions, leading to variation in its capacity. Hence the physical links have to be identified separately for each direction.

8.2.2 PNNI Hierarchy

PNNI routing being source based, all the nodes should be aware of the entire network topology as well as the condition of the network in terms of resource availability, node congestion, and link congestion for ensuring QoS. If a flat routing is used, the scalability will be highly constrained owing to the large amount of information required to be stored at each node, as well as the regular update on the condition of each link and node. The hierarchical topological structure built up using PNNI ensures scalability by reducing overheads and makes routing efficient in an ATM network. PNNI can have a hierarchical level from 1 to 10.

In PNNI routing, the nodes are grouped into peer groups, and there are peer groups at various hierarchical levels. Each peer group elects a peer group leader. In addition to the other activities performed by the peer group leader, it also participates as a representative of the peer group in the next peer group which is in the next higher hierarchical level. The nodes in a peer group share topology information with each other through flooding. All the nodes in a peer group thus have the same topological information.

Figure 8.10 illustrates a network organized in a hierarchy. The network has six lowest-level peer groups (PGs) named PG-A.1, PG-A.2, PG-A.3, PG-B.1, PG-B.2, and PG-C. The nodes, which have been depicted by the node IDs A.1.1, A.1.2, A.1.3, and A.1.4, form the peer group PG-A.1, and the node A.1.1 is the peer group leader of PG-A.1. Similarly, the nodes A.2.1, A.2.2, A.2.3, A.2.4, and A.2.5 form the peer group PG-A.2, and the node A.2.2 is the peer group leader. The other peer group nodes and the peer group leaders are indicated in a similar manner. This notation of the peer group as PG-x.y and a node as a.b.c.d is just for indicating the nodes and peer groups in a simple way during description of the protocol. The actual addressing of the node uses a 20 byte addressing scheme, which indicates clearly the node ID and the peer group. The peer groups PG-A and PG-B are the second-hierarchical-level peer groups formed by the peer group leaders of its lower-level peer groups. PG-A has in it a logical node each from PG-A.1, PG-A.2, and PG-A.3, and similarly PG-B has in it a logical node each from PG-B.1 and PG-B.2. The second-hierarchical-level peer groups also select a peer group leader among them to represent the peer group in the next hierarchical level. The highest-hierarchical-level peer group in the example has a logical node each from PG-A, PG-B, and PG-C. The highest-level peer group does not require any peer group

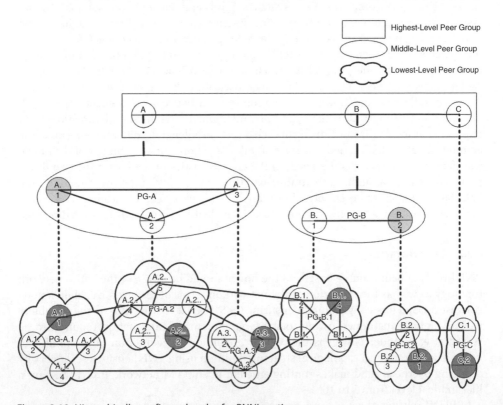

Figure 8.10 Hierarchically configured nodes for PNNI routing.

leader within its logical nodes. However, if the nodes in the ATM network increase by a large number and the hierarchical levels of the PNNI increase further, the peer group that is presently the highest will have another level of peer group above it, and hence a leader will be required to be elected to represent the peer group in the newly formed topmost hierarchical peer group.

8.2.3 Building the Network Topology

The nodes on either side of a link are the neighboring nodes. The neighboring nodes may be in the same peer group or in different peer groups. The neighboring nodes periodically exchange 'hello packets' with each other, indicating their peer group IDs. If the peer group IDs of the neighboring nodes are different, topology information is not exchanged between them. A node with its neighbor in another peer group is known as a 'border node'. If the peer group IDs of the neighboring nodes are the same, they share topology information with each other. Two different peer groups can communicate with each other if both have border nodes that connect to a border node in the other peer group. The connectivity can also be through some other intermediate peer groups that are connected to these peer groups. The peer group is aware of the border nodes available in it, and all the nodes in the peer group communicate with other peer groups through these border nodes. The border nodes also share information about their higher-level peer group and the peer group leader representing the peer group in the higher level with their neighbor node belonging to the other peer group. This helps the border nodes to locate the lowest-level peer group, which is logically connected to both the border nodes. In Figure 8.10, the nodes C.1 and B.2.2 are the border nodes and the highest-level PG in the network is the peer group that logically connects these lowest-level nodes, and the connectivity is: C.1 – highest-level PG [C–B] – PG-B [B.2] – B.2.2. The nodes A.2.2 and A.3.1 are border nodes and the lowest-level peer group that connects these border nodes is PG-A, and the connectivity is: A.2.2 – PG-A [A.2–A.3] – A.3.1.

The peer group leader may not be the border node, but it uses the border node to route its traffic. Thus, the connectivity between A.1, A.2, and A.3 in PG-A depicted in Figure 8.10 is not through direct links but through logical links indicating connectivity between the peer groups through some other nodes in the peer group. At the lowest hierarchical level, the links between the nodes are generally physical links or virtual circuits. The links inside a peer group are called 'horizontal links', while the links between nodes in separate peer groups are called 'outside links'.

As soon as a link becomes operational, the nodes at either side of it start periodic exchange of 'hello packets'. The 'hello packet', in addition to containing the peer group ID of the node, also contains information about its node ID and the port ID. This exchange of information is done over the routing control channel through a virtual circuit setup over the link between the nodes. This regular exchange of 'hello packets' between the neighboring nodes not only gives information about the continuation of connectivity but also provides information such as link delays and processing delays at the neighboring node owing to high resource utilization.

The state information of a node is passed on to its peer group members through messages called PNNI topology state elements (PTSEs). A PTSE contains nodal information and topology state information. Nodal information comprises system capabilities and nodal state parameters, i.e. outgoing resource availability and a nodal information group that has next higher-level binding information. Topology state information contains information about horizontal links, uplinks, internal reachable ATM addresses, and exterior reachable ATM addresses. Some attributes and metrics of the PTSEs are static, while certain others are dynamic with different rates of change in information. However, the dynamic and static parameters are exchanged in a combined way between the nodes through the PTSEs. The dynamic parameters are not exchanged at a greater frequency between the nodes. Each node floods its PTSEs in the peer group, and thus all the nodes in a peer group have the PTSEs of all other nodes, giving them a complete view of the entire peer group. The aggregated topology information is also passed up in the hierarchy through the peer group leader, through which it reaches the other peer groups. Similarly, the peer group leader passes the information it receives from other peer groups down in the hierarchy to the lower-level nodes, giving each node a view of the complete network. The collection of all PTSE information in a node generates the topology database for the node. This topology database can provide path information from the node to any other node in the network. PTSEs are exchanged between nodes using PNNI topology state packets (PTSPs), which also use the routing control channel (RCC) over the virtual circuits between the neighboring nodes.

8.2.4   Peer Group Leader

The selection of the peer group leader (PGL) is based on the value of 'leadership priority'. A node in a peer group with highest 'leadership priority' is selected as the leader. Once the node is selected as a leader, its 'leadership value' is further increased so as to ensure that it continues to remain the peer group leader. This continuity leads to stability in the peer group with respect to aggregations and communication and avoids change of the peer group leader at short intervals. However, in the case of failure of the peer group leader or joining of a new node with a higher 'leadership value', the PGL election algorithm, which keeps running continuously, selects the new PGL.

Every peer group has only one peer group leader. A peer group is sometimes partitioned owing to failure of links or nodes in the peer group. This leads to the creation of two peer groups with the same peer group ID, but each with a separate PGL. The PGL performs three major activities – link aggregation, node aggregation, and representation of the peer group in the higher-level hierarchy by acting as a logical group node (LGN). If a network has only one peer group, a PGL is not required.

Link aggregation is the process of representing multiple links between two peer groups by using a single logical link. Two peer groups may be connected with each other using one or more border nodes. A border node in a peer group may be connected to two or more different border nodes in another peer group. Alternatively, two or more border nodes in a peer group may be connected to two or more border nodes in another peer group. These links are represented in the lowest-level hierarchy. However, while representing the PNNI in the higher-level hierarchies, these multiple links between any two peer groups are represented in the aggregated form of a single logical link. For example, in Figure 8.10 the logical link between A.1 and A.2 is an aggregation of the links (A.1.1–A.2.4) and (A.1.3–A.2.4), and the logical link between A.2 and A.3 is an aggregation of the links (A.2.1–A.3.3) and (A.2.2–A.3.1). Even at the highest level of hierarchy, the logical link between A and B is an aggregation of the links (A.2.5–B.1.2) and (A.3.1–B.1.1).

In a hierarchical representation of a peer group, the peer group leader represents the entire peer group in the next upper hierarchical peer group. Thus, all the nodes in a peer group are aggregated as a single node through its peer group leader in the next higher peer group, a process known as node aggregation.

The peer group leader, which becomes the logical group node (LGN) in the next hierarchical level, is responsible for sending information from its peer group to the other peer groups. Full topology and addressing information is not sent to other peer groups by exchanging PTSEs or topology databases. Aggregated topology information and summarized reachability information are sent through the PGL to the next level in the hierarchy, and from there it goes to the other peer groups through the LGN. The summarized reachability information is in terms of addresses than can be reached through its lower-level peer group.

8.2.5 Advertizing Topology

When a neighbor node comes up at the other side of the link, this node may not have any topology information available with it as the node might be newly joining the network. In such a case, the entire topology database from the node already operating in the network is copied to the neighboring node that has newly joined the network. Once the network is in operation, the nodes in a peer group regularly exchange PTSE information to have a similar and updated topology database. The topology database in a node has detailed topology of all other nodes in the peer group and abstract topology information about the entire PNNI routing network. Before sending the complete PTSE to the neighbor, a node first exchanges the PTSE header with the neighboring node, indicating availability of a new PTSE with it. If the PTSE header indicates a newer version than the one already existing in the neighboring node, this neighboring node sends back a request to the node that has sent it the PTSE header and requests for the complete PTSE. On receiving the complete PTSE, this neighboring node updates the topology database. After updating its own topology database, the header of this new version of PTSE is sent to the other neighboring nodes except the one from which the PTSE was received, and the process is repeated in the other nodes in the peer group. This leads to a hop-by-hop flooding of PTSE information in the peer group. PTSE information, once entered in the topology database, does not remain there forever. The PTSE has a life and is removed from the topology database if an updated version is not received within the predefined life of the PTSE.

The topology database exchange in peer groups at the higher level is similar to that in the peer groups at the lowest level of the hierarchy. In the case of lowest-level peer groups, there are logical nodes with horizontal links, while in the case of higher-level peer groups there are 'logical group nodes' connected by horizontal links. A PTSE never moves up in the hierarchy, but there are PTSEs in the higher-level peer groups that are exchanged between the logical group nodes. A routing control channel is established between the LGN in a peer. Still, a 'hello packet' is exchanged between the peers to confirm this connectivity and membership to the same peer group. Thereafter there is topological database exchange between these LGNs in the peer. However, this topological database is different from the one that the nodes exchange with each other at the lower hierarchical level. This topology database is based only on the PTSEs from the LGNs flooded within the higher-level peer group. As PTSEs can flow downwards, the higher-level peer groups also receive PTSEs from the peer groups above it in the hierarchy, and they too add to the topology database of the LGNs.

8.2.6 Setting up Connection

PNNI routing [4] provides all the switching nodes with complete topological information of the ATM network. This helps the source node to select the path up to the destination node using source routing without dependency on any other node for routing. All the route calculation is done in the source node, and hence there are no chances of any loop formation. Source routing also makes the network free from running the same routing and path selection algorithm in all the intermediate nodes. Setting up of a connection between two nodes for transfer of data using ATM cells comprises two steps – path selection from the source to the destination and setting up of connections between all the nodes in the selected path. The connections should be in the selected sequence of links in the case of multiple links between the nodes. In an ATM network, the user can specify the minimum bandwidth requirement and the other QoS parameters.

It might happen that a node in the selected path is unable to provide the assured QoS parameters at the time of connection setup. Such a situation occurs when an intermediate node has to assign its resources, after transmitting the last update, to some other connection or this update has not yet reached all the other nodes including the source node. In such a scenario, at the time of connection setup through the intermediate nodes, a node that cannot assure resource availability as per the QoS refuses the connection. On refusal of connection by the intermediate node, an alternative route has to be calculated again from the last connected intermediate node with assured QoS. In this case the last connected node with the assured QoS has to take the routing decision, calculating an alternative path to the destination that the path selection algorithm running on the node feels can provide the assured QoS. This process of rolling back the connection to the last node with assured QoS is known as 'crankback'.

If the source and the destination are in the same peer group, the source node calculates the entire route. However, if the source and the destination nodes are in different peer groups, a designated transit list (DTL) is created by the source node, which contains the entire path in the peer group of the source node up to the border node of the peer group. Thereafter, the DTL contains the abstract path in terms of logical group nodes in the higher-level peer groups and optionally the logical links between them. The path is optimized by selecting the LGN in the lowest common peer group. The logical group node at the higher-level peer group is responsible for routing the path through its lowest-level peer group across the two border nodes in the lowest peer group so as to keep the detailed path consistent with the abstract path described by the source node. If the border node detects the destination in its peer group, it calculates the route to the destination. If the border node detects the destination to be out of the peer group, it calculates a suitable route to another border node in the peer group that the entry border node feels is in the path towards the destination and consistent with the initial abstract path. PNNI gives a great degree of interoperability among the ATM switches as the nodes can have different path selection algorithms.

References

1 S. Mueller. *Upgrading & Repairing Networks*, Chapter 16. Que, 4th edition.
2 B. A. Forouzan. *Data Communications and Networking*. Tata McGraw-Hill, 3rd edition, 2004.
3 K. Sumit. *ATM Networks: Concepts and Protocols*. Tata McGraw-Hill, 2nd edition, 2006.
4 The ATM Forum Technical Committee, Private Network–Network Interface Specification Version 1.0. http://www.broadband-forum.org/ftp/pub/approved-specs/af-pnni-0055.000.pdf, March 1996.

Abbreviations/Terminologies

AAL	ATM adaptation layer
ABR	Available Bit Rate
ATM	Asynchronous Transfer Mode
CBR	Constant Bit Rate
CLP	Cell Loss Priority
CS	Convergence Sublayer
DTL	Designated Transit List
GFC	Generic Flow Control
HEC	Header Error Control
IISP	Interim Interface Signaling Protocol
IP	Internet Protocol
ITU	International Telecommunication Union
LAN	Local Area Network
LGN	Logical Group Node
MUX	Multiplexer
NNI	Network-to-Network Interface
nrt-VBR	non-real-time Variable Bit Rate
PG	Peer Group
PGL	Peer Group Leader
PMD	Physical-Medium-Dependent Sublayer
PNNI	Private Network–Node Interface
	Private Network-to-Network Interface
PSTN	Public Switched Telephone Network
PT	Payload Type
PTSE	PNNI Topology State Element
PTSP	PNNI Topology State Packet
PVC	Permanent Virtual Circuit
QoS	Quality of Service
RCC	Routing Control Channel
rt-VBR	real-time Variable Bit Rate
SAR	Segmentation and Reassembly Sublayer
SONET	Synchronous Optical Networking
SVC	Switched Virtual Circuit

TC	Transmission Coverage Sublayer
TP	Transmission Path
UBR	Unspecified Bit Rate
UNI	User-to-Network Interface
UTP	Unshielded Twisted Pair (cable)
VBR	Variable Bit Rate
VC	Virtual Circuit
VCI	Virtual Channel Identifier
VP	Virtual Path
VPI	Virtual Path Identifier
WAN	Wide Area Network

Questions

1   Explain with a diagram how a multiplexer handles variable-size packets and why it is not good for a network to have a mix of very small as well as very large packets.

2   State the difference between the UNI header and the NNI header.

3   Explain all the fields in an NNI header, along with the number of bits occupied by each field.

4   State the three different types of connection channel in ATM and explain their relation with an example.

5   Describe the three-layered ATM architecture. Also mention the sublayers within each layer.

6   Draw the structure of an ATM switching table and explain its entries.

7   Distinguish between real-time and non-real time service categories.

8   Explain the bandwidth distribution among the five service categories.

9   What is the difference between a private network–node interface and a private network-to-network interface?

10   Describe the hierarchical structure of peer groups in PNNI.

11   What is a peer group and how is the peer group leader elected?

12   Explain the following processes:
    A link aggregation,
    B node aggregation,
    C crankback.

13   Explain the process of source routing in PNNI.

14 Differentiate between a group node and a logical group node.

15 Starting from exchange of 'hello packets', explain the process of exchange of PTSEs among the nodes in a peer group.

16 Why is ATM called a connection-oriented protocol?

17 How is QoS ensured in PNNI?

18 1024 bytes of data has to be sent across an ATM network. How many cells will be formed?

19 State whether the following statements are true or false and give reasons for the answer:
A The size of an ATM cell is 48 bytes.
B In an ATM connection, the cells reach the destination in the same order in which they were transmitted.
C A permanent virtual circuit between endpoints can be set up only through the service provider.
D Two different virtual circuits in an ATM network cannot have the same VCI.
E Segmentation and reassembly (SAR) is a sublayer of AAL.
F Transmission convergence is a sublayer of the ATM layer.
G Video conference can be done on an ATM network with 'available bit rate'.
H A peer group can generally have only one peer group leader.
I There can be multiple border nodes in a peer group.
J Two different peer groups in the same hierarchical level exchange PTSEs with each other through the border nodes.

20 For the following, mark all options that are correct:

A The size of an ATM cell is:
- 5 bytes,
- 48 bytes,
- 53 bytes,
- 1024 bytes.

B The interface between ATM switches is called:
- the switch-to-switch interface,
- the logical link interface,
- the network-to-network interface,
- the user-to-network interface.

C The virtual path identifier in UNI is:
- 4 bits,
- 8 bits,
- 12 bits,
- 16 bits.

D Which field is not present in the ATM cell header for NNI?
- header error control,
- cell loss priority,
- payload type,
- generic flow control.

E Which is not a type of connection in ATM?
- transmission path,
- virtual path,
- virtual circuit,
- virtual channel.

F The real-time services are:
- ABR,
- CBR,
- rt-VBR,
- UBR.

G PNNI stands for:
- private network–node interface,
- private node-to-node interface,
- private network-to-network interface,
- private non-network interface.

Exercises

1 100 MB has to be transmitted between two computers connected over ATM. How many cells will be created? How many bytes will remain unutilized in the last cell?

2 Please refer to Figure 8.1. What will be the output of the multiplexer for each of these rules implemented on the multiplexer: 'last in first out', 'shortest job first', 'longest job first'? What will be the output of the multiplexer in Figure 8.2 if it follows 'last in first out'?

3 Please refer to Figure 8.3. Node 2, node 3, node 4, node 5, node 6, node 7, and node 8 have to transmit data to node 1 using ATM. How many virtual circuits and virtual paths will be formed?

4 Please refer to Figure 8.3. Node 1 wants to communicate with node 5. Write down the contents of the ATM cell header for user network interfaces and network node interfaces. Necessary assumptions may be made.

5 Please refer to the network transmission scenario as mentioned in exercise 3 above. Draw the switching tables with the entries for all the ATM switches in the network.

6 Please refer to Figure 8.3. Node 1 wants to communicate with node 5. Mention the layers of the ATM protocol stack that will operate from source to destination between the various devices and endpoints in the connection.

7 Consider the network given below. Organize the network in hierarchy for PNNI routing by taking four nodes (in ascending order of node identifier number) in the peer group at the lowermost hierarchy level. The peer group finally constructed may have less than four nodes.

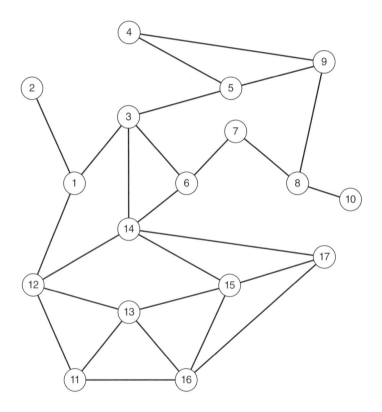

A How many levels of hierarchy are created?
B How many border nodes are there?
C Indicate the route of connectivity between node 1 and node 10 with all the intermediate peer groups in-between.
D If logical links are created between various peer groups, how many logical links will be created and what will be the number of links aggregated for each of these logical links?
E What will be the levels of hierarchy if each peer group has six nodes?

9

Routing in Cellular Wireless Networks

9.1   Introduction

In 1979, the World Administrative Radio Conference (WARC-79) suggested the allocation primarily of the 806–902 MHz frequency band along with the allocation of the 470–512 MHz and 614–806 MHz bands on a secondary basis for land mobiles [1]. The cellular mobile radio became operational for civilian use in the 1980s in various countries across the world. Although the number of users was only in thousands, which is insignificant in terms of today's user base, still it was a time when the cellular network kept improving to overcome various challenges faced by the cellular mobile technology of the age.

European countries were among the first users of the cellular radio network, with a hundred thousand users in Finland, Denmark, Norway, and Sweden where the operations began in 1981. Bahrain has been reported as the first country to have cellular radio operations in mid-1978 through the Bahrain Telephone Company, followed by Tokyo in Japan where the operations started at the end of 1978 through the Nippon Telegraph and Telephone Public Corporation. Other cities in Japan followed soon thereafter. There were about 30 000 mobile radio customers in Tokyo during the early 1980s, and 40 000 users in total in the USA by 1984, where the cellular radio was made operational in about 25 US cities. Chicago followed by Washington DC and Baltimore were the first three cities in the USA, where operations started in 1983. Frequency modulation was used in two-way radio communication.

Network Routing: Fundamentals, Applications, and Emerging Technologies, First Edition.
Sudip Misra and Sumit Goswami.
© 2017 John Wiley & Sons Ltd. Published 2017 by John Wiley & Sons Ltd.
Companion website: www.wiley.com/go/misra2204

Although most countries allocated bands in the 800 MHz region of the spectrum, there were countries operating cellular radio in the 400/450 MHz band also. There was a great variation in the radius of the cell, ranging from 15 to 75 km [2]. The Advanced Mobile Phone Service (AMPS) was developed and used in the USA, while the Total Access Communications System (TACS) was adopted in the UK. Some of the other systems [3] in use during that period were Nordic Mobile Telephone (NMT) in Scandinavia (Sweden, Norway, Finland, Denmark, and Iceland), UNITAX in China, Network C in Germany, the Radio Telephone Mobile System (RTMS) in Italy, the Nippon Automatic Mobile Telephone System (NAMTS) in Japan, Radiocom 2000 (R 2000) in France, and a few other systems.

9.2 Basics of Cellular Wireless Networks

The cellular wireless network is an infrastructure-based wireless network that has emerged as a challenging platform for distributed applications and distributed computing. The cellular network is a terrestrial telecommunication infrastructure. It supports mobile communication with cellular architecture [4] for the coverage area. Coverage areas may or may not overlap. Cellular coverage areas have a telecom infrastructure within them, as well as connecting all other cells so as to enable process calls between various cells. The cellular network can be 'intrasystem', wherein all the cells are under a single administrative control with distributed switching. Alternatively, the cellular network can be 'intersystem', wherein different cellular networks under the administrative control of different operators are networked together for extension of the range, and these support interoperability among themselves. Although it might not be required or feasible to have coverage of the entire geographic area by having overlapping cells, still a cellular network with coverage in terms of overlapping cells is much better for handling applications, security, and call management in mobile communication.

The performance of a cellular network is measured in terms of voice quality, call failures, and spectrum efficiency. The voice quality demanded is equivalent to that of wired telephony. Mobility may lead to variation in voice quality, and increase in distance from the base station can cause voice impairment due to signal fading. Sometimes a call cannot be set up owing to non-availability of channels or signaling errors, leading to ineffective attempts. At times ongoing calls may also be aborted because of signal fading, improper handover, or cochannel interference. The geographic distribution of traffic is highly uneven, leading to call failures in regions with dense traffic or regions with sparse cellular coverage. All mobile communication resources, primarily the bandwidth, are very costly, and hence spectrum efficiency measured in terms of number of subscribers per cell is an important measure of cost [5]. However, increase in spectrum efficiency beyond a limit leads to reduction in service quality.

Each cell has a limited capacity that it can handle, and if the number of users in a cell goes beyond the capacity, blocking of certain calls will occur. However, cellular wireless networks are highly scalable, as they are implemented through a number of small cells, which not only avoids single points of failure but also allows the establishment of microcells to support areas with higher user density. Increase in the number of cells based on demand and use also helps in the dynamic distribution of users across cells. The cell infrastructure can also be established on a mobile platform where the cellular tower is on wheels to support a region with higher call density or during disaster management or in areas that cannot support a permanent infrastructure. The high scalability helps the cellular wireless network to cover huge geographic regions. The smaller cells have to use low-power antennas. Nearness to cellular antennas also saves the transmission power of the mobile devices. Reduced power consumption makes the operations of the cellular wireless network cheaper and affordable. Further, smaller cells help in better reuse of frequency and reduce interference among cells because the adjacent cells have different frequencies.

As the communication in the cellular network is infrastructure based and wireless, it brings along with it associated problems and challenges. The cell towers can generally withstand wind speeds of about 150–200 km/h. Cyclones, hurricanes, and tornadoes can damage the towers. Sandstorms and hail/snow storms can damage the antennas, leading to disruption in service. Compared with wired networks, the transmission of voice and data is not only slow but also costlier in wireless networks. The signal quality or voice quality is also dependent on the location owing to a number of factors such as fading, absorption, attenuation, interference, and handover. Mobile devices inside speeding vehicles or trains gain high mobility, which can be challenging in terms of signal strength, handover, changing topology, and routing. Cellular transmission has to be in the open and in broadcast mode. This not only leads to privacy issues, but the signal can also be hindered by jamming.

Each cell in a cellular wireless network is supported by an infrastructure having an antenna. This antenna with associated infrastructure is known as the base transceiver station (BTS) or the base station. The base station has a transmitter, receiver, and controller with a small range of frequency assigned to it. A number of base stations generally equidistant from each other are installed in the region for uniform coverage of the area. However, in certain cases, special cell sizes are in use, which might be much larger or smaller than the standard cell size, and the antennas are at varying distances in these cells.

A base station controller (BSC) connects a number of base transceiver stations (BTS), and a number of BSCs are connected to a mobile service switching center (MSC). The MSCs are connected to each other on a backhaul public switched telephone network (PSTN). The interconnections between BTS and BSC and between BSC and MSC are also on a wired network or microwave link.

The height of the base station on which the antenna is installed and the power of the antenna are calculated on the basis of location and requirements. Athough the antennas are generally installed on a metallic or concrete tower over the land, they are sometimes installed over rooftops, water tanks, street light poles, or hilltops. Each BSC controls a number of BTSs and is responsible for channel allocation to the BTSs, detection of signal strength of mobile devices, and handovers between the BTSs.

The adjacent cells in the cellular network have different frequencies to avoid interference in overlapped areas. The power level of the transmitter at the base station of the cell has to be optimized to avoid the presence of the signal across neighboring cells, which can lead to interference and cross-talk. The number of cells also has to be increased to increase the geographical coverage. Each cell has a predefined frequency band and hence a fixed capacity in terms of the number of mobile devices within the cell that can simultaneously communicate. Thus, the cell size has to be reduced to increase the number of cells in a region to support a greater number of devices within the same area. The commonly used techniques to increase the capacity of a cell in the cellular

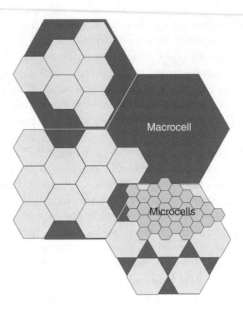

Figure 9.1 Cell splitting.

network are as follows: frequency borrowing from adjacent cells, splitting a cell into smaller cells, sectoring, and zone microcells.

As the number of cells increases by *cell splitting*, it not only increases the capacity of the system, but also reduces the power consumption of the mobile devices and the base station of each cell because the microcell has to operate at a lower radiated power level and signal strength. However, as indicated in Figure 9.1, cell splitting reduces the coverage region of each cell, leading to an increase in the number of handovers for a mobile device that is not stationary.

In *sectoring*, a cell is covered by directional antennas instead of omnidirectional antennas. Depending on the angle of coverage of the directional antenna, the number of sectors in a cell can be calculated. Generally, a cell is divided into sectors of 120° or 60°, as depicted in Figure 9.2. In sectoring, the size of the cell remains the same, but as each sector has its own channel, which is a subset of the channel of the cell, it does not help in reducing the number of handovers. Sectoring is complex to implement as a number of antennas have to be managed at the base station, and there is deterioration in performance owing to reflections from various structures in an urban environment, leading to interference in the sectors. Sometimes, instead of having 3–6 sectors in the cell to cover the entire 360°, a cell might have only one or two unidirectional antennas for wedge-shaped coverage for some specific purpose such as linking towards a valley from a hill or coverage over a bridge or inside a train or tunnel. Such cells are known as 'selective cells'. Figure 9.3 depicts selective cells of 120° to provide coverage only to a highway crossing an uninhabited forest.

Zone microcells help to prevent dead or no-signal zones within a cell. Dead zones can be inside tunnels, basements, in the shadow area of a hill, in areas with reflectors, or in buildings with thick walls. Antennas are fitted in such microcells inside the cell to provide transmission signals to these dead zones. Microcells thus created by using additional antennas use the same frequency as the cell to which they belong. There can be a

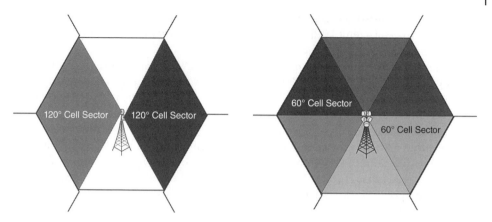

Figure 9.2 Cell sectoring.

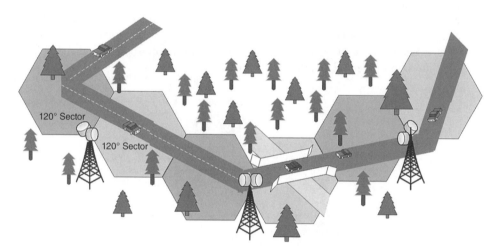

Figure 9.3 Selective cells.

number of microcells within a cell, and all the microcells should have their own antenna. The antennas of the microcells in the cell are generally connected to a base station through a wired network. Increase in the number of antennas in a zone microcell makes the control operation in the base station a complex activity. The handover between microcells is also handled by the base station.

Based on the area of coverage of the cell, it may be termed a macrocell, a microcell, a picocell, or a femtocell. Normally, the size of a cell in a cellular wireless network is 5–15 km in an urban environment, and can reach up to 35 km in rural and obstruction-free regions. Cells with a coverage region greater that the normal cell sizes are referred to as macrocells. A macrocell generally has a coverage of 10 km or more. Macrocells are used in sparsely populated areas and as a means to connect two densely populated areas separated by long distances but with network coverage between them throughout the path.

Microcells have a diameter of about 1–2 km and are used in areas with dense population to increase the capacity of the cell. Picocells, with a coverage of 200–500 m, are used in specific areas such as airports, railway stations, tunnels, basements, bridges, or alleys. The range of a femto cell is about 10 m for personal communication or to provide services in indoor areas or at the edge of the cell.

Frequency reuse. Hundreds of millions of mobile devices have to use the limited frequency range allocated to the cellular wireless network. Each cell is allocated a small frequency range known as a channel. The channels of two adjacent cells cannot be the same, as this could lead to information distortion or packet drop due to mutual interference, which is also known as cochannel interference. Still, frequencies are reused innumerous times with only a constraint that the frequency of the adjacent cells should not be the same. There can be a number of reuse patterns, one of which, commonly referred as the seven-cell reuse pattern, is shown in Figure 9.4. Frequency reuse allows increase in the capacity of the cellular network and makes it highly scalable to enhance the coverage area of the cellular network.

The transmitter at the base station has limited power, leading to a fixed coverage area, which is also known as the 'footprint'. The boundary of a cell is not very well defined or sharp, as the signal strength falls slowly over a region, leading to areas with overlapping signals from neighboring cells. As shown in Figure 9.5, the area of a cell is not perfectly circular, but has an irregular pattern, as the signal in different directions may encounter different types of obstruction and terrain, leading to variation in the signal strength in different directions at equal distance from the antenna.

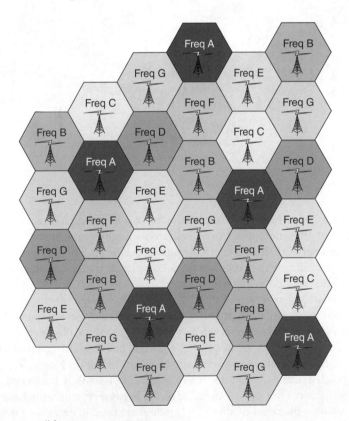

Figure 9.4 A seven-cell frequency reuse pattern.

Fading. In a cellular wireless network, the signals are broadcast in all directions and hence the same signal, in addition to direct reception, may reach a particular receiver at different times on account of a number of factors such as reflection, diffraction, scattering, or propagation at different rates owing to a change in medium [6]. These multiple signals increase the signal-to-noise ratio, cause attenuation, and can cancel each other as a result of the phase difference leading to the weakening of signal strength. Fading can affect all frequencies equally, or different frequencies can be affected in different proportions.

Handover. Owing to mobility, the location of the mobile device keeps changing and may frequently move out of the range of one cell and enter into the range of another. As the cells are overlapping, as shown in Figure 9.6, the transfer of the mobile device from one cell to another is decided by weakening of the signal strength from one base station and strengthening of the signal from the other base station. This leads the base station

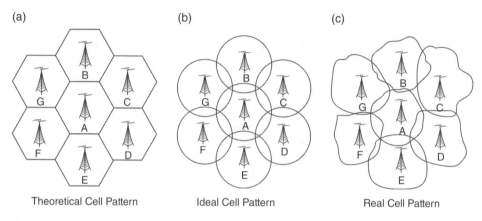

(a) Theoretical Cell Pattern   (b) Ideal Cell Pattern   (c) Real Cell Pattern

Figure 9.5 Cell coverage pattern.

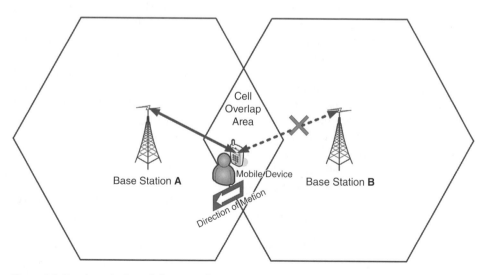

Figure 9.6 Handover in the cellular network.

to inform the MSC, which then transfers the control and reroutes the call of the mobile device to the base station of the new cell which has the mobile device in its range. There is also a change in the communication channel, as the adjacent cells have different channel frequencies. The process of handover is completed in milliseconds, so that the transmission remains uninterrupted and the process remains transparent to the user of the mobile device. Handover is also referred to as 'handoff'. Sometimes the mobile device remains in the overlapping cell area with almost equal signal strength from both base stations. In such scenarios, there might be frequent handover of the mobile device between these two base stations in the overlapping cell, and such a phenomenon is known as 'ping-ponging'.

Two different threshold parameters are conventionally used to trigger handover – the increased power level used in the mobile device to receive the deteriorated signal or the signal strength (or signal-to-interference ratio) [7]. However, the present-day cellular wireless networks use a number of other parameters for effective decision-making with respect to cost, QoS, and performance to trigger handover. Some of these additional parameters are available bandwidth, delay, preferred service provider, signal decay, network interface, link capacity, link cost, power consumption, loss rate, packet delay, jitters, and policies [8-11]. Artificial intelligence and fuzzy logic [12, 13] are also used in decision-making regarding handovers.

A handover can also occur within the same cell when the mobile device changes the channel inside the cell. Apart from this, the handover can be intra-BSC, inter-BSC, or inter-MSC. In intra-BSC handover, the mobile device moves from one cell to another cell where the base stations of the two cells are connected to the same BSC. In the case of inter-BSC handover, the mobile device moves to a new cell from an existing cell such that the base stations of the two cells are connected to two different BSCs. If these two BSCs are connected to the same MSC, this is also known as intra-MSC handover. A special scenario in inter-BSC handover is when these two BSCs are in different MSCs, and in such a case it is known as inter-MSC handover.

The handovers are classified into two categories – hard handover and soft handover. In the case of hard handover, the mobile device releases the channel of the existing cell and thereafter starts using the channel of the new cell. Thus, there is an intermediate break in the signal during changeover from the existing cell to the new cell, hence also referred to as 'break-before-make'. When the mobile device does not release the existing channel but takes on the channel of the new cell, retaining both channels together for some time before releasing the channel of the initial cell, the process is known as soft handover. As there is no instant of time when no channel is available, this phenomenon is also called 'make-before-break' and is used in the CDMA or 3G networks where the neighboring cells operate on the same frequency, enabling soft handover.

Increase in the number of cells in a region enhances the number of mobile devices that can be supported in the region and reduces the cell size. It also leads to an increase in the number of handovers per unit distance travelled, as the cells are smaller in size. The increase in the number of handovers increases the time delay as well as the requirement of buffers in the base station. The increase in the number of handovers also increases the time spent by the base station and the MSC in processing the handovers. The communication between the base station and the mobile switching centers also increases the time to process the handovers. However, movement of the mobile devices across microcells does not require handover, as all the microcells use the same frequency.

Umbrella cells are used to handle mobile devices that frequently change cells. Consider a thickly populated area served by a good transportation system, e.g. port areas, railway stations, or a market place with good road connectivity. As the number of persons per unit space is high in these areas, these areas have to be served by small cells so as to enhance the capacity of the cellular network. However, there are mobile devices inside the transportation system, i.e. vehicles, steamers, and trains, which move at a high speed in these areas, crossing a number of cells in a very small timeframe. This leads to very frequent handovers for these few specific devices travelling at high speed. To cater for the high mobility requirement of such devices in densely populated areas, umbrella cells are implemented. The base station in the umbrella cell uses high-power antennas and hence has a much wider coverage area than a few microcells that lie inside the region of the umbrella cell. As the mobility of a device increases, it is handed over from the microcell to the umbrella cell, and thus the handovers for this high-speed mobile device are reduced, as shown in Figure 9.7.

Multihop cellular network. The basic architecture of a cellular wireless network is based on the single hop, as shown in Figure 9.8a, where the voice or data reaches the mobile device in a single hop from the base station. Two mobile devices in the same cell or in different cells also communicate with each other through the base stations. Unlike an ad hoc network, the mobile devices do not directly communicate with each other. The multihop mobile network [14] optimizes the channel utilization and the number of

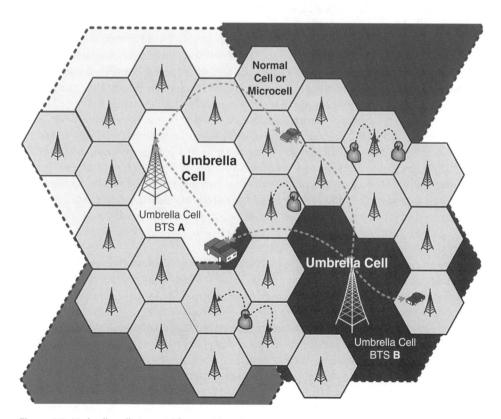

Figure 9.7 Umbrella cells to avoid frequent handovers.

(a)   (b)

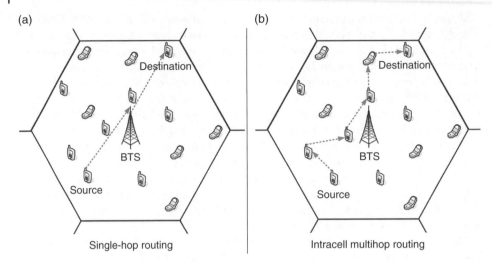

      Single-hop routing   Intracell multihop routing

Figure 9.8 Single-hop and multihop routing.

base stations by incorporating the characteristics of an ad hoc network within the cellular wireless network.

In a multihop cellular network, as shown in Figure 9.8b, two mobile devices within the same cell directly communicate with each other if they are in the transmission range or communicate through other intermediate mobile devices in the cell instead of routing the communication through the base station. If the communicating devices are in different cells as shown in Figure 9.9, then the packet is routed between them through the base stations. However, the packet may not be reachable directly from the source mobile device to its base station, and a few intermediate mobile devices in the cell might be used to forward the packet from the source mobile device to its base station. Similarly, the destination mobile device may not be reachable in a single hop from its base station, and a few intermediate mobile devices in its cell may be used for multihop routing.

A multihop cellular network not only reduces the number of base stations, it also increases the number of simultaneous calls that can be supported within a single cell. Not all calls between the mobile device and base station utilize the channel allocated to the base station. The mobile devices can directly communicate with each other, and there can be a number of simultaneous calls between various mobile devices without utilizing the channel available with the base station.

Communication channel. There are two types of communication channel between the mobile device and the cell infrastructure (the base station) – the traffic channel and the control channel. The traffic channel is used for transmission of voice calls and data transfer between users through the intermediate cell infrastructure. The control channel is used for exchange of information for synchronizing, setting up, maintaining, and disconnecting the call. The control channel handles timing information, phase reference, signal strength, system identification information, and messaging between mobile stations. Interference in the control channel can lead to dropped calls, problems in handovers, or error calls. Error detection and correction information as well as power control information are carried by the traffic channel. The control channel can also be passed through the data channel once the call has been established.

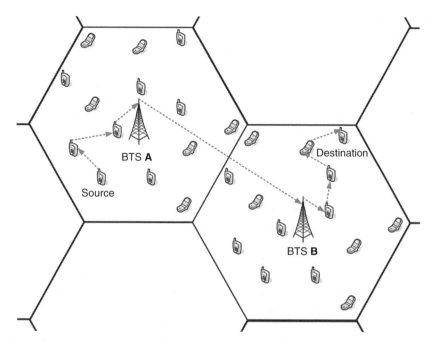

Figure 9.9 Intercell multihop routing.

The frequency used for communication from the cellular base station to the master station or from the cellular base station to the mobile device is known as the forward channel or downlink. The frequency used for communication from the master station to the base station or from the mobile device to the base station is known as the reverse channel or uplink. The frequency of the forward channel is different from the frequency of the reverse channel.

9.3   Resource Allocation

The resources in the wireless cellular network can be classified as radio resources and device resources. Although the base station is a consumer as well as a creator of resources, we assume it to possess sufficient power and other resources and hence not to be constrained. The major radio resources are power, code, and bandwidth. The major device resources are battery power and signal strength (receive as well as transmit). The use of resources is optimised for longer use by least battery consumption with reception of QoS signals, as well as to maximize the revenue of the service provider. The neighboring cells also cooperate with each other for resource sharing, as the resources are equally distributed across all cells, but the density of the users may vary widely across the cells.

Traditionally, the resource allocation has been time aware with traffic-dependent pricing. As the traffic varies with time, with higher demand during peak hours, time-varying resource allocation is implemented. Certain less-priority traffic such as updates can be planned for running during less congested traffic hours when the resource prices and QoS requirements are low.

The resource allocation is generally content based. The cellular wireless traffic contents are of two different types – elastic and inelastic traffic [15]. The inelastic traffic pertains to those applications and services that require real-time flow of voice or data between the source–destination pair. These are generally interactive applications and emergency notifications. A minimum bandwidth, throughput, and QoS have to be ensured for inelastic traffic for the entire duration of the connectivity.

The elastic traffic pertains to those delay-tolerant applications and services for which best-effort delivery is attempted for the recipients of voice and data. This type of traffic generally pertains to applications and services performing firmware updates, messaging services, email delivery and Internet browsing.

The category of application where variable bit rate is acceptable and thus the stringent inelastic traffic pattern may be eased leads to an intermediate or semi-elastic traffic. A video conference held between mobile devices through applications such as Skype, Hangouts or FaceTime is an example of semi-elastic traffic. Although real-time traffic is given priority over delay-tolerant traffic, this does not mean that the elastic traffic gets dropped on account of deprivation of resources.

Resource estimation and allocation become more challenging in a cellular wireless network because it is not a closed system with reference to any cell. A base station has to ensure maximum resource utilization to ensure revenue generation through channel utilization. Bandwidth is the most costly resource in the cellular wireless network. However, the base station has also to keep an optimum level of resources available for handovers from neighboring cells to this base station because a minimum level of success rate for handovers has to be ensured. There may be a large number of active users in the neighboring cell with high resource utilization and inelastic traffic. These users in the neighboring cells may move on to this cell, and the handover of such calls should not be dropped for the lack of availability of sufficient resources as successful handover may be one of the QoS parameters. For the reserved resources, handover calls get preference over any new connection requests because once a mobile device has been granted resources, the rejection is unacceptable [16, 17]. However, the resources cannot be kept reserved for all the ongoing calls in the neighboring cells. For such a case, the requirement will grow manifold and outnumber by a few times the existing resource utilization by the cell as there will be a number of neighboring cells based on the frequency reuse pattern and the radio power. Thus, other parameters such as location of the mobile device, speed, direction, and expected call duration also play a vital role in utilization of reserved resources.

A slow-moving mobile device creates another typical situation when it is in the overlap area of two base stations [18]. Assume it is connected to one of the base stations and may be one of the high-resource-utilization mobile devices. Now, when this mobile device comes into the overlap area, the neighboring cell gets an indication that the mobile device might move from its existing cell to the new cell. Hence, this new cell keeps its resources reserved for this mobile device so that a successful handover can be performed. However, as this mobile device may have very slow mobility speed, it may remain in the overlapped area for a very long time, utilizing the resources of one cell and keeping equivalent resources in the other cell reserved for it.

The requirement becomes more stringent if the QoS is not based on the average of all connections but has to be ensured for each individual connection. The criterion for accepting a mobile device from a neighboring cell is different for voice and data requirements. Voice connections have to ensure a through-and-through non-blocking connectivity, while data connections have to comply with maximum permissible average packet drop and packet delay.

Being in wireless medium, the granted resources are rarely equal to the requested resources. There are losses due to packet drops, attenuation, shadowing, fading, noise, interference, and handovers. Fairness in resource allocation [19] cannot be achieved by equally distributing the resources across all mobile devices, as the nearer devices require less transmit power than the devices at the edge of the cell.

9.4 Routing in GSM Networks

The Global System for Mobiles (GSM) is an international standard for the public land mobile network (PLMN). The standard was published in 1990 and its commercial use started in 1991. The standard belongs to the second generation of cellular networks. It uses digital systems with error detection and correction and encryption for security. The first generation of cellular networks was based on analog systems. The common services provided by the GSM standard are voice telephone, data services, and messaging services. It has also defined supplementary services such as call forwarding, call waiting, conference call, call barring, and caller identification. GSM is based on time division multiple access (TDMA), which operates at a frequency band of around 900 MHz in Asia and Europe and 1900 MHz in America.

In GSM, the user information is contained in the subscriber identity module (SIM). The SIM is small in size and can be swapped across various mobile phones or other GSM devices. The user can change the mobile handset by simply taking out the SIM from the old device and putting it in the new mobile device. The SIM is not only used in handheld mobile devices but also in fixed or portable communication terminals. The power output of the terminal varies with resource availability by virtue of its size and mobility. The power output of a fixed GSM terminal can be 20 W or more, and that of a handheld mobile terminal is generally below 1 W.

The SIM is an integrated circuit that can operate at three different operating voltages: 5, 3, and 1.8 V. There are eight pins in the SIM to provide an interface with the device in which it is inserted. The pin configuration is shown in Table 9.1.

The SIM is available in four different form factors. The first form factor is the full-size SIM, the others being mini-SIM, micro-SIM, and nano-SIM. The embedded SIM, which is much smaller than a nano-SIM, is also in use. The internal architecture of a SIM comprises memory, microprocessor, I/O interface, and internal bus. The memory is in the form of RAM, ROM, and EEPROM. The memory of a SIM varies between 32 and 128 kB. This memory stores a number of pieces of information that can be categorized as user-stored information and GSM or SIM information. The user-stored information comprises the contact details, SMSs, phone settings, and last dialed numbers. The GSM- or SIM-related information stored in the memory comprise the personal identification number (PIN), authentication key, encryption key, available PLMNs, integrated circuit card identifier (ICCID), location identity, and international mobile subscriber identity (IMSI).

The service provider uniquely identifies each SIM by the IMSI that is mapped to the mobile number and thus uniquely identifies the subscriber. IMSI is generally a 15 digit number that can be even smaller, as its size is governed by the regulations in specific

Table 9.1 Pin connection in a SIM.

Pin No.	Signal Name	Function
C1	VCC	Supply voltage
C2	RST	Reset
C3	CLK	Clock
C4	NC	No connect
C5	GRD	Ground
C6	VP	Programming voltage
C7	I/O	Input/output Interface (ISO 7816-3)
C8	NC	No connect

countries. The first three digits represent the mobile country code (MCC), the next two or three digits represent the mobile network code (MNC), and the remaining digits store the mobile subscriber identity number (MSIN). The MCC uniquely identifies a country. In the case of countries with a huge number of mobile service providers, one country may be represented by more than one MCC. MNC uniquely identifies a service provider within a country. Thus, there can be two different service providers in two different countries with the same MNC. The combination of MCC and MNC uniquely identifies the mobile service provider. The 10 digit MSIN is also known as the mobile identification number (MIN) and is the unique 10 digit mobile phone number.

9.4.1 Architecture

The GSM network as depicted in Figure 9.10 comprises three major subsystems – the base station subsystem (BSS), the network and switching subsystem (NSS) and the GPRS core network.

The BSS connects the mobile device to the NSS for access to the telephone network and to the GPRS core network for Internet access. The BSS consists of base transceiver stations (BTSs), the base station controller, and the packet control unit (PCU). Each BTS comprises a radio receiver, a transmitter and an antenna, creating one cell in the cellular network. If a BTS uses a directional antenna, it can create a number of cells. The location of the antenna defines the center of the cell and the transmitting power of the antenna defines the radius of the cell. A BSC controls numerous BTSs or cells. The BSC is responsible for managing the network resources, frequency hopping, allocation of radio channels, and power level to BTSs, managing handovers between the BTSs within the BSS, and connecting to the NSS through the terrestrial channel. The PCU handles the data processing for the BSC and connects it to the GPRS core network.

The NSS interconnects the mobile users among themselves and with the users of the terrestrial fixed-line telephone. The core of the NSS is the mobile switching center (MSC). The NSS also has a few databases to store user information related to subscription and mobility – home location register (HLR), visitor location register (VLR), equipment identity register (EIR), and authentication center (AUC). The databases may be maintained within the MSC or as separate powerful servers in the case of a large subscriber base.

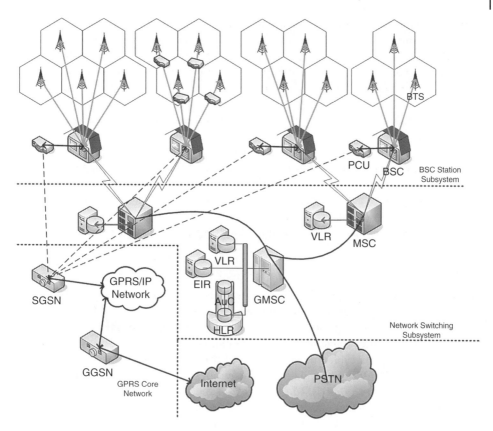

Figure 9.10 Architecture of a GSM network.

The MSC performs mobile registration, authentication of users, call routing for roaming, and inter-BSS call handovers.

HLR contains subscribers' identity record, the list of services hired by the subscriber, and the location updates to support roaming. Each record is semi-permanent in nature and is retained until the subscriber is registered with the service provider.

VLR contains information about the BSC to which a mobile device is connected for all the mobile devices in the range of various cells controlled by the MSC. Whenever a new mobile device enters the range of the MSC, the entry of its location area identity (LAI) is made in the VLR and the same is informed to the HLR of the mobile device to enable update of the location information by the HLR and receive data about the mobile device from the HLR. Some other major data corresponding to the mobile device stored in the VLR are the HLR address of the mobile device, its IMSI, the subscribed services, and authentication data. Whenever a mobile device enters within the area of control of an MSC, the VLR informs the HLR and also keeps track of the location of the mobile device with respect to its cell location even when the device in inactive. When the mobile device moves from the area of one MSC to another MSC, the VLR transfers the data to the other VLR in the new MSC. The VLR database is temporary in nature as the database is reset every day.

Mobile equipment is uniquely identified by its international mobile equipment identity (IMEI). Further, each mobile handset has its make and model, which define various functionalities available in the handset. The EIR contains information about the mobile handset being used by the mobile device. The IMEI is also used to prevent the use of stolen, fake, or unauthorized devices in the network for secured use of the cellular system.

The AUC has encryption keys for providing authentication and secured wireless communication.

The GPRS core network provides connectivity and routing of data packets from the GSM network to the Internet. It also keeps a record of the amount of data transmitted per user. The interconnectivity between the GPRS network and the Internet is provided by the gateway GPRS support node (GGSN). The serving GPRS support node (SGSN) provides GPRS data connectivity to mobile devices and is responsible for authentication, mobility management, and tunneling of data packets from mobile devices to GGSN.

9.4.2 Call Routing

To avail the GSM services, a subscriber has to register with a service provider. On registration with a service provider for the cellular wireless network, a SIM, which also contains the MSIN or the phone number, is activated and given to the subscriber. This SIM uniquely identifies the subscriber or the mobile device used by the subscriber. A subscriber can have multiple mobile devices or SIMs, and a mobile device can also support multiple SIMs. On activation, the phone number and the services registered by the subscriber are added to the HLR in the MSC of the region where the subscriber has registered with the cellular service provider. This coverage region of the service provider is also known as the 'home network'.

When the mobile device is switched on with this service provider's SIM, the mobile device starts detecting the signals being received from the various BTSs in the region. If the channel in the frequency band of the service provider is detected, the mobile device contacts the BSC over the control channel. The mobile device is thus in the coverage area or cell of one of the BTSs of the service provider. If the mobile device finds itself in the overlapping area of two cells, it registers with the cell from which it receives the stronger signal. When a mobile device moves or switches on in a coverage area that is not in the home network, i.e. not under any of the BTSs controlled by an MSC, it is said to be in roaming. This roaming SIM is registered with the VLR of the region, and the VLR in turn contacts the HLR to obtain the details of the SIM in terms of subscriber details and registered services. The VLR also stores the location of the roaming SIM in terms of the cell in which it is present.

When a subscriber dials a phone number, the call setup request goes through the BSC of the nearest BTS to the MSC. The MSC checks the VLR for the services permitted to the mobile device. If the call is permitted, MSC informs BSC for resource allocation to the mobile device for call setup. MSC thereafter forwards the call to the PSTN network or another MSC through the gateway MSC.

At the receiver end, when a call has to reach the mobile device, it first reaches the home MSC of the mobile device, which is always aware of the VLR with which the mobile device is associated at that instant of time. With this information, the call is forwarded to the visiting MSC, which uses the location area identity of the visiting mobile device available in the VLR to forward the call to one of its BSCs. This BSC then forwards the call to the exact BTS in the range of which the mobile device is located.

9.5 Challenges in Mobile Computing

Mobile phones have become so widespread as devices of common utility that the number of mobile phones worldwide is more than double the number of landline phones or the number of personal computers or Internet connections in use in the world, and more than double the TVs or credit cards in use. This indicates the pace of growth and popularity of the mobile phone, which along with its use for voice calls also has computing capabilities through the availability of a processor and memory in the phone. The mobile phone can be used these days for voice and video communication, website access, email communication, and running different utility applications (apps) and various other utility applications that may use the processing capabilities and memory of the mobile phone with limited or no requirement for mobile connectivity. This breed of computing devices brings a new set of rules and restrictions, common among which are the restricted user interface, cheaper computation and storage, but limited and costly bandwidth. Application development for mobile computing is governed by the communication requirement and driven by innovation and consumer requirements [20]. The new areas of applications introduced by mobile computing include mobile commerce, location-based services, and dual-mode operations.

Storage. Data stored in a mobile phone can be categorized into communication-related data (contacts, SMSs, call logs), configuration data, Internet data (browser cache, email), sound (ringtones, music), calendar (reminder, schedules, tasks), and files (photos, ebooks). The data is stored either in file systems or databases, both of which have been designed specifically for use and storage in the memory of mobile devices which is primarily NAND flash. The flash memory used in mobile phones is non-volatile and similar to that used in digital cameras, memory cards (SD cards), USB drives/storages, and ultrathin laptops. The flash drives may be based on NAND gates or NOR gates and can be termed as a category of electrical erasable and programmable read only memory (EEPROM), which is non-volatile. Although flash memory is compact, energy efficient and inexpensive, it has a finite number of write/erase cycles and writes/erases in blocks, as the write operation can only change 1s to 0s and hence the complete block has to be changed from 0s to 1s during erase. Such erase cycles of converting 0s to 1s is longer for NOR flash than for NAND flash, which rules out NOR flash for a lot of applications. The difference between NAND flash and NOR flash is indicated in Table 9.2.

As the software is not updated frequently in a mobile device, the code requires random access, and the storage requirement is much less than the data, these requirements make the costlier NOR flash suitable for storing the code.

The number of write/erase cycles is finite in flash memory and restricted to ten thousand to a million cycles, depending on the type of memory. Thus, suitable writing mechanisms have to be devised for traditional file system and relational databases (RDBMS) to distribute write across the sectors for wear leveling. To overcome these challenges, file systems for use in mobile computing have been devised, such as TrueFFS (True Flash File System), YAFFS (Yet Another Flash File System), JFFS (Journaling Flash

Table 9.2 NOR flash vs NAND flash.

	NOR Flash	NAND Flash
Access	Random	Sequential
Write/erase speed	Slow	Faster
Substitute	EEPROM	HDD
Usage	Suitable for code	Suitable for data
Reliability	Relatively higher	Single bit errors

File System), and JFFS2, some of which are proprietary while many are open source. With regard to databases for use in mobile devices, either the traditional RDBMS can be retrofitted for mobile devices or new databases optimized for use in flash can be built from scratch to reduce serialization. Excessive modification in the traditional databases to optimize for use in mobile computing may bloat the code and make the database inefficient. Some of the common RDBMSs used in mobile devices are Apache Derby, Sybase iAnywhere, SQLite, and Berkley DB.

Synchronization. Based on update characteristics, the mobile data can be put into two categories – the data that is created and deleted only in the mobile device such as SMS, logs, emails, ringtones, photos, videos, and music, and the data that is not only created and deleted but also updated in the mobile device such as contacts, codes, configuration data, calendar, operating system, and apps data. Some of this second category of data requires synchronization between the data stored in the mobile device and that stored in the servers of the service provider. The data sync can either be required for maintaining the same copy of the data at both ends for codes, configuration, contacts, or calendar, or just version control at the server end and the entire data or software in the mobile device as in the case of the operating system or applications. Synchronization is a complex process in mobile computing owing to unreliable and costly communication channels, and it is supported by change logs to detect the amount of change, date of change, and hash functions to trace similarity.

The sync can either be for the files or for an application. It can either be from the server to the mobile device or bidirectional. The logs have to be maintained in the servers as well as the mobile devices. The logs may contain a variety of information pertaining to creation, deletion, or change in terms of size, timestamps, and hash. The process of sync is termed fast if it takes place between a mobile device and a server, while it is termed slow if the mobile device has to sync with a number of servers. As the mobile devices are power and resource constrained, the sync process confines maximum processing at the server end. In slow syncs, too, the mobile device syncs to a server and the server in turn syncs with other clients and servers.

The hash-based sync does not require any historical logs and can also be used for detection of substring similarity using rolling hash. Synchronization of code is more complex than data synchronization. Patches should be available for differences in version. However, the patches should be small and reusable for execution over a number of devices. The patches should be highly compressed to save on bandwidth. Further, there is no resource constraint at the server end to compress the patch, but it becomes challenging to decompress the patch at the mobile device with limited resources.

The patches should be installable in the limited disk space of the mobile device and capable of fast, flawless, and trace-free rollback on failure.

Updates. The mechanism for sending updates and notification to the mobile phone should be highly scalable as it has to reach millions of devices. The mechanism generally does not require user intervention, and the service provider or any other server asynchronously sends the update or notification to the mobile device. Some common examples of such notifications and updates are emails, SMSs and IP calls. These notifications should be small to save radio battery but less complex to decode in order to save computational requirements. The latency should be low and scalability should be high. The mechanism should be secure to avoid impersonation, obfuscation, and channel safety and to ensure policy-compliant firewall passage. The version management should be error free with provision for roll-back. There are three different mechanisms through which updates are received in mobile devices:

- *Round-the-clock connectivity.* The devices use supporting protocols such as Hypertext Transfer Protocol (HTTP) or Transmission Control Protocol (TCP) to remain connected to the server on a 24×7 basis or may connect-disconnect frequently to receive any data from the server. As the number of devices is very large, this leads to a high number of connections between the server and the devices, which becomes challenging for the protocol to support, and most of the connections remain idle.
- *Polling.* The device checks for any update from the server after every 't' seconds. By changing the value of 't', the frequency of the polling can be varied to provide high scalability or reduce unwanted polls in the case of rare updates.
- *Queuing on inbound port.* There are three variants to this methodology of receiving updates on a dedicated port. The device gets a dedicated server socket to which it connects periodically to get an update. The server may also contact the client using SMS or UDP connection. For bulky updates, SMS and UDP may not be suitable, and hence the server uses the 'poke-n-pull' mechanism to send updates to the device. The server, when it has an update to send to the device, pokes the device using SMS or UDP, following which the device establishes a TCP connection to receive the actual update. These mechanisms ensure highest scalability.

Broadcast. The communication from server to mobile devices is a one-to-many communication that is in broadcast mode, while the communication from mobile devices to server is a one-to-one communication. Hence, the broadcasts are asymmetric communications with a further complexity of time synchronization, as all the mobiles may not be ON or in the communication range during the broadcast by the server. Thus, broadcast-on-demand may be more suitable than live broadcast for version upgrade and code update in mobile devices. The live broadcast may, however, be used for certain information-related applications in the mobile device, such as streaming information related to news, weather, or traffic. A few other challenges with broadcast, some of which are typical of a mobile computing network, are lack of acknowledgement from the device, each device starts listening to the broadcast at different points in time, and the packets dropped by each device may be different owing to interruption, coverage, jamming, or resource constraint. This requires repeated broadcast of the data in cycles by the server to enable mobile devices to pick up again the relevant data that was dropped or a block of which was deleted in the storage. The retransmission of updates consumes 'n' times the bandwidth than a one-time broadcast, where 'n' is the number of times the broadcast has to be looped. Further, the mobile device has to wait for one complete cycle to get the dropped data. Waiting by the mobile device until the last unwanted packet also leads to loss of power owing to active radio listening to the broadcast.

To ensure redundancy in the broadcast data and save waiting time of the devices, multiple tracks may be broadcast by using broadcast disks or data carousel. Erasure codes may also be used where the server pads the broadcast with redundant data, which helps the devices to reconstruct the dropped data from the padded data. The forward error correcting codes encode the message from 'n' blocks to 'n' + 'x' blocks, where $x << n$ to help to reconstruct the erased data. The value of x decides the bandwidth efficiency of the erasure code. Some of the commonly used erasure codes are Reed Solomon code [21, 22], Tornado codes, and LT codes [23–25].

Device management. Mobile devices are the newest and most popular type of computing equipment that is nearest to the human attention span. The applications in the devices find everyday use, and hence users drive innovation leading to regular updates over the air. The data received by the devices relates to software version upgrade, firmware upgrade, application data, new installations, access codes, and diagnostic data. For effective device management, it is essential to have information about the protocol stack running in the device, details of the device, and details of all the enablers running in the device, such as security overlay, authorization framework, browser, data synchronization, management schedule, domain name system, navigation, identity management, location awareness, services interface, and APIs.

The authorization for device management is held by a number of stakeholders, such as the service provider, equipment manufacturer, application provider, operating system/firmware developer, processor manufacturer, Government, employer/parent, and the user. Each of the stakeholders has to manage its own part of the equipment or service in the device, such as incorporation of new features, bug fixing, monitoring, and service provisioning, and hence wants access control in device management. The device management updates should be atomic and are performed in three phases: (i) identification and authentication, (ii) downloading, and (iii) update. In the case of a failed device update, it should be capable of rolling back to the stable state previous to the failure. The device update is challenging as it should ensure sufficient battery availability before the start of the process, availability of the device in the home network (and not on roaming), memory limitations, and variation in the previous version number in the devices that are undergoing updates.

References

1 WARC-79 overview, actions, and impacts, Chapter 4, in *Radiofrequency Use and Management: Impacts from the World Administrative Radio Conference of 1979*, Congress of the US Office of Technology Assessment, 1982.

2 T. E. Bell. Communications: Highlights include the AT&T divestiture, a new long-distance record in fiber-optic communications, local-area networks, and cellular radio. *IEEE Spectrum*, 22(1):56–59, 1985.

3 D. M. Barnes. The introduction of cellular radio in the United Kingdom. *35th IEEE Vehicular Technology Conference, Volume 35*, pp. 147–152, 1985.

4 J. J. McCarthy and G. E. Marco. Cellular networking functionality & application. *37th IEEE Vehicular Technology Conference, Volume 37*, pp. 305–311, 1987.

5 J. Whitehead. Cellular system design: an emerging engineering discipline. *IEEE Communications Magazine*, 24:8–15, 1986.

6 Cellular wireless networks, Chapter 14, in *Data and Computer Communications*, 8th edition, William Stallings.

7 J. Kim, E. Serpedin, D. R. Shin, and K. Qaraqe. Handoff triggering and network selection algorithms for load-balancing handoff in CDMA-WLAN integrated networks. *EURASIP Journal on Wireless Communications and Networking*, 2008(136939):14, 2008.

8 Q. Zhang, C. Guo, Z. Guo, and W. Zhu. Efficient mobility management for vertical handoff between WWAN and WLAN. *IEEE Communications Magazine*, 41(11): 102–108, 2003.

9 H. J. Wang, R. H. Katz, and J. Giese. Policy-enabled handoffs across heterogeneous wireless networks. *2nd IEEE Workshop on Mobile Computing Systems and Applications (WMCSA '99)*, pp. 51–60, New Orleans, LA, USA, 1999.

10 L. J. Chen, T. Sun, B. Chen, V. Rajendran, and M. Gerla. A smart decision model for vertical handoff. 4th ANWIRE International Workshop on Wireless Internet and Reconfigurability (ANWIRE '04), Athens, Greece, 2004.

11 S. Balasubramaniam and J. Indulska. Vertical handover supporting pervasive computing in future wireless networks. *Computer Communications*, 27(8):708–719, 2004.

12 Y. Nkansah-Gyekye and J. I. Agbinya. Vertical handoff decision algorithm for UMTS-WLAN. *2nd International Conference on Wireless Broadband and Ultra Wideband Communications (AusWireless '07)*, Sydney, Australia, 2007.

13 H. Liao, L. Tie, and Z. Du. A vertical handover decision algorithm based on fuzzy control theory. *1st IEEE International Multi-Symposiums on Computer and Computational Sciences (IMSCCS '06)*, pp. 309–313, Hangzhou, China, 2006.

14 Y. D. Lin and Y. C. Hsu. Multihop cellular: a new architecture for wireless communications. *19th Annual Joint Conference of the IEEE Computer and Communications Societies, Volume 3*, pp. 1273–1282, 2000.

15 R. Madan, S. P. Boyd, and S. Lall. Fast algorithms for resource allocation in wireless cellular networks. *Transactions on Networking, IEEE/ACM*, 18(3):973–984, 2010.

16 D. A. Levine, I. F. Akyildiz, and M. Naghshineh. A resource estimation and call admission algorithm for wireless multimedia networks using the shadow cluster concept. *IEEE/ACM Transactions on Networking*, 5(1):1–12, 1997.

17 J. Tajima and K. Imamura. A strategy for flexible channel assignment in mobile communication systems. *IEEE Transactions on Vehicular Technology*, 37:92–103, 1988.

18 S. G. Choi, O. S. Yang, and J. K. Choi. An efficient resource allocation scheme during handoff in mobile wireless networks. *7th International Conference on Advanced Communication Technology, ICACT, Volume 2*, pp. 988–992, 2005.

19 L. Wayne and Y. Pan (eds). *Resource Allocation in Next Generation Wireless Networks, Volume 5*, Nova Publishers, 2006.

20 S. Roy. *Lecture Notes in Mobile Computing*. IIT, Kharagpur, 2008.

21 I. S. Reed and G. Solomon. Polynomial codes over certain finite fields. *SIAM Journal of Applied Mathematics*, 8:300–304, 1960.

22 B. Sklar. *Digital Communications: Fundamentals and Applications*, Prentice Hall, 2nd edition, 2001.

23 J. W. Byers, M. Luby, M. Mitzenmachert, and A. Rege. A digital fountain approach to reliable distribution of bulk data. *ACM SIGCOMM Computer Communication Review*, 28(4):56–67, 1998.

24 J. W. Byers, M. Luby, and M. Mitzenmachert, A digital fountain approach to asynchronous reliable multicast. *IEEE Journal on Selected Areas in Communications*, 20(8):1528–1540, 2002.

25 M. Luby and L. T. Codes, Foundations of Computer Science. *43rd Annual IEEE Symposium*, pp. 271–280, 2002.

Abbreviations/Terminologies

3G	Third-Generation (networks)
AMPS	Advanced Mobile Phone Service
apps	(mobile) Applications
AUC	Authentication Center
BSC	Base Station Controller
BSS	Base Station Subsystem
BTS	Base Transceiver Station
CDMA	Code Division Multiple Access
EEPROM	Electrically Erasable Programmable Read-Only Memory
EIR	Equipment Identity Register
GGSN	Gateway GPRS Support Node
GPRS	General Packet Radio Service
GSM	Global System for Mobiles
HDD	Hard Disk Drive
HLR	Home Location Register
HTTP	Hypertext Transfer Protocol
ICCID	Integrated Circuit Card Identifier
IEEE	Institute of Electrical and Electronics Engineers
IMEI	International Mobile Equipment Identity
IMSI	International Mobile Subscriber Identity
JFFS	Journaling Flash File System
LAI	Location Area Identity
MCC	Mobile Country Code
MIN	Mobile Identification Number
MNC	Mobile Network Code
MSC	Mobile Switching Center
MSIN	Mobile Subscriber Identity Number
NAMTS	Nippon Automatic Mobile Telephone System
NMT	Nordic Mobile Telephone
NSS	Network and Switching Subsystem
PCU	Packet Control Unit
PIN	Personal Identification Number
PLMN	Public Land Mobile Network
PSTN	Public Switched Telephone Network

QoS	Quality of Service
RAM	Random Access Memory
RDBMS	Relational Database Management System
ROM	Read-only Memory
RTMS	Radio Telephone Mobile System
SD card	Secure Digital Card (memory)
SGSN	Serving GPRS Support Node
SIM	Subscriber Identity Module
SMS	Short Message Service
TACS	Total Access Communications System
TCP	Transmission Control Protocol
TDMA	Time Division Multiple Access
TrueFFS	True Flash File System
VLR	Visitor Location Register
WARC	World Administrative Radio Conference
YAFFS	Yet Another Flash File System

Questions

1   Why is the cellular wireless network termed an 'infrastructure-based network'?

2   Mention the advantages and disadvantages of a cellular wireless network.

3   Why does the high user density in a region require cell splitting?

4   With the help of a diagram, explain the concept of various cell sizes.

5   Describe the process of handover in a cellular wireless network. With reference to movement between the same or different BSCs or MSCs., what are the different types of handover?

6   Why is a software handover also termed 'make-before-break'?

7   Explain the process of handling high-mobility mobile devices using umbrella cells.

8   Draw the broad architecture of a GSM network, properly marking all the components and subsystems.

9   How is a voice call routed in a GSM-based cellular network?

10   Explain the different mechanisms for receiving updates in a mobile device.

11   Differentiating between NAND flash and NOR flash, explain the applications where each can be used.

12 In the area of a cellular wireless network, 'Communication is the primary driver, not computation'. Please justify this statement.

13 Write short notes on the following:
A cochannel interference,
B call handover,
C selective cells,
D frequency reuse,
E network and switching subsystems,
F international mobile subscriber identity,
G device databases,
H device management,
I flash memory,
J data synchronization.

14 Differentiate between the following:
A intrasystem and intersystem cellular networks,
B base transceiver station and base station controller,
C 60° cell sectoring and 120° cell sectoring,
D cell splitting and cell sectoring,
E inter-BSC handover and intra-BSC handover,
F hard handover and soft handover,
G single-hop and multihop cellular networks,
H traffic channel and control channel,
I forward channel and reverse channel,
J elastic traffic and inelastic traffic.

15 For the following, mark the option that is correct:
i The performance of a cellular network can be measured in terms of:
(a) voice quality, (b) call failure, (c), spectrum efficiency, (d) all of these.

ii The capacity of a cell in a cellular network cannot be increased by:
(a) sectoring, (b) frequency borrowing, (c) scattering, (d) cell splitting.

iii A parameter that can trigger handover is:
(a) speed of the mobile device, (b) signal strength, (c) movement of the mobile device into a cell overlap area, (d) moving out of line-of-sight of the BTS.

iv The same broadcast signal reaches a point at different times owing to:
(a) diffraction, (b) scattering, (c) reflection, (d) any of these.

v The area or shape of a real cell is:
(a) irregular, (b) hexagonal, (c) circular, (d) square.

Exercises

1 A hexagon cell has a radius of 20 km. The cell has to be split into equisized smaller hexagonal microcells. The minimum number of microcells should be created to cover the cell without any overlapping. What will the radius of each microcell be? The microcell has to be further split to one more level to create picocells under the same splitting conditions as for microcells. What will the radius of each picocell be?

2 When did the cellular radio operations start in your city, who was the service provider, what was the radius of each cell, how many voice call charges were there per minute, and what was the approximate number of users? As of today, who are the service providers in your city, what is the average radius of each cell, how many voice call charges are there per minute, and what is the approximate number of users?

3 Present-day humans carry a number of interconnected devices with them, and these devices communicate with each other through a personal area network. The transmission range of the devices is 5 m. A number of similar devices used by different people operate on the same frequency and hence interference should be avoided. In a convention hall of 50 m × 100 m, how many persons can be accommodated if the transmission frequency of their devices is not known and collision/interference has to be avoided?

4 In a convention hall of 50 m × 100 m with a height of 4 m, transmission antennas can be fitted on the walls and roof. Mounting of antennas on walls is preferred, and hence a roof-mounted antenna should be used only if a wall-mounted antenna is not feasible. Each antenna has a range of 5 m. What is the minimum number of antennas required if the entire convention hall has to be under cellular wireless network cover without any non-covered area? How many of the antennas will be on the roofs and how many of the antennas will be on the walls? The reflection of the transmission signals may be ignored.

5 Please refer to exercise 4 above. Assume that it has been said that there should be no areas with overlapping signals, even if certain areas remain outside the coverage of the cellular network. However, such uncovered areas should be minimized. All the other conditions of mounting the antennas remain the same. How many of the antennas will be on the roofs and how many of the antennas will be on the walls?

6 Attempt a frequency reuse pattern with less than seven cells as well as a frequency reuse pattern with more than seven cells. Any other uniform shape of the cell other than a hexagon may also be assumed if required.

7 Try to identify all the digits of the IMSI with respect to your mobile phone number. Also identify the IMEI of the mobile device and the form factor of the SIM used.

8 A cell switches on in 'roaming' and then starts moving and reaches its home network. Indicate the initial value and changes in the values of various registers used in the GSM network. Necessary assumptions may be made.

10

Routing in Wireless Ad Hoc Networks

10.1   Introduction

Gordon Moore, who was the cofounder of Intel, predicted 'Transistor density of minimum-cost semiconductor chips would double roughly every 18 months'. The prediction was made in 1965 and it has been more than five decades since then, but still the predicted trend holds its ground. This has led to its being called Moore's law. The density of transistors is directly related to the processing speed and inversely related to the cost of the microprocessor. The effect of Moore's law is that the performance of a processor is doubling up every 2 years, and the capacity of memory for the same covered area doubles up every 18 months. This has led to miniaturization of the computers and other computing systems over the past few years, as well as reduction in their cost. The advancements in microprocessor and computing technology have made available laptops, mobile phones, and iPADs at affordable cost at the grassroots level. This has led to the application of computing systems in all imaginable applications, scenarios, and geographic domains.

The size of the modern day computing device and its application has provided mobility to these devices. The ease of operations, fast deployment possibility, freedom from wired mesh, and support for mobility have led to invention and advancements in the field of wireless technology, and the area is growing rapidly [1, 2]. The wireless mode of

Network Routing: Fundamentals, Applications, and Emerging Technologies, First Edition.
Sudip Misra and Sumit Goswami.
© 2017 John Wiley & Sons Ltd. Published 2017 by John Wiley & Sons Ltd.
Companion website: www.wiley.com/go/misra2204

communication between two nodes, either mobile or static, can be supported by two different methodologies:

- infrastructure-based mobile wireless communication, i.e. a wireless cellular network;
- infrastructureless mobile wireless communication, i.e. a wireless ad hoc network.

The difference between a wireless cellular network and a wireless ad hoc network has been tabulated in Table 10.1.

An ad hoc network is made on-the-fly by the participating nodes without requirement of any centralized infrastructure. 'Ad hoc' is a Latin phrase that means 'for this', and this phrase defines well the characteristic of an ad hoc network, which is created by its participating nodes for this (any) particular application [3]. An ad hoc network is formed by two or more participating nodes that want to communicate with each other and do not possess the time, resource, or scenario to set up a traditional network in terms of wired connectivity or routing and administrative infrastructures in terms of access points, switches, and routers. The nodes in an ad hoc network may

Table 10.1 Comparison of a cellular wireless network and an ad hoc wireless network.

Comparison Parameter	Cellular Wireless Network	Ad Hoc Wireless Network
Centralized infrastructure and management	Required.	Not required.
Time for setup	High.	Low.
Networking cost	High.	Low.
Network designing and planning	Required for establishing the base stations.	Little or even not required.
Reliability and stability	Comparatively higher than an ad hoc network.	Comparatively lower than a cellular network.
Resource constraint	The nodes might have resource constraint, but the routers and access points have sufficient resources.	The nodes act as hosts as well as routing devices, and hence resource constraint.
Reachability	The range of communication is pre-fixed. The network can be operational only at those locations where the cellular infrastructure exists.	New nodes can automatically be connected to the network, changing the range of connectivity. The network can be made operational at any place, including underwater, above the land, or in the sky.
Range of communication and connectivity	The range may even extend to communication connectivity across a country.	The spread of the network is generally confined to its application theater, which may be limited to an operational area, building, campus, or a very small geographic region for some specific defense or civilian application.
Topology	The topology of the backbone is static.	The entire network topology is dynamic.

communicate continuously, regularly, or on a requirement basis. The lack of network support infrastructure necessitates the participating network nodes of the ad hoc network to help each other in communicating in the network. This makes them play the dual role of host as well as the networking device. The 'network device' role of the node is responsible for one or more of the activities such as signal amplification, packet forwarding, detecting neighbor nodes, participating in routing, detecting network attacks, and ensuring secured communication. The reach or the size of an ad hoc network completely depends on the number of participating nodes, as well as the willingness of these ad hoc nodes to participate in the process of routing information to and from their neighbors.

The precursor of the ad hoc network [4] was the packet radio network (PRNET) designed and developed by the US Department of Defence (DoD). The research on the packet radio network was initiated in 1972 to provide reliable communication between computers for military applications. The PRNET, which used ALOHA and Carrier Sense Multiple Access (CSMA) protocols, was continuously improved during the period 1973–1987 when a number of network devices and protocols came up to improve the network. The PRNET used radio frequency to provide half-duplex communication at 100 and 400 kbps. In the early 1980s, the survivable adaptive radio network (SURAN) evolved to support networking between battlefield equipment, platforms, and soldiers in a mobile, infrastructureless, and hostile terrain. SURAN provided packet switching over improved radios, which were smaller and cheaper. The PRNET used a distance vector algorithm, while SURAN was based on a hierarchical link state algorithm. The era of PRNET is called the 'first generation' of ad hoc networks, while SURAN initiated the beginning of the 'second generation' of ad hoc networks, which continued until 1993.

The second generation of ad hoc networks also saw the advent of global mobile (GloMo) information systems and near-term digital radio (NTDR) systems from DARPA. The GloMo information system used flat as well hierarchical routing algorithms and brought in the concept of self-configuring and self-healing networks. The GloMo network was meant mainly for providing anytime-anywhere access to multimedia-based defense information systems in the hand-held devices of the mobile forces under deployment. The NTDR was a self-organizing and self-healing radio network based on clustering and link state routing that was used to provide connectivity at a brigade level of the defense forces. A brigade comprises 3–5 regiments. Each regiment has about a thousand soldiers, along with small and medium arms, equipment, guns, vehicles – tracked or wheeled or other platforms such as armored vehicles, depending on the role and formation of the regiment. NTDR had a two-tier architecture and was meant to provide data communication between the brigade-level command and control center and the regiment and other formations further in the regiment.

The third generation of ad hoc networks marked its beginning from 1990 and saw a shift in ad hoc network applications from defense use to commercial applications. Miniaturization of computers led to the invention of laptops and notebooks and the introduction of 802.11 PCMCI cards enhancing mobility. Research in the area of ad hoc networks gained momentum in the academic institutes during this period. IETF created the MANET working group. Bluetooth, HIPERLAN, and mobile ad hoc sensor networks were also introduced in this period.

10.1.1 Basics of Wireless Ad Hoc Networks

The nodes participating in the creation of a wireless ad hoc network should possess a network adapter capable of wireless communication. The nodes may be mobile or static. There are certain special types of node in an ad hoc network that are generally static, e.g. access points, which may provide connectivity between the ad hoc network and the Internet or some other backbone network, either wired or wireless. Two nodes in a wireless ad hoc network can communicate only if they are in the direct transmission range and reception range of each other. They can also communicate with each other through one or more intermediate nodes if routing of packets from the source to the destination is allowed through the intermediate nodes. Hence, the source node as well as the destination node should individually have an overlapping transmission and reception region with at least one intermediate node each nearer to it. Each of the intermediate routing nodes should also have an overlapping reception–transmission region with at least two other intermediate node or source node or destination node as indicated in Figure 10.1.

As can be understood from Figure 10.1, ad hoc networks are also capable of forming multiple paths between a particular source and destination pair. Furthermore, as there is no centralized node for managing or administrating the network, there is no single point of failure. This leads to enhancement of the reliability and fault tolerance of a wireless ad hoc network. There are frequent topological changes in an ad hoc network because the nodes are mobile, toggle between sleep mode and active mode, come into the transmission zone, or move out of it. The routing protocols for an ad hoc network are designed to accommodate these basic characteristics of the network. These routing protocols accommodate topological changes not only due to movement of nodes but also due to

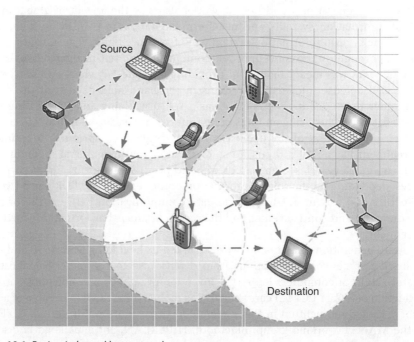

Figure 10.1 Basic wireless ad hoc network.

frequent link disruptions and new link availability. The link instability is due to sudden availability or non-availability of intermediate nodes for routing. The wireless nature of an ad hoc network further enhances the need for such routing protocols, as the nodes are free from the constraint of limited or restricted mobility as in wired networks.

The nodes participating in the formation of a wireless ad hoc network can be laptops, mobile phones, smart phones, personal digital assistants, and any other ad-hoc-network-enabled device used in applications for defense, health, industry, home, or transportation. The main characteristic of an ad hoc network, which has opened up the entire spectrum of its application, is that the network is 'infrastructureless'. Hence, it is the most preferred type of network in a disaster-hit area. A disaster, natural or man induced, e.g. an earthquake, flash floods, landslides, a tsunami, an avalanche, explosions, or a fire, may wipe off all the available network infrastructure, rendering the wired communication as well as the wireless cellular communication inactive. In such scenarios, an ad hoc network is the cheapest and fastest way to set up a communication network for information exchange among the participating nodes. The network provides sufficient bandwidth and is established immediately on demand and is reliable. Similar requirements also exist in the case of events that lead to a sudden influx of people in a particular region that may not have any prior networking infrastructure or may lack sufficient infrastructure to cater for a sudden increase in node connectivity requirement. These events may be a collection of people at conferences, jamborees, carnivals, melas, exhibitions, trials, and cross-country events, or defense applications such as fleet reviews and land or offshore exercises, or some political event leading to a mass exodus of citizens. In the case of these scenarios, an ad hoc network can help to set up, without exploiting any pre-existing network infrastructure available in the region, communication between the enabled devices held by the people.

Ad hoc networks are designed to be self-configuring, self-organizing, and self-healing in nature [5]. The nodes in them are able not only to interconnect with each other but also to continue to retain their connectivity by adapting to various changing parameters in the network, including variation in signal strength, changing mobility rate, variation in topology of nodes, node failure, or link failure. They discover their neighbors on booting up, and thereafter they are capable of computing the route from themselves to every other connected node in the network and selecting the optimum route. The optimum route can be in terms of number of hops, total energy consumption, number of packets transmitted, or control messages exchanged.

A node in an ad hoc network collaborates with the neighboring nodes to identify the network and continuously adapts to the changing network parameters to route data packets to the destination. The node shares with all its neighboring nodes on a continuous or on-demand basis the information gathered relating to its neighbor connectivity, network topology, and route to destination nodes using multihop over other intermediate nodes. This enables all the nodes to self-heal themselves to the changing network with the help of a routing algorithm being run by it. This characteristic of the ad hoc network is also called 'fault tolerance'.

The network architecture, routing capabilities, and operational requirements of an ad hoc network lead to the possession of the following characteristics by a wireless ad hoc network:

- Distributed processing across nodes for path determination and packet forwarding.
- Variation in network topology and number of nodes.

- Dynamic topology due to mobility of nodes.
- Energy constrained, low power nodes.
- Unreliable links and variable link capacity.
- Heterogeneous link characteristics such as unidirectional links, omnidirectional links, and point-to-point links.
- Capability of neighbor discovery for peer-to-peer routing.
- Autonomous mode of operation by each node owing to lack of a centralized administrator or management.
- Self-configuring and self-healing.

In an ad hoc network, a link may be unidirectional not only because of its manufacturing or configuration setting but also because of a number of other operational, environmental, or technical factors. A few factors that lead to unidirectional behavior of the links are as follows [6]:

1) The reception hardware and the transmitting radio may not be of equal power and range in the node.
2) A node may decide to maintain electromagnetic silence owing to some operational threat in the area or to protect itself from being detected in an enemy territory. In such a scenario, the node, even though capable of bidirectional communication, will use a unidirectional link only to listen to the network transmissions or receive packets from the network and impede its ability of packet transmission.
3) The toggling between bidirectional and unidirectional states of the link may vary with time owing to environmental effects and traffic parameters.
4) The link in any one direction may become unavailable owing to heavy signal interference around one of the nodes. The node density may be very high around one of the nodes, reducing channel availability towards this node from some other node. A node willing to transmit may be from a less dense area and waiting very long to transmit data to a node in a dense area. The waiting period may be long enough effectively to make it an unavailable transmission link and turn it into a unidirectional link. A scenario depicting this situation can be understood from Figure 10.2. Node 1 wants to transmit a packet to node 2. However, as node 2 has a huge number of active nodes in its reception range, which may be involved in packet exchange, node 2 may not be able to receive a data packet from node 1. Contrary to this, whenever node 2 wants to send a packet to node 1, it can do so with relative ease as there will be no interference around node 1.

The hidden terminal problem and exposed terminal problem are also two typical technical issues faced in a wireless network, and hence solutions for them have to exist in mobile ad hoc networks. Hidden nodes, with respect to a particular node or a collection of 'N' nodes in a wireless network, refer to the set of nodes that are outside the range of 'N'. To present the problem in a simple manner, consider three nodes, node 1, node 2, and node 3, such that node 1 is in the range of node 2 as well as node 3, but node 2 is not in the range of node 3, as depicted in Figure 10.3. Hence, node 1 can transmit to (referred to as 'see') node 2 and node 3, but node 2 cannot see node 3, and vice versa. Now, at a particular instant of time 't', node 2 plans to communicate with node 1, and so it senses the wireless medium, does not detect any interference, and therefore transmits. At the same time instant 't', node 3 also plans to communicate with

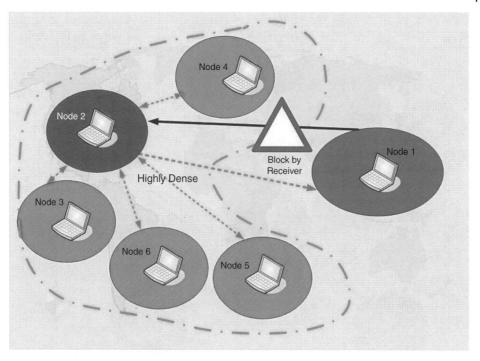

Figure 10.2 Unavailable node due to high interference.

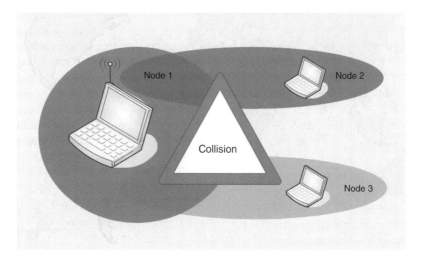

Figure 10.3 Hidden node.

node 1, senses the wireless channel to be free, and thus transmits. In this particular scenario, Carrier Sense Multiple Access/Collision Avoidance (CSMA/CA) also fails to work. This leads to collision of signals at node 1, leading to loss of data, and node 1 does not receive any transmission either from node 2 or from node 3. This is referred to as the hidden terminal problem.

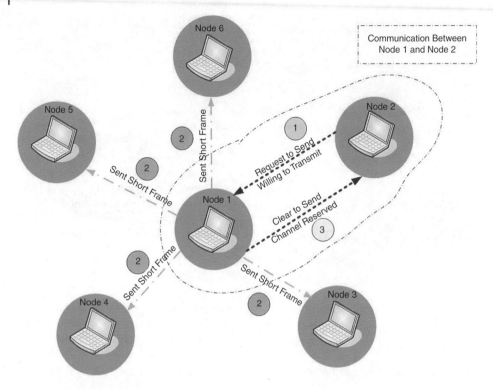

Figure 10.4 Resolving the hidden node problem.

The hidden terminal problem is resolved by reserving the channel and notifying others before transmission of data, as depicted in Figure 10.4. Node 2 sends a 'request to send' (RTS) to node 1 that it is willing to transmit. On receiving the RTS signal from node 2, node 1 sends a short frame to its neighbors to indicate that shortly a transmission involving node 1 will occur, a 'clear to send' (CTS) signal is sent to node 2, and thus the channel is presumed to be reserved as the neighbors restrain from transmission until the channel passes off a large frame. The RTS/CTS signaling works only in those cases where the packets to be transmitted are relatively larger than the RTS/CTS packet size. If the packet to be transmitted from one node to another node is comparable with the packet size of RTS/CTS, the packet is transmitted directly over the wireless channel without invoking RTS/CTS signaling to reduce time and complexity. Furthermore, RTS packets from node 2 and node 3 can also collide. Hence, an optimization has to be done for using RTS/CTS signaling, and the threshold level has to be determined to invoke it.

The solution to the hidden node problem leads to another problem, the 'exposed node' problem, which can be understood from the same example as in Figure 10.3 but with another node, node 4, added to the network, as depicted in Figure 10.5, such that node 4 is in the range of node 3 but is not in the range of node 1 or node 2. Now, if node 1 is communicating with node 2 and at the same time node 4 wants to communicate with node 3, this cannot be done because node 3 will not grant a CTS signal to node 4 because node 3 has already received a short frame from node 2 and hence assumes that it cannot send a clear signal to node 4, even though the point-to-point communication

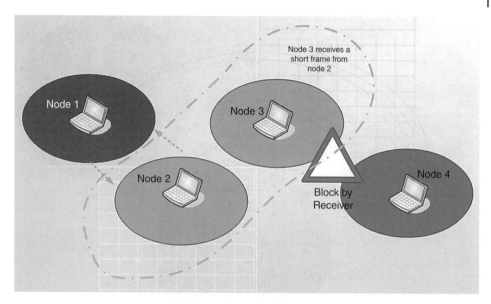

Figure 10.5 The problem of an exposed node.

between node 1 and node 2 and between node 3 and node 4 could happen simultaneously. If such a transmission were permitted, none of the nodes would face any collision as the area of collision would be in the region between node 2 and node 3.

The typical characteristics of an ad hoc network raises a number of challenges in successful and efficient routing of packets between any two nodes. The dynamic topology leads to a time-dependent change in the optimal path between any two nodes. The major technical challenges faced in ad hoc routing are listed below:

- *Communication channel related* – hidden terminal problem, exposed terminal problem, signal fading effect, low bandwidth, packet loss, variation in channel quality, environmental effects.
- *Data traffic related* – variation in traffic characteristics and profile.
- *Device related* – heterogeneous nodes, limited wireless range, unidirectional communication, i.e. either receiving or sending in certain nodes, mobility of nodes, variation in mobility pattern and speed, dual role of host as well as intermediate routing device.
- *Power related* – constrained battery power, power planning, and rationing cannot be done owing to unpredictable packet forwarding requirements.

Based on the interconnectivity among various independent wireless ad hoc networks, these networks are classified [7, 8] into two categories:

- closed mobile ad hoc network,
- open mobile ad hoc network.

In a closed mobile ad hoc network, all the nodes belong to the same network, and hence the nodes participate in networking to cater for the same application. As the nodes belong to the same network, there is a high probability that the nodes are

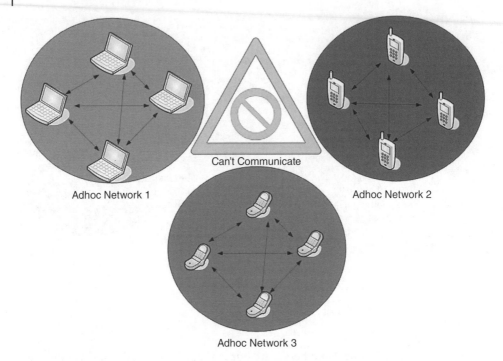

Figure 10.6 Closed ad hoc network.

homogeneous. These nodes do not communicate either with any other mobile ad hoc network or with any other type of network such as a cellular network or wired network. All the neighbor discovery, route determination, and packet forwarding are confined within the network, and there is no exchange of information with other networks. These networks lack gateways for connectivity to other networks, as shown in Figure 10.6.

An open mobile ad hoc network, as shown in Figure 10.7, is formed by interlinking of different mobile ad hoc networks. They are also capable of communicating with other types of network, i.e. a cellular network, a sensor network, and a wired network, through gateways available in the open mobile ad hoc networks. As the network is formed by different types of participating mobile ad hoc network, there may be different types of node in the networks, leading to a high degree of heterogeneity of nodes in the network.

A closed mobile ad hoc network is generally formed for accomplishment of a single mission by an identified and known group of participating nodes. The mission accomplishment holds priority in the routing over resource conservation in the network. An ad hoc network of laptops formed by a group of researchers conducting a soil study in a desert, an ad hoc network of mobile phones of rescuers to share location and mobility information among each other at a disaster site, an ad hoc network formed among radars for surveillance of a designated area – these are examples of closed mobile ad hoc networks.

In an open ad hoc network, different mobile ad hoc networks, each with its own mission objective, collaborate with each other for resource sharing in terms of packet forwarding to a large extent and information exchange to a lesser extent. As different networks collaborate with each other, this leads to an increase in the spread of the

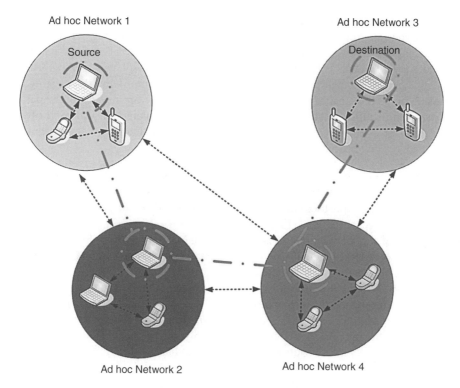

Figure 10.7 Open ad hoc networks.

network. The transmission and reception range of the network can increase many fold as intermediate nodes from other networks contribute to packet forwarding, enhancing the transmission range.

As shown in Figure 10.7, some of the participating networks may be located at the periphery of the network, while a few may be located in the central region of the network connecting the networks at the periphery. There might even be networks that have a wider spread and have nodes across the entire open ad hoc network. The nodes in the central region of the network receive more requests for packet forwarding, as they bridge the nodes at the periphery and are in the routing path of most of the source destination pairs. These nodes will therefore consume more resources in terms of processing power and wireless transmission, and hence consume their battery at a faster pace. As each node will attempt to maximize its benefit by participating in the open network, this might lead to selfish behavior by the node.

Each node attempts to use its resource optimally by ensuring its availability for information detection and forwarding related to its own mission. It also attempts to restrict its resource usage when it participates in the packet forwarding related to the mission of some other network forming a part of the open network. The most common example of selfish behavior by a node is that it does not participate in packet forwarding for other participating nodes to save its own battery power consumed for processing the packet and forwarding it, as transmission consumes considerable power in an ad hoc node.

The ad hoc network has a variety of applications and various deployment scenarios. The same routing protocol will not be able to cater for different ad hoc network requirements. Hence, there exist a number of routing protocols, each optimizing one or more parameters on the basis of the priority of the requirements. Still, new routing protocols are evolving to cater for the changing requirements and improving the existing algorithms. A single protocol cannot have all the functionalities optimizing all the parameters required in different scenarios, as they may be complementary to each other. A cumulative list of properties [3, 9-11] that should be considered before designing or analyzing a routing protocol is given below:

- The protocol should be able to do *multicasting*, i.e. send packets from one node to multiple nodes simultaneously. The property of multicasting should not be assumed to be present if the protocol has the capability of broadcasting packets. Multicasting consumes less resources than broadcasting.
- The protocol should *converge rapidly* in the dynamic network topology.
- The protocol should be *highly scalable*, enabling it to connect to a large number of nodes that may even be heterogeneous.
- The protocol should be small and simple for reduced memory requirements and ease of path prediction and performance analysis.
- The protocol should discover a *loop-free path*, i.e. it should not visit the same link twice unless the network has been reconfigured while the data packet was in transit, leading to changes in the routing table of intermediate nodes, necessitating a revisit to particular links.
- To recover from link failures and congestion avoidance and to save time spent rediscovering routes in the case of some change in network topology, the routing protocol should discover *redundant paths* from source to destination before transmission of packets.
- The protocol should *avoid a single point of failure* in terms of its presence in a single node for any administrative, management, or control function. The protocol should be distributed in nature, so that it is unaffected in the case of any random failure of nodes.
- The protocol should be capable of *effective link utilization*. The ad hoc network can have unidirectional as well as bidirectional links. The protocol should not be designed to support only one kind of link for particular operations or in a particular region of the network.
- The routing protocol should be designed to *minimize resource consumption* in terms of processing, channel utilization, and exchange of packets, as all these lead to power consumption, which can be the most constrained resource in an ad hoc network during its usage in certain application scenarios.
- The protocol should be *adaptive to misbehaving nodes*, sleeping nodes, and captured nodes, as these may lead to unpredictable behavior.
- The protocol should be *secure*. It should be able to avoid eavesdropping, snooping, disruptions, modifications, repudiation, and cryptanalysis.

10.1.2 Issues with Existing Protocols

The characteristics that make an ad hoc network different from the traditional networks lead to a different set of requirements in wireless ad hoc routing protocols [3]. Mobility is the prime factor driving the wireless ad hoc protocols. The distance as well as the orientation of the nodes may keep changing. To make it further complex, the rate of change in the distance and orientation may also vary. These complexities are then driven by the typical behavior of an ad hoc network, such as processing limitation, power constraint, bandwidth restriction, heterogeneity of the nodes, high scalability, change in the transmission and reception range of a node due to variation in climatic condition, or variation in the number and distance of neighbor nodes. It is not only these characteristics of the ad hoc nodes that drive the routing protocol, but also unpredictable behavior of the participating nodes, such as selfish behavior, byzantine or captured nodes, sleep nodes, and dumb nodes, which makes the ad hoc routing protocol one of the most researched areas for enhancements and improvements.

In order to effectively handle the ad hoc networking characteristics and constraints, the routing protocol should be highly scalable, adaptive to rapidly changing topology, demand low processing capacity, generate least traffic, and save battery power. However, all these routing characteristics may not be required simultaneously by all types of ad hoc network. Each ad hoc network will require one or more of these routing characteristics, based on the devices used, the application, and the deployment scenario.

An ad hoc network of vehicles on a highway will require the routing protocol to handle high mobility, but it will not be constrained in terms of power owing to the uninterrupted and sufficient availability of battery power from the vehicle. A different scenario would be an ad hoc network of laptops and mobile phones of delegates in a conference hall where scalability of the network will be of utmost concern for the network. Mobility and power conservation may not be the driving characteristics in their routing protocol. Another scenario is the ad hoc network formed by buoys floating in the sea for regular measurement of the oceanographic and sea conditions and processing of the data for weather prediction. In such a network, power saving will be of prime concern, failing which there will be a frequent requirement to retrieve the buoys to change the power source. The power constraint will automatically govern constraints over processor requirement, frequency of transmission, periodicity of updates, and minimization of control traffic. However, in this deployment, mobility is of less concern and scalability will rarely come into play for a particular deployment. Hence, a single routing protocol cannot cater for all the types of ad hoc network.

A particular ad hoc routing protocol that performs best in a particular deployment scenario may not be able to perform optimally in a number of other deployment scenarios. Hence, there are a variety of ad hoc routing protocols, each having its own advantages, disadvantages, and specific application scenario, and each performing optimally only if used after an intelligent judgment.

A routing protocol is designed with two prime objectives – discovery of optimum routes between all source–destination pairs and efficient delivery of packets between any source–destination pair. The objective of the routing protocol remains the same for the conventional routing protocols as well as the routing protocols required for wireless ad hoc networks. Still, there are problems in porting the routing protocols used in wired networks directly to wireless ad hoc networks [3, 12]. The following are the major problems:

Protocol convergence. The conventional protocols were designed for fixed network topology. As the topology of an ad hoc network is dynamic, it becomes difficult for the conventional protocols to converge.

Route discovery. The conventional protocols were designed to have a complete view of the network and discover routes for all source–destination pairs. This leads to a high volume of data exchange between the nodes in the form of control signals, which is not acceptable for an ad hoc network as it is resource constrained. Furthermore, in an ad hoc network it is not necessary for each node to know the details of the entire network, and nor does an ad hoc network necessitate the discovery of all source–destination pairs.

Link characteristics. The conventional protocols were designed for bidirectional links, which were assumed to be reliable in terms of QoS, while in ad hoc networks, not only may the links be unidirectional, they may be turned on and off by the intermediate nodes, and the link parameters are dynamic and unpredictable owing to environmental effects.

Unusable class of protocols. The conventional protocols can be differentiated into certain categories, and there are particular categories that are downright unsuitable for direct porting from static wired networks to mobile ad hoc networks [13, 14]. A few such categories are as follows:

1) *Centralized routing protocol.* There is a class of conventional protocols that are based on centralized routing where the route discovery is made by a single node in the network. This is not an acceptable practice in ad hoc networks. This therefore renders the entire class of conventional centralized routing protocols in their existing form unusable in ad hoc networks.

2) *Static routing protocol.* The class of static routing protocols for conventional networks is ineffective in its existing form in ad hoc networks with node mobility and dynamic topology.

3) *Multicast routing protocols.* The traditional multicast routing protocols, which are based on source of origin and reverse path forwarding, are not able to perform in the mobile ad hoc environment as the source node may move to a new position after transmitting the packet, rendering the routing protocol ineffective and inefficient. The movement of nodes that are members of a multicast group and the formation of a transient loop during reconfiguration of the spanning tree are two other problem areas when using conventional multicast routing protocols.

10.2 Table-Driven (Proactive) Routing Protocols

The table-driven protocols are the family of ad hoc network routing protocols that are closest to wired routing protocols, as most of the algorithms are inherited from the wired network with least variation to fit into the wireless community. Proactive routing protocols keep the network nodes ready for routing the packets from source to destination. The proactive routing protocols are also known as table-driven protocols, as they maintain in the nodes a routing table with information about connectivity to other nodes in the network. The routing table is regularly updated through topology information exchange between neighboring nodes with the intention to maintain the route information for every other node in the network. The advantage of this class of protocol is that the shortest distance between any source–destination pair can be found in the least time. Another common name given to the class of table-driven routing protocols is 'global' routing protocols, as the nodes have global (complete) information about the entire network.

Traditionally, the proactive routing protocols are classified into two categories – distance-vector-based protocols and link-state-based protocols. In distance-vector-based proactive routing, each node maintains a routing table that has two major pieces of information – the distance to each node and the next hop node through which the packet should be forwarded to that node. To maintain an updated distance vector table, each node periodically sends its distance vector table to all its neighboring nodes. A node, on receiving a table from its neighboring node, matches the entries in the received table with its own entries. If it finds that the distance to any node is less through its neighbor, then the distance already available in it for that particular destination node is decreased accordingly and the route to that destination node is made via this particular neighboring node from which it received the updated table. The protocol also effectively supports loop avoidance, neighbor discovery, and changes in network topology, though not at a rapid rate.

In the link-state-based routing protocol, each node in the network maintains complete information about the network topology. This information is in terms of which node is connected to which other nodes and the condition of the links among them. To maintain updated information, each node regularly broadcasts the status of its links to its neighbors. Dijkstra's algorithm is generally used by a node to create the routing table. Each node can independently calculate the best path from itself to all other nodes in the network, as they all possess the link state information of the network.

The disadvantages of the table-driven proactive routing protocol are as follows:

- As the routing table is regularly updated, the resource utilization in terms of bandwidth, processing power, memory, and battery consumption is high.
- The scalability is less, as the routing table stored in each node becomes too large to store and maintain. Furthermore, as all nodes broadcast information, this leads to congestion and interference.
- This routing protocol reacts slowly to the dynamic network topology and may lead to routing failures in the case of rapid node movements.
- There may be certain routes over which no data packet has travelled and also by which there is a high probability that no packet will travel in the future. Still, the protocol utilizes resources to establish and regularly update such paths.

A few table-driven proactive routing protocols are listed below:

- Ad Hoc Wireless Distribution Service (AWDS),
- BABEL,
- Better Approach to Mobile Ad Hoc Networking (BATMAN),
- Clusterhead Gateway Switch Routing Protocol (CGSR),
- Destination Sequence Distance Vector (DSDV),
- Direction Forward Routing (DFR),
- Distributed Bellman–Ford Routing Protocol (DBF),
- Fisheye State Routing (FSR),
- Hierarchical State Routing Protocol (HSR),
- Optimized Link State Routing Protocol (OLSR),
- Wireless Routing Protocol (WRP).

10.3 On-Demand (Reactive) Routing Protocols

In a reactive ad hoc routing protocol, the path from a source to a destination is determined when a packet is to be routed between the source–destination pair. As the path is determined on a requirement basis, it is also known as 'on-demand' routing. This routing strategy reduces the routine bandwidth utilization by avoiding exchange of update information between the nodes. However, it adds to the delay in routing a packet to the destination as the path to the destination has to be determined before the packet can be routed. This routing strategy is effective in the case of large networks with relatively stable nodes, as the strategy adapts well to a dynamic network. Hence, the performance of a proactive routing protocol as well as a reactive routing protocol depends on the size of the network, the mobility of the nodes, and the frequency of communication between the nodes, and one routing protocol will perform better than another, depending on these parameters.

The source node initiates the process of route discovery when it is required to forward a packet to a destination. The family of reactive protocols generally relies on a variety of classical flooding algorithms and their improvements for route discovery. The reactive routing protocol has two primary operations to perform – route discovery and thereafter maintenance of the discovered routes. Dynamic Source Routing (DSR) is one of the initial on-demand routing protocols that is explained here with an example in Figure 10.8 to give an overview of an on-demand routing protocol.

In a network running DSR [3], the source node embeds in the header of the packet the entire route in terms of the hopping sequence on intermediate nodes before forwarding the packet towards the destination. If the source node (SN) is not aware of the path to the destination node (DN), then SN initiates route discovery. SN broadcasts a route request (RREQ). Each intermediate node (IN), when it detects that it is not the DN for the RREQ, adds its identifier before forwarding the RREQ further. These added identifiers of the INs build up the required hop sequence. The IN forwards the RREQ only if it finds that the route to the DN is not available in its route cache, otherwise it sends a route reply (RREP) back to the SN through the same hop of intermediate nodes through which this IN received the RREQ. As the RREQ already contains the list of node identifiers through which it came, the RREP need not be flooded, but it is unicast back.

The network may not have any INs with information of the route to the DN in its route cache or the RREQ may reach the desired DN before reaching any IN that has the cached route to the DN. When the DN detects that the RREQ is meant for it, it sends back the RREP in a similar way to the IN with the cached route would have done. To reduce flooding of the RREQ, any node before broadcasting the RREQ sends the RREQ to its neighbor with an instruction not to broadcast it and return back the route information if it has the route to the DN in its cached route. If the neighbors do not possess the information, then only the broadcast mode is opted. A node may also operate in promiscuous mode to receive all the packets and use the header information in the packet to gain information about the source route it is carrying.

An RREP can be sent back through the same route from which the RREQ packet came if all the links are bidirectional. If some of the intermediate links are not bidirectional, the RREP may rely on the route cache of the nodes to get the path back to the SN. Alternatively, the RREP is piggybacked on a RREQ packet and the route back to the SN from the DN is again discovered.

(a)

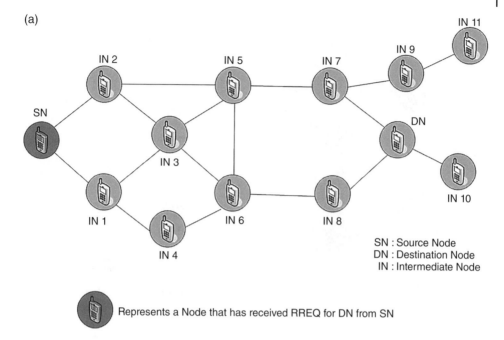

SN : Source Node
DN : Destination Node
IN : Intermediate Node

Represents a Node that has received RREQ for DN from SN

(b)

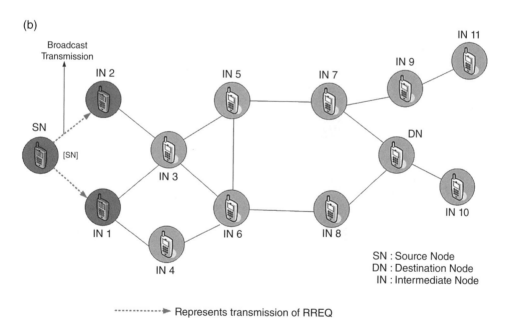

SN : Source Node
DN : Destination Node
IN : Intermediate Node

········▶ Represents transmission of RREQ

Figure 10.8 Sequence of steps in dynamic source routing.

(c)

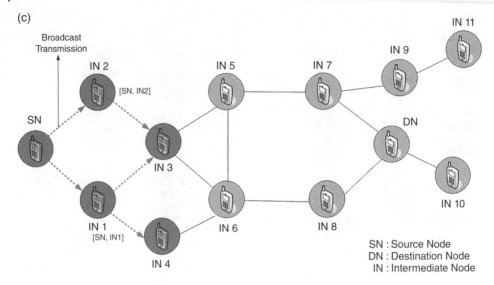

IN 3 receives packet RREQ from two neighbours: **Potential for collision**

(d)

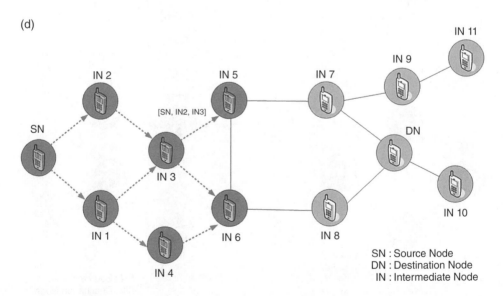

Figure 10.8 (*Continued*)

The reactive routing strategy also has the task to maintain the discovered route. A route once detected by the source towards the destination may not be available for the entire time duration required to transmit the complete data available at the source to the particular destination. Some of the intermediate nodes in the selected routing path may move out of the transmission range or may run out of energy. These missing intermediate nodes leading to failed links can be detected actively or passively. Active

(e)

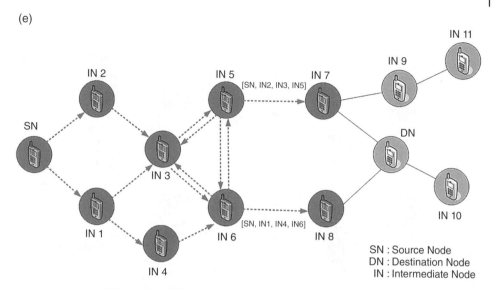

IN 3 receives RREQ from IN 5 and IN 6, but does not forward
it again, because node IN 3 has **already forwarded RREQ** once

(f)

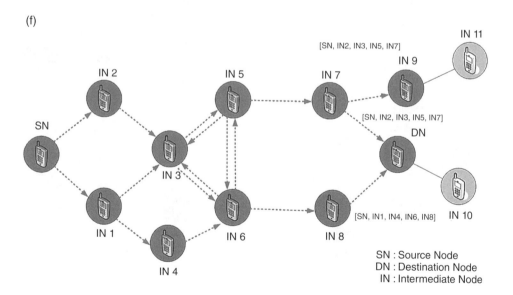

• Nodes IN 7 and IN 8 both broadcast RREQ to node DN
• Since nodes IN 7 and IN 8 are hidden from each other, their
  transmissions may collide

Figure 10.8 (*Continued*)

(g)

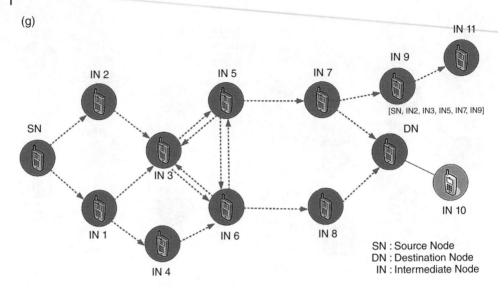

Node DN **does not forward** RREQ, because node DN is the **intended target** of the route discovery

(h)

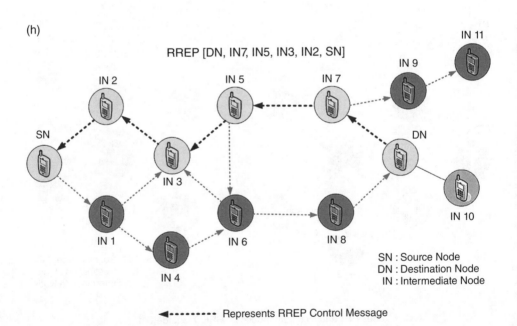

Figure 10.8 (*Continued*)

detection is by receiving back acknowledgement from all the previous hops. In passive detection, each node in the routing path listens to its neighbor by entering into promiscuous mode. Information about any detected link failure is transmitted back up to the source node. All the intermediate nodes also learn about this node failure and truncate the path beyond the failed node.

During the process of route discovery from source to destination, the intermediate nodes involved in the route discovery process also learn a number of routes, which each of the nodes stores in its route cache. The source node in the course of discovering the destination node learns about the path to all other intermediate nodes that are on the path to the destination. All the intermediate nodes that receive an RREQ packet from the source learn the path from themselves to the source as well as from themselves to all the other intermediate nodes on the path to the source node. Similarly, the RREP packet makes all the intermediate nodes aware of the route to the destination node and to all the intermediate nodes up to the destination node. The nodes also learn paths while forwarding error messages related to node failure. A node may also operate in promiscuous mode, learning new routes. Thus, reactive routing is associated with a number of opportunities through which the nodes learn about routes to a number of other nodes during the process without initiating any new process specifically for route discovery. The routes learned in this process are stored in the route cache of the nodes, which they use to help other nodes in faster route discovery.

Advantages. The major advantages of the reactive routing protocol are as follows:

- The reactive routing strategy is bandwidth savvy as it not only initiates route discovery on demand but also searches by an incremental method, stopping with the discovery of the destination node.
- Multiple routes are traced from the source to the destination during the process of route discovery. This is due to fresh route discovery as well as replies from intermediate nodes aware of the destination through locally cached route information.
- During the process of route discovery from source to destination, the source also gains information about the routes to the intermediate nodes. The intermediate nodes involved in the process of node discovery also learn about the routes to the source, destination, and intermediate nodes involved in the route discovery.
- There will be certain routes in the network that will never be discovered by the routing strategy. Only those routes through which a packet can possibly be routed efficiently will be discovered.
- Caching of route information by the participating nodes reduces resource utilization during route discovery.

Disadvantages. Some disadvantages of the reactive routing protocol are as follows:

- When a path to a destination has to be determined, generally flooding is used so as to speed up the route discovery process. This results in congestion and collision in the network and also enhances the resource utilization in a burst mode, leading to sudden decrease in the lifetime of the network.
- It takes time to determine a route from source to destination.
- The header size of the packet increases with increase in intermediate nodes between source and destination, as the header carries the route information.
- The nodes cache route information, which they generally do not update automatically. If a node replies to a route request from its cache, it may lead to incorrect information owing to the dynamic network topology and high mobility of the nodes.
- During the process of route discovery, a node may receive a request from more than one of its neighboring nodes at the same time, as all the nodes might be participating in the flooding of requests for route discovery. This increases collision and delay in route discovery.

Example.

- Ad Hoc On-Demand Multipath Distance Vector (AOMDV),
- Ad Hoc On-Demand Distance Vector (AODV),
- Admission-Control-Enabled On-Demand Routing (ACOR),
- Ant-Based Routing Algorithm for Mobile Ad Hoc Networks (ARA),
- Backup Source Routing (BSR),
- Caching and Multipath Routing (CHAMP),
- Cluster-Based Routing Protocol (CBRP),
- Dynamic Manet On-Demand Routing (DYMO),
- Dynamic Source Routing (DSR),
- Flow State in Dynamic Source Routing,
- Multivariate Ad Hoc On-Demand Distance Vector Routing Protocol,
- Power-Aware DSR-Based,
- Reliable Ad Hoc On-Demand Distance Vector Routing Protocol (RAODV),
- SENCAST,
- Temporally Ordered Routing Algorithm (TORA).

10.4   Hybrid Routing Protocols

The hybrid routing protocols are designed to maximize the advantages of the proactive routing protocols as well as the reactive routing protocols. The hybrid routing protocol uses both proactive and reactive routing either in different stages of routing or in different areas of the ad hoc network to trade off effectively between the advantages and disadvantages of proactive and reactive routing. This type of routing is commonly used in hierarchical routing, location-based routing, and cluster-based routing where proactive routing is preferred in one region, primarily in the lower hierarchy or within a cluster, and reactive routing is preferred in another region, mainly in the upper hierarchies or between clusters. Generally, a proactive approach is taken near the source and a reactive approach is taken near the destination so as to discover as well as to maintain the route near the destination even in the case of high mobility. A proactive approach is also taken to maintain the route to nearby nodes, and a reactive approach is adopted to discover routes to distant nodes. A few hybrid routing protocols are as follows:

- Ad Hoc Routing Protocol for Aeronautical Mobile Ad Hoc Networks (ARPAM),
- Core Extraction Distributed Ad Hoc Routing (CEDAR),
- Distributed Dynamic Routing (DDR),
- Distributed Spanning Tree (DST)-Based Routing Protocol,
- Hazy Sighted Link State Routing Protocol (HSLS),
- Hybrid Ad Hoc Routing Protocol (HARP),
- Hybrid Routing Protocol for Large-Scale Mobile Ad Hoc Networks with Mobile Backbones (HRPLS),
- Hybrid Wireless Mesh Protocol (HWMP),
- Order One Routing Protocol (OORP),
- Scalable Location Update Routing Protocol (SLURP),
- Scalable Source Routing (SSR),
- Zone-Based Hierarchal Link State Routing Protocol (ZHLS),
- Zone Routing Protocol (ZRP).

10.5   Hierarchical Routing Protocols

Hierarchical routing protocols enable high scalability of the ad hoc networks. As the number of nodes increases in a mobile ad hoc network, it increases the memory requirement for storing the routing table. It also increases the resource requirement for establishment as well as reconfiguration of routes. After a certain breakeven point, the performance of the network routing may deteriorate with further increase in the number of nodes in the network owing to the extensive memory and processing requirement. Hierarchical routing best fits into such scenarios by dividing the network into regions and/or clusters. The nodes within a region are aware of other nodes only in their region and can route the packets only to them. For routing packets to any other node outside the region, the node forwards the data packet to a cluster coordinator. The cluster coordinators are capable of communicating with each other and forwarding the packet to that cluster coordinator for delivery that has the destination node in its region. Each cluster coordinator requires awareness of the topology only of its region. As the complexity of the network increases, the levels of hierarchy may not be restricted to two and may be further raised to retain flexibility and ease in routing.

Clustering is the distribution of nodes into groups called 'clusters'. The clustering can be physical or logical. In the case of physical clustering, the cluster is formed from nodes that are in physical proximity, preferably within a single-hop communication link with each other. Logical clustering may be based on a certain predefined set of relationships among the nodes. Each cluster generally selects one of the nodes within it as the cluster head or the cluster coordinator. All the nodes in a cluster maintain information about all its neighbors, the topology of the cluster, and the link status within the cluster. The communication between two nodes in a separate cluster is only through their cluster coordinator node. In a generalized manner it can be said that when a packet is routed from the source node to the destination node, as shown in Figure 10.9, it first rises through single or multiple hops to the node highest in the hierarchy towards the source, then moves to the node highest in the hierarchy of the destination node, and thereafter moves down to the destination node through single or multiple hops in intermediate hierarchical levels.

The exact routing strategy used in a hierarchical routing differs with the level of hierarchy in which the node exists. For example, within a cluster the nodes may depend on proactive routing or flooding, while reactive routing or position-based routing may be in use between the cluster heads. The hierarchical level at which a node exists can also be dynamic, as it may be given the role of a cluster head or removed from it for a number of reasons such as mobility, energy constraints, or regular polling for cluster head.

The disadvantage of a hierarchical routing strategy is that it becomes very complicated to establish a channel between two nodes after passing through a number of intermediate nodes at various hierarchical levels. Although the protocol is extremely efficient for huge networks, its performance decreases and becomes similar to any other proactive or reactive protocol for a small-sized network. As every cluster has a cluster coordinator, the process of selection of the cluster coordinator adds to the complexities in the network, resource utilization, point of failure, and energy depletion. Examples of a few hierarchical routing protocols are as follows:

- Cluster-Based Routing Protocol (CBRP),
- Clusterhead Gateway Switch Routing (CGSR),

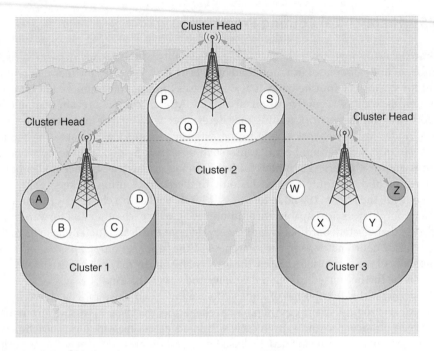

Figure 10.9 Hierarchical routing.

- Distributed Spanning Tree (DST)-Based Routing Protocol,
- Fisheye State Routing (FSR),
- Hierarchical Optimized Link State Routing (HOLSR),
- Hierarchical Star Routing (HSR),
- Hierarchical State Routing (HSR),
- Hybrid Ad Hoc Routing Protocol (HARP),
- Landmark Ad Hoc Routing (LANMAR),
- Multimedia Support in Mobile Wireless Networks (MMWN),
- Scalable Location Update Routing Protocol (SLURP),
- Zone-Based Hierarchical Link State (ZHLS) Routing.

10.6 Geographic Routing Protocols

The routing protocols can be broadly divided into two categories –topology-based routing, which includes the proactive, reactive, and hybrid routing protocols, and position-based routing [15, 16]. The position-based routing is also referred to as geographic routing or georouting. In the case of geographic routing, every node is aware of its position in space as well as the position of its neighboring node. The routing protocol determines the route on the basis of the known position of the source and the destination node, and the packet is forwarded on the basis of the geographic position of the nodes instead of the network address. Like the topology-based protocols, the georouting protocol may also operate in unique path mode, redundant path mode, or through flooding.

The geographic routing handles huge as well as dense networks exceptionally well. The topology-based class of protocols consumes a huge amount of resource for route determination, periodic updates, and reconfiguration in terms of exchange of table information as compared with the georouting for such networks. The memory requirement is also less for the georouting protocols, as huge routing tables do not have to be retained in the memory.

As the packets are forwarded on the basis of the position of the destination and not its network address, the source node has to know the destination address (position) for inclusion in the header. The position of each device in terms of latitude, longitude, or some other reference parameter is required [17-19]. This position parameter is generally provided by devices such as a global positioning system (GPS), with which each position-aware node should be fitted. Another approach is based on the relative coordinates in the network, wherein the distance between two nodes is calculated on the basis of the received signal strength or the time delay between transmission and reception.

The georouting ad hoc network has to run a localization service on a requirement basis for exchange of position information among nodes. The localization service in any network can either be centralized or distributed. The centralized localization service, though popular in cellular networks, is generally not preferred in an ad hoc network as it has a single point of congestion, failure, and resource utilization. As the localization server will be resource hungry, it might be required to have connectivity to a wired backbone, which the ad hoc network has to access through a gateway, leading to complexities. The alternative to a centralized localization service is the distributed localization scheme. The distributed localization scheme may be partial, wherein a few nodes in the ad hoc network run the localization services catering for the requirements of all their neighbors or the zone. It can also be a fully distributed service in which all the nodes run the localization service. The position of the nodes is registered in the nodes running the localization service, and whenever any node wants to locate the position of the destination node, it seeks the location from the localization service.

The localization service [20-22] can be proactive or reactive, i.e. the nodes may either voluntarily forward their location information at regular intervals or the location information is shared only on demand. Furthermore, there are a number of techniques for distribution of the location information within the network, such as flooding, quorum-based location service, grid location service, home-zone-based location service, or movement-based location service. In the case of flooding, a node broadcasts its position information in the network proactively or reactively to enable other nodes to update the position information of this particular node. In certain georouting protocols based on flooding, for e.g LAR, the flooding may be restricted to a specific geographic region. Flooding is effective and reliable. However, it increases the resource utilization and communication overhead. In the quorum-based approach, the position information is not available with all the nodes of the network, but with certain specific nodes. In the case of a home-zone-based approach, the network is divided into zones, and each node has a location service provider in its home zone that provides the location information of the node. In the case of a movement-based location service, the location information is not disseminated across the entire network over multiple hops. A node shares its location information locally with its neighbors. Now, while a node is mobile, it continues to share its time-stamped location information with all the other nodes that come in its communication range during its movement. Compass routing may further help to increase the efficiency of georouting by providing restricted directional flow.

Although a number of protocols have been developed for geographic routing and most of them are being improved with further research, the georouting protocols can mainly be classified into the following strategies:

a. Greedy routing

In the greedy strategy of geographic information routing, the source node and the intermediate nodes apply the greedy principle in forwarding the packet to a next-hop node that is closer to the destination. This is depicted in Figure 10.10. The greedy principle may be based on advancement, distance, or direction. The traditional greedy geographic information routing does not guarantee delivery of the packet to the destination, even though a path may exist, as the packet may land up, as shown in Figure 10.11, at an intermediate node beyond which there may be no node that satisfies the greedy principle towards the destination, i.e. no one-hop node exists after this dead-end node that reduces the distance to the destination, and this dead-end node does not have a communication link with the destination. However, a number of recovery strategies to avoid or cope up with such dead-end situations have been proposed by researchers.

b. Restricted directional routing

In the case of restricted directional routing, the packets are not forwarded omnidirectionally. The packets are forwarded only to those intermediate nodes that are in the zone in the direction of the destination. The limited zone towards which the packets are forwarded is created on the basis of the specific restricted-direction-strategy-based routing protocol in use. For example, in DREAM a circle, known as the expected zone, is created around the last known position of the target, and a cone is created using this circle as the base and the source as the vertex of the cone. The packet is forwarded only by the next-hop nodes within this area, as depicted in Figure 10.12. In the case of

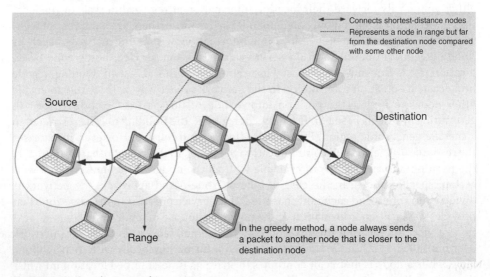

Figure 10.10 Greedy geographic routing.

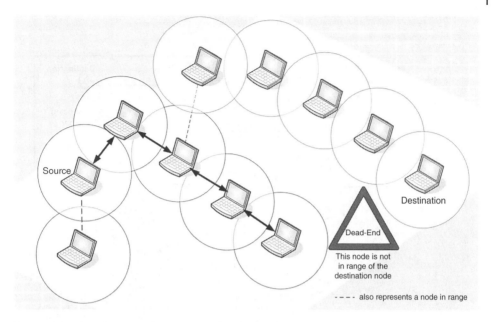

Figure 10.11 Greedy geographic routing with a dead end.

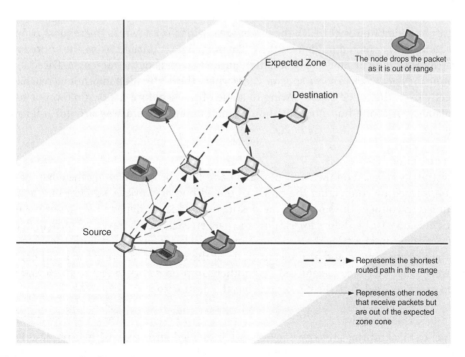

Figure 10.12 Packet forwarding in DREAM.

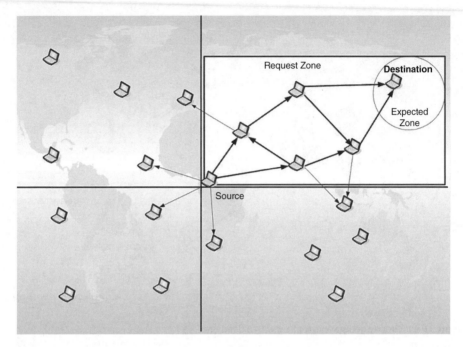

Figure 10.13 Location-Aided Routing.

Location-Aided Routing (LAR), the target area, which is known as the request zone, is a rectangle, as depicted in Figure 10.13. The rectangle is created using the source and the expected zone around the destination at the two ends of the rectangle. Directional routing may not work in some specific cases where the destination may not be reachable directly from the request zone owing to some obstacle before the destination or non-availability of a one-hop node to the destination, and thus a way around might be required.

c. Face routing

The strategy is based on the concept that a packet can be routed along the edges of the graph formed by connecting the nodes of the ad hoc network. A straight line 's–d' is assumed to connect the source to the destination, which is depicted for a sample network in Figure 10.14. Before forwarding the packet, each face is traversed once to determine the edge intersecting the s–d line. Now the packet is forwarded along the edges on the face of the source and at the start node of the edge intersecting the s–d line, the packet changes the face as depicted for another sample network in Figure 10.15, and the process is repeated until the packet reaches the destination.

d. Hierarchical routing

In hierarchical routing strategy, the network is divided into at least a two-tier architecture. This leads to reduction in the complexity of the network to be handled by each node and thus reduces the resource utilization. A hierarchical routing is highly scalable and is capable of efficient routing even in the case of a drastic increase in the number of nodes in the network. The hierarchical routing also remains less affected by the high

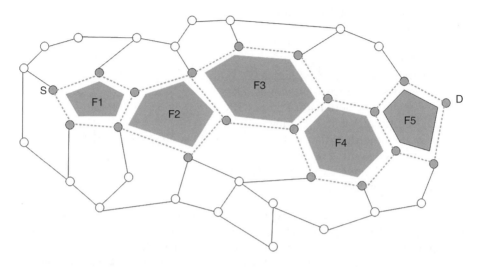

Figure 10.14 A graph formed by connecting the nodes of the ad hoc network for face routing.

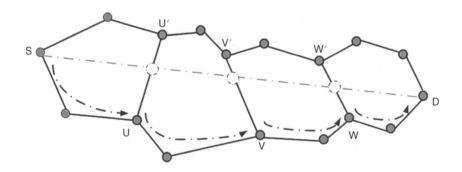

Figure 10.15 Depiction of a face change taking place at nodes *u, v, w*.

degree of mobility of the node, as node mobility is handled in one of the tiers, keeping the nodes of other tiers free from processing mobility information. The collaboration and redundancy requirement is also reduced in a hierarchical routing. In one of the traditional position-based hierarchical routings, the network is divided into a two-tier architecture [23]. The source forwards the data packet to the destination using greedy geographical routing. Once the packet reaches the vicinity of the destination node, a local, non-position-based routing protocol is used for guaranteed delivery. If the source and destination nodes are nearby, the positioning routing may not be used at all, and the packet is routed directly using a proactive routing protocol.

A few geographic routing protocols [24] are listed below:

- Anchor-Based Street and Traffic-Aware Routing (A-STAR),
- Boundary State Routing (BSR),
- Distance Routing Effect Algorithm for Mobility (DREAM),

- Energy-Aware Geographic Routing (EGR),
- Geographic Distance Routing (GEDIR),
- Geographic Routing Protocol (GRP),
- Geographic Source Routing (GSR),
- Greedy Other Adaptive Face Routing (GOAFR),
- Greedy Path Vector Face Routing (GPVFR),
- Greedy Perimeter Coordinator Router (GPCR),
- Greedy Perimeter Stateless Routing (GPSR),
- Location Routing Algorithm with Cluster-Based Flooding (LORA_CBF),
- Location-Aided Routing (LAR),
- Oblivious Path Vector Face Routing (OPVFR),
- Priority-Based Stateless Georouting (PSGR).

Some drawbacks have also been reported in geographical routing [24], such as failed or unreliable packet delivery in the case of an inaccurate positioning service. Moreover, geographical routing leads to a high energy demand in certain nodes, which falls in the assumed center of gravity of the ad hoc network as most of the packets will be crossing through it. The position-based protocols generally believe that, if two nodes are geographically close to each other, they will be in each other's transmission range and able to transmit to/receive from each other, which may not always be true. The greedy approach of geographical routing does not always guarantee delivery of the packets, especially in the case of high node mobility.

10.7 Power-Aware Routing Protocols

A wireless mobile ad hoc network remains operational when a minimal number of nodes in the network are alive. Failure of nodes beyond this threshold value may make communication between the nodes unreliable and difficult without any guarantee of packet transmission. As the battery power in the node is fixed and the battery is generally not rechargeable, power is one of the major constraints in a mobile ad hoc network. This necessitates awareness of the power consumed for each transaction and its associated activities to enable efficient power rationing. To avoid network breakdown due to power failure in specific nodes, it is preferred to distribute the power consumption uniformly across all the nodes by distributing processing and packet transfers accordingly to avoid too much communication with a single node or a group of nodes. As shown in Figure 10.16, node N9 is in the shortest path of communication between any pair of nodes, with one each from {N1, N2, N3, N4} and {N5, N6, N7, N8}. Hence, shortest-path routing will deplete N9 of battery power, leading to network failure after breakdown of N9. To avoid such points of failure, the battery power in each node should be known and the remaining battery power should be taken into consideration as one of the essential parameters while taking routing decisions. Such a routing strategy is known as power-aware routing [25].

However, mobility of nodes makes it difficult to implement this strategy as a depleted power node may move from its area having neighbors with sufficient battery power to an area with low-battery-power neighbors making the zone susceptible to network disintegration. The research in the area of battery technology is not as fast as in the

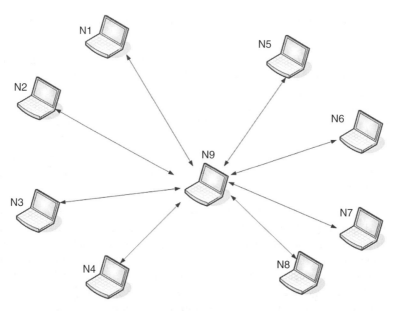

Figure 10.16 A sample ad hoc network depicting power depletion of an intermediate node.

areas of computing, communication, networks, and devices. There has been a tremendous increase in the capability of devices, increase in processing capability, miniaturization of size, and increase in density. All this increases the demand on power, but research on batteries cannot keep pace with the research in the field of devices. It is also difficult to provide enhanced power with lesser space requirement. The battery size and weight has to be optimized to ensure mobility of the nodes. This leads to a very high demand for energy-aware and energy-efficient routing protocols in the field of ad hoc networks.

The two major operations in an ad hoc network that consume power are communication and computation [26]. The communication power is consumed by the receiver–transmitter module installed in the ad hoc device. The computation power is used to process data received from the reception module as well as to prepare data for transmission through the transmission module. Exchange of data over transmission links requires data preparation such as compression/decompression, encoding/decoding, or encrypting/decrypting, which also consumes computational power. An ad hoc device continues to draw battery power whether it is in active mode or in sleep mode. While in active mode, the communication module consumes relatively more power when transmitting than when receiving. An ad hoc node in idle mode also consumes power, as it keeps listening to the transmissions from other nodes. All these processes have associated processing requirements which, too, consume power. However, when the node is in sleep mode, it may switch off its communication module but still continue to drain the battery power by a small amount, as some essential components of the node remain operational and hence continue to consume power. In any routing protocol, generally there is a tradeoff between the power consumption in computation and in communication, increasing one and reducing the other, even though it may not be in the same ratio, depending on the efficiency of the protocol.

The ad hoc network can be divided into two categories based on the strength of transmission:

- *Fixed-power-network transmission.* The transmission strength of each node is fixed. The power consumed in such a network is directly proportional to the size of the data packet received or transmitted and the cost of acquiring the transmission channel. It is independent of the distance between the transmitting and receiving node.
- *Variable-power-network transmission.* The node can vary the transmission strength according to the distance between the source and the destination. The power consumed is directly proportional to the distance between the two nodes exchanging the data packet and the environmental parameters. However, this strategy requires awareness of the relative position of the nodes in the network.

A few power-aware metrics [27] that are extensively used in devising energy-efficient routes are the energy consumed per packet, the time remaining for network partition and the difference in power level of the nodes. In a low-density network, the difference in power savings by calculated routes using an energy-aware routing and some non-energy-aware routing strategy, such as shortest path or minimum hop, may not be significant. However, in a dense network, the power saving in the energy-aware route may be significantly different from the minimum-hop route. The power-aware routing can also lead to awareness about the remaining time that the network communication will be sustained before the network is partitioned owing to failure of a group of nodes. The energy-aware routing also attempts to achieve uniform depletion of energy across the nodes.

A few power-aware routing protocols are listed below:

- Conditional Max-Min Battery Capacity Routing (CMMBCR),
- Geographical and Energy-Aware Routing (GEAR),
- Localized Energy-Aware Routing (LEAR),
- Low-Energy Adaptive Clustering Hierarchy (LEACH),
- Minimum Energy Routing (MER),
- Power-Aware Localized Routing (PLR),
- Power Source Routing (PSR),
- Power-Aware Multiple-Access Protocol with Signaling (PAMAS),
- Sensor Protocols for Information via Negotiation (SPIN).

References

1 Moore's law and Intel innovation, Intel. www.intel.com/content/www/us/en/history/museum-gordon-moore-law.html.
2 J. L. Hennessy and D. A. Patterson. *Computer Architecture: A Quantitative Approach.* Morgan Kaufmann Publishers, 3rd edition, 2003.
3 T. Larson and N. Hedman. Routing protocols in wireless ad hoc networks – a simulation study, Lulea University of Technology, Stockholm. www6.ietf.org/proceedings/44/slides/manet-thesis-99mar.pdf, 1998.
4 J. A. Freebersyser and B. Leiner. *A DoD Perspective on Mobile Ad Hoc Networks, Ad Hoc Networking.* Addison-Wesley, Boston, MA, USA, 2001.
5 S. Misra, I. Woungang, and S. C. Misra (eds). *Guide to Wireless Ad Hoc Networks.* Springer, 2009.

6 S. K. Sarkar, T. G. Basavaraju, and C. Puttamadappa. Routing protocols for ad hoc wireless networks. *Ad-hoc Mobile Wireless Networks: Principles, Protocols and Applications*, pp. 59–67. CRC Press, 2008.

7 Z. Ismail and R. Hassan. Effects of packet size on AODV routing protocol implementation in homogeneous and heterogeneous MANET. *Third International Conference on Computational Intelligence, Modelling and Simulation (CIMSiM), 2011*, pp. 351–356, IEEE Computer Society, Washington, DC, USA, 2011.

8 H. Miranda and L. Rodrigues. Friends and foes: preventing selfishness in open mobile ad hoc networks. *23rd International Conference on Distributed Computing Systems Workshops*, pp. 440–445, 19–22 May 2003.

9 T. Santhamurthy. A comparative study of multi-hop wireless ad hoc network routing protocols in MANET. *International Journal of Computer Science Issues (IJCSI)*, **8**(3):176–184, 2011.

10 M. S. Corson and J. Macker. Mobile ad hoc networking (MANET) routing protocol performance issues and evaluation consideration. *IETF RFC 2501*. www.ietf.org/rfc/rfc2501.txt, 1999.

11 G. Aggelou. *Mobile Ad Hoc Networks*. McGraw-Hill, 2004.

12 A. G. Patil, D. P. Durgade, and D. D. Ghorpade. Routing in mobile ad hoc networks. *National Conference on Advances in Recent Trends in Communication and Networks ARTCON 2010*. Allied Publishers, 2010.

13 C. C. Chiang and M. Gerla. Routing and multicast in multihop, mobile wireless networks. *IEEE 6th International Conference on Universal Personal Communications Record, 1997, Volume 2*, pp. 546–551, 12–16 October 1997.

14 C. Mala, S. S. Kumar, and N. P. Gopalan. Pervasive service discovery protocol for power optimization in an ad hoc network. *Selected Topics in Communication Networks and Distributed Systems*, pp. 261–275. World Scientific, 2010.

15 A. Vineela and K. V. Rao. Secure geographic routing protocol in MANETs. *International Conference on Emerging Trends in Electrical and Computer Technology (ICETECT)*, pp. 1063–1069, 23–24 March 2011.

16 J. C. Navas and T. Imielinski. GeoCast – geographic addressing and routing. *3rd Annual ACM/IEEE International Conference on Mobile Computing and Networking (MobiCom '97)*, pp. 66–76, New York, USA, 1997.

17 I. Stojmenovic. Position based routing in ad-hoc networks. *Communications Magazine, IEEE*, **40**(7):128–134, 2002.

18 M. Mauve, J. Widmer, and H. Hartenstein. A survey on position based routing in ad-hoc networks. *IEEE, Network*, **15**(6):30–39, 2001.

19 J. Widmer, M. Mauve, H. Hartenstein, and H. Fubler. Position-based routing in ad-hoc wireless networks. *The Handbook of Ad Hoc Wireless Networks*, pp. 12-1–12-14. CRC Press, Boca Raton, FL, USA, December 2002.

20 S. Ruhrup. Theory and practice of geographic routing. *Ad Hoc and Sensor Wireless Networks: Architectures, Algorithms and Protocols*, pp. 69–88. Bentham Science, 2009.

21 D. Liu, X. Jia, and I. Stojmenovic. Quorum and connected dominating sets based location service in wireless ad hoc, sensor and actuator networks. *Computer Communications*, **30**(18):3627–3643, 2007.

22 E. Kranakis, H. Singh, and J. Urrutia. Compass routing on geometric networks. *11th Canadian Conference on Computational Geometry*, pp. 51–54, 1999.

23 M. Mauve, H. Fubler, J. Widmer, and T. Lang. Position-based multicast routing for mobile ad-hoc networks. *Mobile Computing and Communications Review*, 7(3):53–55, 2003.

24 G. Wang and G. Wang. An energy aware geographic routing protocol for mobile ad hoc networks. *International Journal of Software Informatics*, **4**(2):183–196, 2010.

25 J. Chokhawala and A. M. K. Cheng. Optimizing power aware routing in mobile ad-hoc networks. *International Journal of Computer Applications*, **1**(1):1–4, 2011.

26 V. Rishiwal, M. Yadav, S. Verma, and S. K. Bajapai. Power aware routing in ad hoc wireless networks. *Journal of Computer Science and Technology*, **9**(2):101–109, 2009.

27 S. Singh, M. Woo, and C. S. Raghavendra. Power-aware routing in mobile ad hoc networks. *4th Annual ACM/IEEE International Conference on Mobile Computing and Networking, MobiCom '98*, pp. 181–190, New York, USA, 1998.

Abbreviations/Terminologies

ACOR	Admission-Control-Enabled On-Demand Routing
AODV	Ad Hoc On-Demand Distance Vector
AOMDV	Ad Hoc On-Demand Multiple Distance Vector
ARA	Ant-Based Routing Algorithm
ARPAM	Ad Hoc Routing Protocol for Aeronautical Mobile Ad Hoc Network
AWDS	Ad Hoc Wireless Distribution Service
A-STAR	Anchor-Based Street and Traffic-Aware Routing
BATMAN	Better Approach to Mobile Networking
BSR	Backup Source Routing/Boundary State Routing
CBRP	Cluster-Based Routing Protocol
CEDAR	Core Extension Distributed Ad Hoc Routing
CGSR	Clusterhead Gateway Switch Routing
CHAMP	Caching and Multipath Routing
CMMBCR	Conditional Max-Min Battery Capacity Routing
CSMA/CA	Carrier Sense Multiple Access/Collision Avoidance
CTS	Clear to Send
DARPA	Defense Advanced Research Projects Agency
DBF	Distributed Bellman–Ford Routing
DDR	Distributed Dynamic Routing
DFR	Direction Forward Routing
DN	Destination Node
DoD	Department of Defense
DREAM	Distance Routing Effect Algorithm for Mobility
DSDV	Destination Sequence Distance Vector
DSR	Dynamic Source Routing
DST	Distributed Spanning Trees
DYMO	Dynamic Manet On-Demand Routing
EGR	Energy-Aware Geographic Routing
FSR	Fisheye State Routing
GEAR	Geographical and Energy-Aware Routing
GEDIR	Geographic Distance Routing
GloMo	Global Mobile
GOAFR	Greedy Other Adaptive Face Routing

GPCR	Greedy Perimeter Coordinator Router
GPS	Global Positioning System
GPSR	Greedy Perimeter Stateless Routing
GPVFR	Greedy Path Vector Face Routing
GRP	Geographic Routing Protocol
GSR	Geographic Source Routing
HARP	Hybrid Ad Hoc Routing Protocol
HIPERLAN	High-Performance Radio LAN
HOLSR	Hierarchical Optimized Link State Routing
HRPLS	Hybrid Routing Protocol for Large Scale
HSLS	Hazy Sighted Link State
HSR	Hierarchical State Routing/Hierarchical Star Routing
HWMP	Hybrid Wireless Mesh Protocol
IETF	Internet Engineering Task Force
IN	Intermediate Node
LANMAR	Landmark Ad Hoc Routing
LAR	Location-Aided Routing
LEACH	Low-Energy Adaptive Clustering Hierarchy
LEAR	Localized Energy-Aware Routing
LORA_CBF	Location Routing Algorithm with Cluster-Based Flooding
MANET	Mobile Ad Hoc Network
MER	Minimum Energy Routing
MMWN	Multimedia Support in Mobile Wireless Networks
NTDR	Near-Term Digital Radio
OLSR	Optimized Link State Routing
OORP	Order One Routing Protocol
OPVFR	Oblivious Path Vector Face Routing
PAMAS	Power-Aware Multiple Access Protocol with Signaling
PCMCI	Personal Computer Memory Card International Association
PLR	Power-Aware Localized Routing
PRNET	Packet Radio Network
PSGR	Priority-Based Stateless Georouting
PSR	Power Source Routing
QoS	Quality of Service
RAODV	Reliable Ad Hoc On-Demand Distance Vector (Routing)
RREP	Route Reply
RREQ	Route Request
RTS	Request to Send
SLURP	Scalable Location Update Routing Protocol
SN	Source Node
SPIN	Sensor Protocols for Information via Negotiation
SSR	Scalable Source Routing
SURAN	Survivable Adaptive Radio Network
TORA	Temporally Ordered Routing Algorithm
WRP	Wireless Routing Protocol
ZHLS	Zone-Based Hierarchal Link State Routing Protocol
ZRP	Zone Routing Protocol

Questions

1 Why is a wireless ad hoc network known as an infrastructureless network? Name two other infrastructureless networks.

2 An ad hoc network can be set up between laptops without the requirement of any router or switch. Try to establish a wireless ad hoc network among at least three laptops of your friends and transfer a file from one laptop to another.

3 State the characteristics of a wireless ad hoc network.

4 Explain the process of solving the hidden node problem.

5 What is the requirement of hybrid routing protocols when reactive and the proactive routing protocols exist?

6 The routing protocols used in wired networks cannot be used directly for a wireless mobile ad hoc network. Justify this statement, giving reasons.

7 Explain the difference between the following:
   A a closed mobile ad hoc network and an open mobile ad hoc network,
   B DREAM and LAR,
   C proactive routing and reactive routing,
   D problems in multicast routing and problems in flooding,
   E distance-vector-based protocols and link-state-based protocols.

8 Explain the following:
   A the hidden terminal problem,
   B an exposed node,
   C face routing,
   D the problem of a dead end in greedy geographical routing,
   E greedy geographic routing.

9 Write the names of at least three protocols from each of the following categories:
   A proactive routing protocols,
   B reactive routing protocols,
   C hybrid routing protocols,
   D hierarchical routing protocols,
   E geographic routing protocols,
   F power-aware routing protocols.

10 State whether the following statements are true or false and give reasons for the answer:
   A In some application scenarios, the node of an ad hoc network may have uninterrupted power supply or a replaceable battery.
   B Geological routing protocols cannot work in the absence of a GPS.
   C LAR and DREAM work by the greedy approach.

D Hierarchical routing can have a maximum of three hierarchical levels.

E The class of protocols with a mix of reactive routing and proactive routing are termed hierarchical routing protocols.

F DSR is an on-demand routing protocol.

G The proactive routing protocols are all link-state-based protocols.

H An ad hoc network can have selfish nodes, sleeping nodes, misbehaving nodes or captured nodes.

I An ad hoc network in a particular area or for a particular application should have homogeneous nodes.

J The commercial application of ad hoc networks started with the third generation of ad hoc networks.

Exercises

1 Draw a UML sequence diagram to indicate how the problem of a hidden terminal is handled using RTS and CTS signals.

2 A number of ad hoc network nodes have been positioned in various patterns, as indicated in the various diagrams below. It may be assumed that the range of each node is just sufficient to reach the neighbor node. If more than one neighbor node is at almost equal distance from the node, they are all within range. Indicate in each topology if it is prone to the hidden terminal problem or exposed terminal problem or both, and circle the nodes that will face the problem.

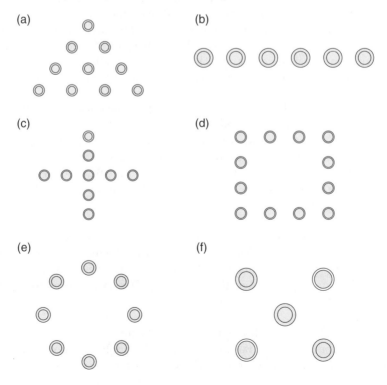

3   Assume the existence of an ad hoc network as depicted in Figure 10.4 that uses table-driven routing. Node 1 is in the communication range of all the other nodes. Additionally, node 2 is in the communication range of node 3 and node 6 and node 5 is in the communication range of node 6 and node 4. Communication range indicates the capabilities of receiving as well as transmitting in full duplex. Make an indicative table that will be stored at each node once the network has converged. Now, node 1 goes into sleep mode, starts moving towards the left for some time, and then stops and becomes active. On becoming active, it is detected that node 1 is now in the communication range only of node 5. What data will be shared among the nodes during the mobility of node 1, and what will now be the indicative routing table in each node after convergence again. Do the exercise separately for distance vector routing and link state routing.

4   It has been mentioned that one of the disadvantages of proactive routing protocol is that 'There may be certain routes over which no data packet has travelled and also by which there is a high probability that no packet will travel in the future. Still, the protocol utilizes resources to establish and regularly update such paths'. Design a network and the associated routing tables to prove that this statement is correct for certain networks.

5   Consider the network given in Figure 10.8a, which runs DSR. The nodes that are in communication range of each other are connected by lines. The links are bidirectional. Now the node SN wants to transmit some data to node IN3 (the destination node in this case). How many nodes in total will receive the RREQ packet during the process of path discovery? Apart from the source and destination nodes, how many nodes will come to know about the entire path from SN to IN3? If IN 10 wants to communicate with IN 11, calculate the number of nodes receiving RREQ and the number of nodes that will gain knowledge of the route between IN 10 and IN 11. Assume the presence of no routes of any intermediate nodes in the cache at the beginning of the process.

6   Consider the network given in Figure 10.8a where the lines indicate the communication range of the node. Node SN wants to transmit 2 MB of data to node IN 3. Use any ad hoc network simulation algorithm to compare the performance in terms of power consumption and bandwidth utilization to perform this activity for the following three categories of routing protocols (any specific protocol within the category may be used): distance-vector-based table-driven routing, link-state-based table-driven routing, and reactive routing.

7   It has been said that 'Directional routing may not work in some specific cases where the destination may not be reachable directly from the request zone owing to some obstacle before the destination or non-availability of a one-hop node to the destination, and thus a way around might be required'. Design a network to prove the correctness of this statement.

8   Consider an ad hoc network as indicated in the figure on the following page. The lines indicate the communication range of the nodes. Geographic/location-based routing has been implemented in the network. Node 2 wants to transmit data to node 4. Which will

be the intermediate nodes involved in forwarding the packet in the case of greedy routing, DREAM, LAR, and face routing? Necessary assumptions may be made regarding the expected zone for the destination node. Will the packet reach the destination with all four routing techniques?

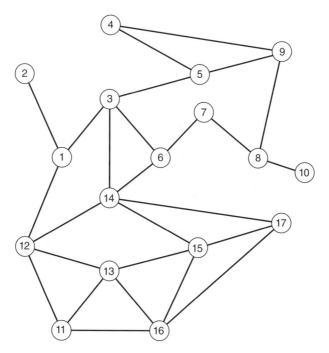

11

Routing in Wireless Sensor Networks

11.1 Basics of Wireless Sensor Networks

Wireless sensor networks (WSNs) are being adopted in various applications that require deployment for continuous operations in inhospitable terrains, constrained space, or challenging environments. The features that make the WSN the preferred choice for such applications are its low cost, ad hoc deployment, rapid installation, low power consumption, ease of use, distributed operation, and small sensor size [1]. The number of devices deployed in a typical WSN for any particular application may vary from a few hundred to thousands. The advancements in wireless technology have been a boon for advancement in the field of sensor networks. The mesh of wires required to connect a few thousand devices, if not operational in wireless but in wired mode of

Network Routing: Fundamentals, Applications, and Emerging Technologies, First Edition.
Sudip Misra and Sumit Goswami.
© 2017 John Wiley & Sons Ltd. Published 2017 by John Wiley & Sons Ltd.
Companion website: www.wiley.com/go/misra2204

communication, can well be imagined in terms of deployment time, space requirement, number of interconnections, essentiality of manual deployment, and finally the inability to conceal operation. Advancements in the area of application-specific integrated circuits (ASICs), microelectromechanical systems (MEMSs), and nanotechnology has led to the design of compact and energy-efficient sensor devices. The research in the area of battery technology is also at a pace to catch up with the advancement in computing technology and miniaturization of devices, and this has led to prolonged life of WSN devices. The algorithms being used for WSNs are getting optimized, power aware, distributed, and secured, enhancing the performance of WSNs from the available underlying device hardware.

The size and deployment pattern of nodes of the WSN lead to **typical characteristics** of resource constraint, limited processing capability, limited power availability, and limited view of its environment. This calls for distributed sensing, coordinated processing, and routing of information to achieve its goal over an unreliable network, as well as to prolong the lifetime of the WSN. Depending on the criticality of the network, one of the requirements – such as power saving, QoS, performance, bandwidth, reliability, or security – may get preference over others. With the use of proper routing protocols, the life of a WSN can be extended much beyond the life of any of its individual nodes by putting the nodes in sleep mode when not in use and activating a single node at a time in a particular coverage area with multiple node deployment. Effective bandwidth as well as data redundancy can also be increased by using multiple paths.

The characteristics of a WSN generally necessitate deployment of a number of sensor nodes to monitor an object, environment, or event. This is due to the fact that the range of event detection and message transmission of the nodes is limited and requires a collaborative effort to forward information. Being tiny and power constrained, all the nodes may not be operational at all times, which adds to the unreliability of a single node. The terrain of deployment may not permit line-of-sight communication between far-off nodes. The dense deployment of nodes not only leads to robustness and fault tolerance but also gives highly accurate information about the geographic point of action, also known as localization. The nodes may be uniformly deployed over an area or randomly spread.

The applications of WSNs are varied, and some of the common areas where these are being used include healthcare and patient monitoring, industrial applications and plant safety, structural monitoring and catastrophe warning, environmental and wildlife monitoring, traffic regulation and vehicle monitoring, manufacturing and inventory tracking, and smart and automated home appliances. The variety of defense applications is huge, and the deployments, applications, spread, and topology are unknown, as WSNs provides a high degree of concealment. It is also possible to deploy sensors inside concrete structures, under the skin of animals, or inside medical implants, which opens up another area for sensor network applications.

In defense parlance and a few other specific fields of application, the size of the sensor may typically not be of the order of a few millimeters or centimeters – surveillance aircraft, submarines, autonomous underwater or surface vehicles, UAVs, aerostats, or communication and detection systems such as radars are also referred to as sensors. The evolution of sensor networks can be traced to the era of the Cold War, when the

USA deployed a network of interconnected radars for aerial surveillance and acoustic sensors in the seabed to detect Soviet submarines. These sensors had wired deployment, relieving them from the bandwidth and power constraints faced by wireless sensors. DARPA started the distributed sensor network program in the 1980s to interconnect for collaborative activity low-cost independent sensing nodes spread over an area. Under this project, the Massachusetts Institute of Technology (MIT) realized a sensor network for tracking aircraft. It was based on acoustic sensors that captured and analyzed the sound of aircraft, processed it, and transmitted the data using microwave radio. Subsequent research in this area led to low-cost, energy-efficient, and miniaturized hardware. In the present day, a typical sensor node consumes about $10\,\mu W$ power and has a life of about 10 years.

A sensing and communicating device connected to a wireless sensor network is generally referred to as a 'node'. There might also be devices in the WSN that do not have sensing capability but participate only in the routing. The area where the sensor nodes are deployed is known as the 'sensor field'. The wireless sensor network consists of a number of sensor nodes that may be connected to a gateway or a sink node. The gateway or the sink node is generally assumed to have sufficient computational capability, memory, and power. This gateway sensor node receives data from all the nodes, stores it, and processes the data to derive a decision about an event from the sensed data. The sink or the gateway is also commonly referred to as an 'aggregation point'. The sink node is connected to the Internet or to any other reliable wired network. The connectivity between the sink node and the wired network can either be wired or through a high-bandwidth reliable wireless link. There may be more than one gateway or sink node in a wireless sensor network.

11.1.1 Hardware Architecture of Sensor Node

The major components of a sensor node are the sensing system, processing unit, communication system, power supply, and optionally an actuator. The architecture of a typical wireless sensor node is depicted in Figure 11.1.

The **sensing system** gives the input to the sensor nodes in terms of parameters sensed from the external environment. The sensors may vary in size, the sensing technology used, and the parameters being sensed. The sensor can be for the detection of sound, temperature, weight, vibration, humidity, motion, chemical contents, biological agents, explosive vapors, electromagnetic signals, or any other parameter for which the WSN has been deployed. The sensors may be continuously ON, detecting the environment, or may be activated at regular time intervals.

The **processing unit** comprises the microprocessor and memory. It has interfaces to connect to the sensing system, communication system, power source, and actuators. An analog-to-digital converter may be used between the processing unit and the sensing system if the output of the sensing system is in analog form. The processor unit runs an operating system and the routing algorithm. A real-time operating system configured for a given application is generally deployed in the sensor nodes. A few examples of operating systems used in sensor networks are TinyOS, MANTIS OS, and Smart Card OS. The processing system controls the communication system as well as the sensors and actuators for power saving and maintaining electromagnetic silence. Depending on

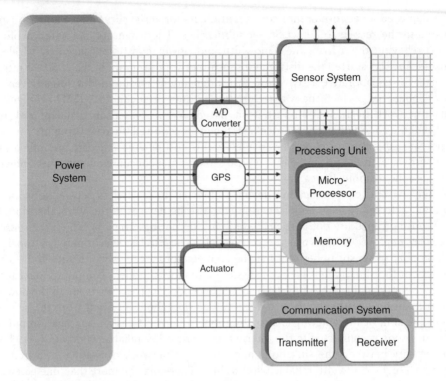

Figure 11.1 Block diagram of a typical wireless sensor node.

the application, the processing unit may have a GPS attached to it. The entire processing unit is designed as a system on chip (SoC).

The **communication system** comprises a short-range radio receiver and transmitter. There may be nodes without receivers but only transmitters to support one-way communication. The transmitter has various modes of operation, namely active, idle, or sleep. The wireless radio circuits may be proprietary or standard based. The standard-based radios help in interoperability. Although 802.11.x has a high data transfer rate and transmission range, it is not suitable for use in WSNs owing to the high power requirements. Bluetooth, IEEE 802.15.4, Zigbee, and IEEE 1451.5 are the preferred interfaces for communication in WSNs.

The **power source** should essentially feed energy to the processing unit and communication system. The power requirement of the sensing system varies with the design and technology used in the sensor. While electronic sensors essentially require power, the mechanical sensors may work without power. The power requirement of the radio can be controlled not only by switching off the radio at frequent intervals but also by a number of other power-saving strategies. Some of the strategies that can reduce the power consumption of radios are data compression to reduce the amount of transmitted data, activation of radio only on event detection, reduction in packet overhead, avoidance of collisions and retransmission, optimization of the time interval for exchange of control signals, and lowering of the frequency of data transmission.

11.1.2 Network Topology

Although there are a number of available topologies for wireless networks, radio networks, or ad hoc networks, the topologies for a WSN [2] are very limited and are as follows:

Star network. This can be viewed as a point-to-multipoint communication, where a number of nodes can send or receive data from a single node, which may be termed the coordinator, base station, leader node, or central node. The nodes communicating with the coordinator cannot communicate directly among themselves. The advantages of such a topology are the simplicity of the architecture, the avoidance of duplicate data packets, the reduction in processing at the nodes, the reduction in flow of data, lower latency, and power saving. The disadvantages of the topology are that there is a single point of failure, the power of the coordinator is depleted faster than other nodes, and all other nodes must be in the reception range of the coordinator node. This type of topology is used in hierarchical routing.

Mesh network. In a mesh topology, any node of the WSN can communicate with any other node in its range. Flat routing uses this topology, as it supports multihop communication. Participation from all intermediate nodes helps in passing the data packet from one node to another and enables the packet to reach the destination, which may be out of communication range of the source. The topology has no single point of failure and can have multiple data paths, leading to redundancy. Any number of new nodes can be added to the network as the topology is highly scalable. The spread of the network can also be huge, limited by the distance between the two extreme nodes under the condition that there are intermediate nodes between these extreme nodes to participate in multihop routing. The disadvantage of the network is that there are duplicate data packets flowing in the network, leading to congestion. As there are a number of hops between the source and destination, the latency for message delivery increases. A node has to check every data packet it receives so as to forward it to other nodes as well as to avoid forwarding duplicate packets. This requires all the nodes to keep the radio on and process the data packets. Both these activities consume battery. As all the nodes are active and operating at the same pace, the power of all the nodes is consumed almost at the same pace. All the nodes fail almost at the same time, which first partitions the network and thereafter makes the complete network fail.

Hybrid network. Topology that is a hybrid of the star network and the mesh network takes the advantages from both to provide an underlying design for a robust and energy-efficient routing. The power-constrained nodes communicate only with their coordinators, forming a star topology, while the coordinators can communicate with each other in a mesh pattern. Thus, the data packet from a low-power node reaches the coordinator in a single hop over star topology and thereafter reaches its destination using multihops across various coordinators in a mesh topology. Cluster-based routing generally relies on this topology.

Figures 11.2 to 11.4 present typical star, mesh, and hybrid network topologies in a WSN.

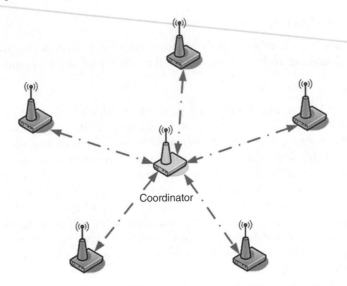

Figure 11.2 Star network topology in a WSN.

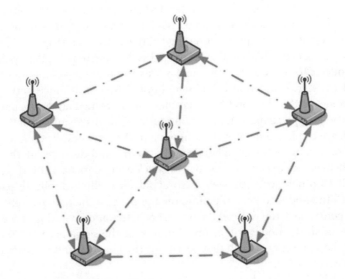

Figure 11.3 Mesh network topology in a WSN.

11.1.3 Design Factors

The designs, applications, and devices used in a WSN are varied. Still, the design factors that are considered equally important across all types of WSN are almost the same because all WSNs have similar constraints and similar underlying technologies. The major design factors in a WSN [3] are as follows:

Reliability. Some of the factors that lead to failed nodes or dropped packets are deployment in a harsh environment, heavy interference due to high node density as well as

Figure 11.4 Hybrid network topology in a WSN.

noise, frequently being out of transmission range, and energy depletion. The routing protocol should be reliable enough to handle such failures.

Scalability. The number of nodes deployed in a WSN may vary according to the application, coverage area, and redundancy. The routing protocol should be able to support a few hundred nodes to a few million nodes. New nodes may be added when the WSN is already operational, and the routing protocol should be able seamlessly to integrate the new nodes. The protocol should also be able to handle situations of nodes going to sleep and coming back to active state after sleep.

Hardware cost. As the WSN nodes are one-off deployed in their thousands without being collected back, the hardware cost per node should be very low.

Hardware design. A sensor node comprises a processing unit, a sensing unit, a power source, and a communication system. It might also optionally have actuators, a positioning system, and a mobilizer to support node movement. All these systems should be fitted in a small space, preferably on a single chip.

Network topology. The WSN protocol should be able to discover the topology at the time of deployment and support topological changes post-deployment. At the time of deployment, the nodes may be introduced in the network one at a time, allowing the WSN to keep discovering each newly introduced node and its nearest neighbors. Alternatively, all the nodes might be deployed en masse by spraying, aerial dropping, or exploding a container of sensors in proximity to the site. In such a scenario, all the nodes start discovering their neighbors simultaneously, leading to the formation of a WSN. Post-deployment, the nodes may be mobile all by themselves, or their motion can be dictated by the environment. This leads to topological changes, which the WSN should be capable to handle. Topological changes may also occur owing to deployment of additional nodes or nodes becoming inactive.

Operating environment. The nodes can be deployed in extreme and varied environments. The extremity of the environment can be in terms of height, temperature, pressure, humidity, wind speed, or electromagnetic interference. They can be deployed on land, on the seabed, floating in the sea, and even in space. The deployment environment can be a home, office, hospital, industry, forest, or battlefield. Thus, the WSN should be designed to cater for such controlled as well as harsh operating environments.

Mode of communication. The nodes in a WSN communicate with each other using wireless transmission. The wireless transmission can be on radio frequency, infrared, electromagnetic, or light waves at any frequency. Use of optical media is also being explored. The transmission media in use should preferably not be proprietary, to enable interoperability and global operations.

Energy requirements. The nodes have a non-replaceable, non-chargeable, and limited power battery. This should be used judiciously to power all the components of the node to participate in sensing, processing, and communicating. The life of the node depends on its own power requirements and consumption pattern. The nodes consume energy not only in sensing their own environment and in processing and forwarding their data but also in rerouting data for other nodes. It may be preferable for the energy of all the nodes in the WSN not to be depleted almost at the same time, but only a few nodes keep failing with time owing to energy depletion. This will lead to graceful degradation in the performance of the WSN, but at the same time enhance the life of the WSN by rerouting the packets through paths across available nodes.

11.1.4 Classification of Routing Protocol

The routing protocols may be classified on the basis of various parameters [3-5]. Based on the network topology in which the WSN routing takes place, the routing protocol can be classified as flat, hierarchical, location based, or direct. On the basis of the time when the path is determined between the source and destination, i.e. prior to the need to route the packet or after the need arises to route between the pair, the routing protocols are classified as proactive, reactive, or hybrid. For determination of the route between the source–destination pair, the communication can be initiated either from the source or the destination, leading to these two categories based on the initiator of communication. On the basis of protocol operation, the routing has been classified into negotiation-based routing, multipath routing, query-based routing, QoS-based routing, and coherent routing.

In proactive WSN routing protocols, the routes between all source–destination pairs are computed even before any particular route is required to be discovered for actual transmission of a data packet. Any change in the network topology is also incorporated to keep the routing table updated, even if no data packet might be routed through these changed routes. As a WSN generally has thousands of sensor nodes, discovering such a huge number of routes and storing them in the memory of each sensor node is not feasible, and hence proactive routing protocols are generally not suitable for WSNs. In reactive routing protocols, the routes are discovered on a requirement basis from the source to the destination whenever a packet reaches an intermediate node for transmission. Hybrid routing protocols are a combination of reactive and proactive routing whereby the routes to neighbors or other nearby nodes may be known proactively and the complete end-to-end path is discovered reactively.

The communication in a WSN can be initiated by the source or the destination. A source initiates communication when it wants to transmit data to the sink, gateway, or any other destination about an interesting event or some other data. The source may initiate the communication at regular time intervals, at predefined times, or only on detection of some event of interest or on measuring a parameter above the threshold level. The communications that are originated by the destination are generally query driven. When the sink node or any other destination wants some information from the source, it sends a request/query to a node for data. The source-initiated communication is generally a one-lap communication from source to destination, while the destination-initiated communication is a two-lap communication in the form of a query from destination to source and the data as a reply from the source to the destination.

The network topology of a WSN can be hierarchical (star) or flat (mesh). In a flat structure, all the nodes are equivalent and any node can communicate with any other node, while in the hierarchical structure there are a few special nodes known as coordinators or cluster heads, and the other normal nodes can communicate only with their coordinators and not with each other. These two network structures lead to two different classes of routing protocols – flat and hierarchical. In a hierarchical routing protocol, the nodes are separated into zones called clusters, and there is a cluster head in each zone. The nodes in a cluster can communicate only with their cluster head and not with the other nodes in the same or different clusters. In a flat routing protocol, generally multihop routing is in practice, where any node can communicate with any other node in the WSN.

Two other routing protocols that are also based on network topology are direct routing and location-based routing. In direct routing, the source should be able to send the data packet directly to the destination, i.e. the source and destination should be connected over a single hop. In a typical WSN implementation, this indicates that all the sensor nodes are a single hop distance away from the sink, which generally is a rare architecture and may be in use only in some specific applications. In location-based routing, each node is aware of its location. The location can be in terms of a common frame of reference for the entire WSN, the relative position to each other, or based on a GPS fitted with each node. The broad classification of the routing protocols in a WSN is indicated in Figure 11.5.

11.2 Routing Challenges in Wireless Sensor Networks

Although a WSN appears to be very similar to a wireless ad hoc network in its characteristics and working paradigm, it greatly differs from a wireless ad hoc network. Some of the major differences [3, 6] between a wireless sensor network and a wireless ad hoc network are depicted in Table 11.1.

The node of a WSN is not generally deployed manually or preplanned. This leads to a random distribution of nodes over an area. Once deployed, the nodes cannot be upgraded or configured from outside. All this further adds to the pre-existing

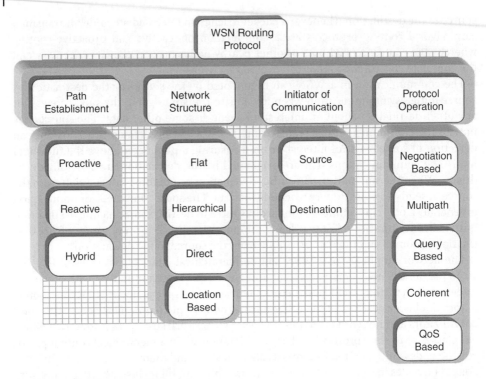

Figure 11.5 Classification of WSN routing protocols.

challenges of wireless communication. The major constraints challenging the development of a robust, optimized, and common routing protocol for a WSN are as follows:

Deployment pattern. The sensor nodes are randomly deployed with no predefined layout. The separation between the nodes has wide variations. The deployment of nodes
by aerial dropping is a common scenario in the field of WSN. This not only requires
the nodes to be environmentally hardened but also requires the routing algorithms to
support a wide variety of network topologies.

Autoconfiguration. The WSN nodes, once deployed, have to be self-configurable as
there is no external configuration mechanism available individually to configure each
node by intimating about the neighboring nodes, routing pattern, and path to be
selected. The routing algorithm should be robust and intelligent enough regularly to
detect all the available nodes and create a network out of these available nodes for
data communication.

Mobility. Some of the nodes may be mobile by themselves, or the location of the nodes
may keep changing owing to the mobility of the deployment environment or platform, such as sand dumes, wildlife, glaciers, cultivable land, or vehicles. The routing
algorithm should support change in network topologies to cater for such scenarios.

Energy constraint. Owing to deployment in inhospitable, unreachable terrains, the
WSN nodes once deployed may not be retractable for repairs or change of battery.
Intractability may also be due to their small size. Furthermore, the devices are so
small and cheap that it may not be economical to retrieve them. Hence, the battery of

Table 11.1 Difference between a wireless ad hoc network and a wireless sensor network.

Characteristic	Wireless ad hoc network	Wireless sensor network
Number of nodes in the network	It varies from a few nodes to a few hundred nodes	It varies from a few hundred nodes to thousands of nodes
Transmission standard	Wireless LAN (802.11.x), Bluetooth	Bluetooth, 802.15.4, ZigBee, Wireless HART, ISA SP100
Devices involved	Laptops, notebooks, palmtops, cellular phones, smart phones, etc.	Sensor nodes, motes, etc.
Density of deployed nodes	Low	High
Rate of failure of nodes	Low	High
Retractable (taking back) nodes	Possible	Not possible/not economical
Change/recharge battery	Possible	Not possible/not economical
Frequency of topology change	Low, of the order of minutes	High (owing to node failures, addition of nodes, nodes going into sleep mode or coming into active mode from sleep mode), of the order of milliseconds to seconds
Resource (bandwidth, memory, processing, power) limitation	Less constrained	Constrained
Transmission	Point-to-point or broadcast	Broadcast
Unique node identification	Yes. Generally through the IP address	In deployments with a large number, global unique identification may not be possible

a WSN node decides the life of the node. Radio transmission is an activity that consumes huge energy. This cannot be done away with in the WSN nodes, as radio transmission is generally the only way of communication in a typical WSN. Thus, the transmitter has to be used optimally so as to use it for the minimum possible time and range. The other parts of a sensor node that consume power are the onboard processing, the sensing activity, and the actuators, if any. Thus, the routing protocol has to be designed to use all the resources in the node to an optimum level of energy conservation.

11.2.1 Self-Healing Networks

Wireless sensor networks are infrastructureless and comprise a huge number of nodes, which may be homogeneous or heterogeneous. This leads to an increase in the probability of failure of nodes or links within the network as a result of malfunctioning, routing strategy, energy saving, misbehavior of a node, or an external attack. In order to ensure seamless operation of the wireless sensor network, it is imperative to have a scheme to detect and self-heal such failures. With the advancement of technology, new areas and technologies such as nanotechnology, microelectromechanical systems (MEMSs), smart dust, and microsensors have evolved, which are highly complex systems with reduced physical dimensions. The complexity of the system increases its points of failure, and the size of the systems prohibits the recovery or repair of the system, leading to further strengthening of the requirement for healing capabilities of the sensor network.

A self-healing network is able to detect by itself any failure in the network. The detection capability of the sensor network is further linked with the process of fault modeling and identification of the cause of the fault so as to decide on suitable intervention. For example, it should be able to detect whether a node has failed, a link has failed, a bidirectional link has become unidirectional (owing to failure of a particular receiver or transmitter), a node has gone into energy-saving mode, or an attack has been launched on the network. Another advantage of fault modeling is that it helps in damage mitigation studies as well as degradation modeling in the case of failures. After detection of the cause of the failure, a self-healing network is capable of launching a corrective action to bring the network or bring that particular portion of the network back into action. Generally, it is assumed that a network is self-healing if it can autoconfigure itself to retain network connectivity and recover the point or zone of failure. However, self-healing should actually ensure that, after the healing process, the network is equally reliable, with at least the same degree of connectivity and quality of service as in the network prior to the failure. This is to ensure that the recovered points are not the weakest links in the network. Certain systems ensure that the post-healing strength is greater than the design strength.

The self-healing of a network can be based on network realignment so that operational network components take over the load of the failed nodes and failed links. Alternatively, there can be redundant nodes, which can either move locally or throughout the network to replace the points of failure. There can also be reinforcement nodes, which act *a priori* on the basis of fault modeling. These nodes move to a network section detected with a high probability of failure and so reinforce that network section. Redundant nodes can be detected using sensor optimization, and these redundant nodes can be used as healing nodes. Sensor network health monitoring and damage algorithms can also be used to gain an insight into the extent of harm caused, its effects, and the remaining time for which the network can keep operating with graceful degradation.

Wireless sensor networks with memory about the shape of the deployed sensor network are a promising development in improving the damage resistance property of a sensor network in a highly turbulent deployment environment or in an area under surveillance of the sensor network and being attacked.

11.2.2 Security Threats

The nodes of the WSN transmit data omnidirectionally and in open medium, leading to an imminent security threat. Authentication, integrity, confidentiality, and availability are the major security concerns in WSNs as well. Denial of service (DoS) attack is the most common threat in a WSN, wherein an attacking node can place itself in the sensor field and create interference in the communication medium by continuously generating noise. If the attacking node is strategically placed in a dense part of the network, it can lead to disruption of multiple routes. The attack can be made more severe if this attacking node continuously sends request messages to the neighboring nodes. This would initiate a continuous conversation with the nodes and engage the communication channel as well as draining the power source of the neighboring nodes. Packet sniffing is an inherent characteristic of a wireless sensor network as the nodes generally accept all the packets, analyze them, process them, and retransmit the packets, leading to a major concern related to confidentiality.

Selfish nodes, inactive nodes, dumb nodes, and sleeping nodes add to the difficulty in distinguishing these nodes from a node planted by an adversary in the sensor field or any node of the WSN captured by an adversary. It is difficult to distinguish between the failure of a node as a result of malicious action and natural death due to hardware malfunction or power exhaustion. As the sensor nodes communicate in wireless mode, they are also prone to jamming, spoofing, tampering, and replaying. Some common attacks on a WSN are sybil attack, black hole attack, wormhole attack, and hello flood attack.

A wireless network in general, and a WSN in particular, is exposed to all the threats and attacks that a wired network faces. However, a WSN, being infrastructureless, cannot have specific security nodes performing the activities of a firewall, intrusion detection system, intrusion prevention system, information security gateway, and gateway encryptor device. Hence, the traditional intrusion detection and prevention techniques used in a wired network cannot be implemented in a WSN.

Secure routing protocols used in WSNs are Security Protocols for Sensor Network (SPINS), Authentication Confidentiality (AC), cluster-based secure routing protocol, etc.

11.3 Flat Routing Protocols

In a flat routing protocol [7], all the sensor nodes are considered to be equivalent in terms of transmission and reception capabilities. Any node can communicate with any other node in the WSN, thus giving the name 'flat' to this category of protocols for the unstructured network topology where there are no hierarchies within nodes. The level and the amount of information possessed by all the nodes are similar. All the nodes are equally responsible for sensing the sensor field, processing data, forwarding and routing data, as well as ensuring the security of the network. As all the nodes are considered to be at an equal level, it is not necessary to have a unique identifier for each sensor node. Moreover, by doing away with a unique ID for each sensor node, the scalability of the WSN is also enhanced. Flat routing saves the resources and time involved in leader election, which is done in hierarchical routing. Most of the communication in flat routing is based on broadcast mode or by using wave propagation algorithms. In a flat-routing-based WSN, selective sensor nodes that lie in the path from the source to the sink are involved in forwarding the packet to the sink. The frequency of traffic flow and the traffic pattern dictate the power consumption across nodes.

One disadvantage of flat routing is the asymmetric energy consumption across nodes. The nodes participating in flat routing protocol communicate with the base station by direct communication or multihop communication. Depending on the

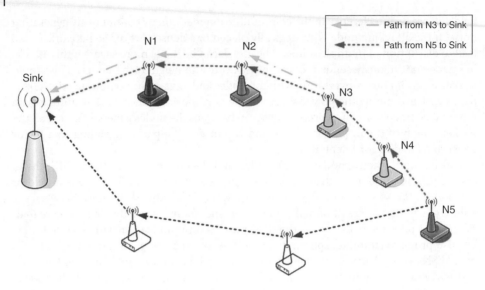

Figure 11.6 Multihop transmission in an indicative WSN.

distribution of the nodes with respect to the sink node and the location of the transmitting sensor node in the WSN, direct communication may be more efficient than multihop communication, or vice versa. Consider a WSN with nodes N1 to Nn (where $n = 2$ to 5), as shown in Figure 11.6. In a multihop communication, when Nn wants to send data to the sink, it will move over N$n-1$, ..., N2, and N1. When N$n-1$ wants to send data to the sink, it will move over N$n-2$, ..., N2, and N1, and so on. Thus, all the data intended for the sink will hop over N1, with maximum energy depletion of the sink's nearest node. The distant nodes from the sink have minimum energy depletion. Alternatively, if the nodes were involved in direct communication with the sink, the node farthest from the sink would have to use the highest-strength signal, while the nodes nearer to the sink would use the minimum-strength signal. This would lead to minimum energy depletion in the nearby nodes and to maximum energy depletion in the farthest nodes.

The flat routing protocols vary from very simple techniques such as flooding and gossiping to various application-specific and resource optimization protocols, including energy-aware routing. Some of the commonly used flat routing protocols and their brief description are as follows:

Flooding. Flooding is the simplest routing protocol and needs neither topological information nor a routing algorithm. It can easily accommodate mobility, topological changes, and scalability. However, with increase in the number of nodes, the number of packets, including the redundant packets, increases manifold in the network. In flooding, as indicated in Figure 11.7, a sensor that receives a packet broadcasts the packet to all its neighbors. Thereafter, all the neighbors that receive this packet broadcast it further, and the process continues. Only a node that has already transmitted the packet does not broadcast it again when it receives the same packet from its neighbor's broadcast. The process continues until the packet reaches the

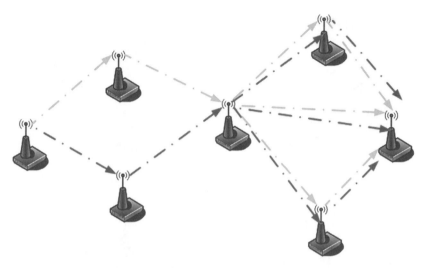

Figure 11.7 Data implosion during flooding.

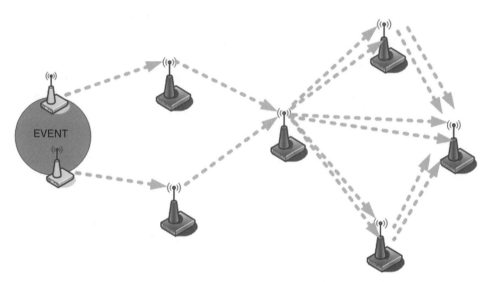

Figure 11.8 Interesting event covered by two sensors in the overlap region and flooded through the neighbors by both sensors.

destination or has completed the predefined number of hops [8]. The disadvantages of the flooding technique are as follows:

- It leads to 'data implosion', as a sensor receives the same packet from a number of its neighbors.
- An interesting event might be covered by two or more sensors, as indicated in Figure 11.8 and all these might transmit the same data to their neighbors. Thus, two or more sensor nodes might sense a region independently, but send the same sensed data from the 'overlap' region to a common intermediate node.

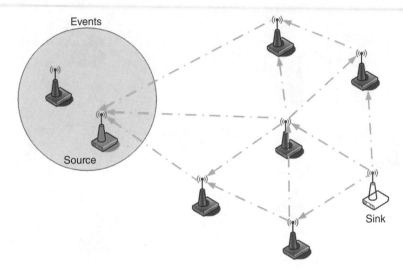

Figure 11.9 (a) Diffusion of interest from the sink node in the WSN.

- The technique does not try to optimize resource utilization in terms of processing, power, and buffer memory. This is termed 'resource blindness.'

Gossiping. This is an optimized version of flooding to reduce data implosion, whereby each sensor node sends the packet to one or a selected few neighbors instead of to all its neighbors. The neighbors receiving the data packet send it further, again only to one or a selected few of its neighbors. The process continues until the data packet reaches the destination or exceeds the maximum number of hops. This type of routing may lead to propagation delay, as the selected sensor nodes in the neighbor may not be in the optimum path.

Direct diffusion. This is a destination-initiated reactive routing protocol and operates in three stages: diffusion of interest, gradient setup, and data delivery along a reinforced path [9].

During diffusion of interest, depicted in Figure 11.9 (a), the sink node floods the query in the WSN. The query contains the attribute the sink is interested in, the particular value or a range of values that the sink desires to receive, and the frequency at which it intends to receive the sensor data. The geographical region of interest may also be transmitted. For example, the sink of a battlefield sensor network may send a query that it is interested in weight data and the value should be more than 50 000 kg (to enable it to detect a tank or launchers). This keeps the sink away from receiving data from all those sensor nodes that are not sensing the attribute of interest and even those sensors measuring the attribute of interest but that are presently not sensing data within the range of interest. The frequency of transmission requested at this stage is generally low, to prevent flooding of the network by the eligible sensing nodes. All the nodes maintain an 'interest cache', where they record the attributes of interest, as they might have to participate in the multihop routing of the data from the qualifying sensors to the sink, and during that stage will use this attribute in their interest cache for data aggregation.

In the gradient setup stage, shown in Figure 11.9 (b), the qualifying sensors start sending the values of the attributes of interest (which are in the specified range) at specified

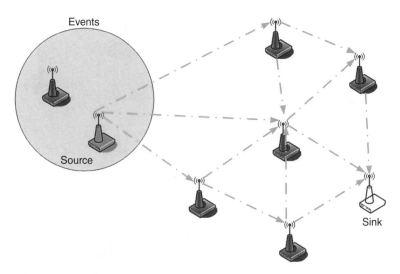

Figure 11.9 (b) Gradient setup from source node to sink node.

intervals by flooding their neighbors with the data of interest. All the intermediate nodes that receive this data of interest perform data aggregation by checking whether the received data matches the entries in the 'interest cache' of the sensor nodes. If the data exists in the 'interest cache', it is flooded to neighbors and an entry is made in 'data cache'. If the same data is received again, the data cache is used to verify that it has already been forwarded and hence the packet is dropped. When the data packet reaches the sink, the sink may decide to search for more paths from certain nodes. It sends another query for the same attribute of interest and range for one or more specific destination nodes. This initiates the generation of paths to the sink from these nodes also.

In the final stage, as shown in Figure 11.9 (c), the qualified sensors start sending data at the specified rate towards the sink using the discovered paths. Once the path is established, the sink may resend the original query to the destination with a change in the frequency at which it intends to receive the sensor data. The frequency is generally increased so as to receive the data of interest more frequently.

The protocol supports topological changes and path repairs by selecting from the other paths that have sent data at a slower rate from the sensor to the sink in the gradient setup stage. It also does not require an addressing scheme or unique identifier for each node, as the communication process is neighbor to neighbor. The disadvantage of this routing is that there is flooding of data, which has associated resource utilization during the first two stages – diffusion of interest and gradient setup. The memory requirement is also high for each sensor, as it has to maintain an 'interest cache' and a 'data cache'.

Rumor routing. A diffused routing has a major overhead if there are a very few sensors sensing data of interest. Rumor routing [10] is a variation of diffused routing where the data query is not sent across the entire network but to specific sensor nodes that have indicated that they have captured the event of interest. When a sensor node detects an event of interest, it adds it to its event table and creates a packet called an 'agent', which contains the event table and floods it in the network with certain hops-to-live. All the

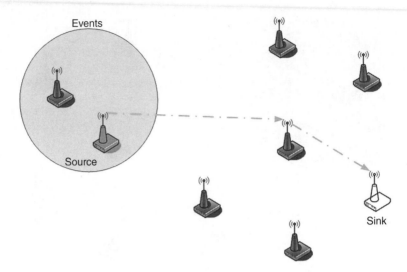

Figure 11.9 (c) Data delivery from source to sink through the discovered path.

nodes receiving the agent use the event table being carried by the agent to update the sensor nodes' own event table. New events can be added in the event table of the agent, and it is propagated to the nodes. The agent also maintains a list of visited nodes so as to ensure visiting a new node on every hop and reduces its hop-to-live count by 1. When a sink is interested in an event, it sends a request in a random path until it reaches a node that has information about the sensor node that has recorded the event of interest and the path to that sensor node.

The disadvantage of the protocol is that the size of the agent may become large, as it maintains the event table as well as the list of visited nodes. As the agent gives only one route to the sensor that has recorded the event of interest, a request from the sink to the sensor for data at frequent intervals may deplete the energy of the nodes in the path.

Gradient-based routing. This is a destination-initiated routing protocol, a variation of directed diffusion, where the distance in terms of hop count to the node capturing the event of interest is used as a gradient [11]. The minimum number of hops between any node and sink is termed the height of the node here. When a node has more than one neighbor at the same height, one of the neighbors is selected randomly. The protocol has an energy-saving scheme incorporated in it, where any node with decreasing energy levels can increase its height parameter, thereby refraining its neighbors from forwarding packets through it.

Sensor Protocol for Information via Negotiation (SPIN). The protocol uses data descriptors or metadata to send information about the data to neighbors before sending the actual data to the sensor nodes interested in it. Those neighbors that do not have the data will be interested in it, and hence, on receiving metadata related to unreceived data from a neighbor, the sensor node sends a request for the actual data. On receiving the request, the actual data is sent to the interested node by the sensor node that had initially sent the metadata. This does away with the requirement of data aggregation and sensing of events by overlapped sensors, and prevents flooding. The format of the data descriptor is application specific. The protocol has to use three types of message, each

Figure 11.10 (a) SPIN – advertisement of metadata to all neighbors.

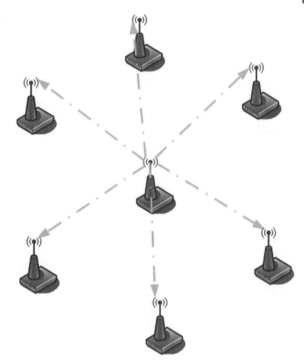

of which is used during the process, depicted in Figures 11.10 (a)–(c) respectively: (1) 'advertisement' of metadata, (2) a 'request' message for actual data, and (3) transmission of actual 'data'. However, SPIN does not guarantee delivery of data between any two nodes because, if the intermediate nodes are not interested in the data, the data cannot cross over the region of uninterested nodes and reach the destination that is interested in the data [12].

Some of the other flat routing protocols are as follows:

- energy-aware routing [13],
- Constrained Anisotropic Diffusion Routing (CADR) [14],
- cougar approach [15],
- Active Query Forwarding in Sensor Networks (ACQUIRE) [16],
- Minimum-Cost Forwarding Algorithm (MCFA) [17],
- Minimum-Energy Communication Network (MECN) [18].

11.4 Hierarchical Routing Protocols

The hierarchical routing protocols preferably use faster and better sensor nodes as we move up in the hierarchy of the sensor nodes. These nodes may have higher power and better resources. Alternatively, if all the nodes in the WSN are of similar capabilities, the nodes that operate at higher levels of hierarchy are replaced regularly, and some other nodes are selected to perform the task of the cluster head or coordinator, as these roles demand more processing requirements and drain the power of the sensor node very

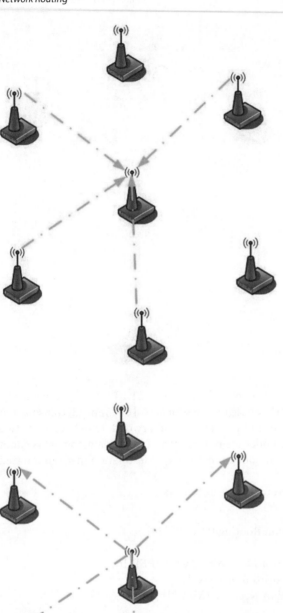

Figure 11.10 (b) SPIN – after checking of metadata, a request for actual data by nodes not having the data.

Figure 11.10 (c) SPIN – transmission of actual data to the interested nodes.

rapidly. Hence, the responsibility is generally rotated across all the available nodes to avoid any particular node being exhausted of power.

Hierarchical routing keeps the flow of data packets restricted within the WSN by avoiding duplicate packets as well as broadcast of packets to all the nodes. Data broadcasts and communication between sensors are generally confined within small regions of the sensor field, and these regions are generally termed 'clusters'. Only selective traffic moves up in the hierarchy. Data communication between two far-off sensor nodes is through the nodes at a higher level of the hierarchy forming a high-reliability and minimum-congestion 'backbone' of communication in the WSN. In a hierarchical routing protocol, the role of sensing an event of interest is generally assumed to be performed by the sensors at the lowest level of the hierarchy, which can be the low-energy, low-processing-capability, and reduced-communication-capability (only transmission is required) sensor devices. The nodes at the higher level perform the tasks of data processing, aggregation, and forwarding and are generally high-energy nodes. This enhances the life of the WSN by saving energy requirements, and avoids resource blindness. The energy dissipation is very similar in all the nodes at the same hierarchical level, and the energy dissipation cannot be controlled on the basis of traffic pattern because all the nodes at the higher hierarchical levels have to exchange data even if a single sensor node forwards data of interest.

Cluster routing is the most common hierarchical routing in a WSN, with two layers of hierarchy. The WSN is divided into regions called clusters, comprising low-end devices at the lower level, and each cluster can have one or more cluster heads at the upper level. The nodes in the cluster can communicate only with their cluster head, and the cluster heads can communicate with all other cluster heads in the WSN.

The major disadvantage of hierarchical routing is the resource overhead in selection of the cluster heads. Furthermore, there might be regions in the WSN where no data of interest is being generated, but still these regions have to utilize resources in creating and maintaining the cluster structure. This involves creating cluster heads and coordinating with sensor nodes in the cluster to inform them about the cluster heads or the discovery of cluster heads by the lowest-level sensor nodes. Resources are also used by the sensors in deciding their association with any particular cluster and thereafter change of cluster owing to mobility, signal strengths, and QoS issues.

Routing is generally very simple in a hierarchical routing – the sensor node sends the data to the cluster head. In the case of a two-level hierarchy, the cluster heads communicate with each other, similarly to a flat routing protocol. In the case of a multilevel hierarchy, the data moves up to the highest level in the hierarchy and then comes down in the hierarchy up to the destination node. Most of the hierarchical routing protocols therefore handle the following two problems in hierarchical routing:

- selection of a cluster head,
- selection of a cluster by a sensor node to get associated with.

Low-Energy Adaptive Clustering Hierarchy (LEACH). In LEACH [19], the cluster head is selected randomly based on election. The nodes can send data only to their cluster head, and the cluster head in turn sends the data to the sink after processing and aggregating the data. Although the scheme reduces the power requirement of an individual sensor node, it enhances the power consumption of the cluster heads by a huge amount. As the cluster heads also do not have a fixed power source, the protocol rotates the role of the cluster head among the sensor nodes based on two factors – the energy left in the node and the number of times the node has performed the role of cluster head. A node elects itself as a cluster head based on random number selection between 0 and 1. If the selected number is below a threshold value, it elects itself as the cluster head. The threshold value is determined on the basis of an equation with the parameters: desired percentage of cluster heads, current round, and set of nodes that have not been cluster heads in certain previous rounds. One disadvantage of this technique of cluster head election is that the elected cluster heads may be concentrated in a certain region of the WSN instead of being equally distributed across the WSN.

The membership of a node in a cluster is based on the signal strength of the cluster head being received by the node. A node is a member of the cluster from which it receives the strongest cluster-head signal strength. Proximity to the cluster head and signal-to-noise ratio may also be used as criteria for selection of cluster membership. The protocol operates in rounds, with a different cluster setup and cluster head in each round. The following are the operations in each round:

- A node elects itself as the cluster head with a certain probability and advertises its election.
- The other nodes select their clusters based on the received signal strength of the cluster head. The cluster head is informed about their membership.
- The cluster head manages transmission from the nodes in its cluster by using TDMA, where each node is informed about its slot so as to enable it to go into sleep mode while it does not have a transmission slot.
- A node sends data to the cluster head in its allocated time slot.
- Once the cluster head receives the data from all the nodes, it performs data processing and analysis and data aggregation and compression, and then sends it to the sink.

Collision between the nodes in a cluster is avoided by using TDMA, and contention between cluster heads for transmitting data to the sink is avoided by CDMA.

PEGASIS is a hierarchical routing protocol that works on 'chaining' [20] instead of clusters as in LEACH. It is more energy efficient than LEACH, as it saves on processing for cluster formation, election of cluster heads, and processing of data from all the sensors at the cluster head. PEGASIS forms a chain of sensor nodes, and one of the nodes from the chain is selected for transmission of processed data to the sink. As the data moves in the chain, it keeps being aggregated along with the data of the node that is forwarding it in the chain. A token passing technique is used in the chain to avoid collisions, as only the node possessing the token is capable of transmitting data in the chain. The working of the routing algorithm can be explained using the deployment example shown in the Figure 11.11. Nodes N1 to N5 are the sensor nodes in a small region within the sensor field. These nodes form a chain, and the data moves along the chain. Initially, node N1 has the token and sends the data to N2. N2 aggregates the data received from N1 along with its own data and sends it to N3. N3 passes on the token to N5, which on receiving the token sends its data to N4. Like N2, N4 also aggregates its data along with the data from N5 and sends it to N3. N3 aggregates the data received from N2 and N4 to the sink, which effectively is the aggregated data of the complete chain.

The disadvantage of PEGASIS is that the central node in the chain, which sends the data to the sink, becomes the point of delay and energy consumption. Furthermore, as

Power-Efficient GAthering in Sensor Information Systems (PEGASIS):

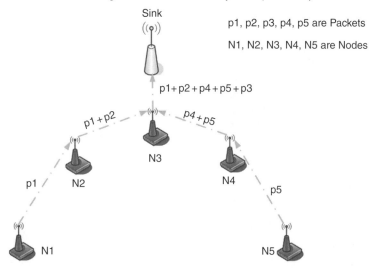

Figure 11.11 Creation of a chain in a WSN running PEGASIS.

the size of the chain increases, the process of data aggregation and transmission is delayed, as it requires time for the data from the farthest link in the chain to move towards the center of the chain.

TEEN is based on clustering with three levels in the cluster hierarchy [21]. As depicted in Figure 11.12, the nodes transmit to the first-level cluster heads. The first-level cluster heads communicate with the second-level cluster heads, and finally the second-level cluster heads communicate with the sink. This is a reactive routing protocol that is used to detect sudden change in the value of the sensed parameter. The routing algorithm defines two parameters: the hard threshold and the soft threshold. The hard threshold is that level beyond which, if the data is sensed by the sensor, it should be transmitted to the sink. If the value has crossed the hard threshold, this does not mean that the data will now be continuously transmitted to the sink. It also does not now necessitate data transmission at regular intervals. The data will be transmitted again only when it is beyond the hard threshold and fluctuates beyond the soft threshold limits. Thus, the soft threshold further reduces the data transmission, and only interesting and significant changes in the values of the sensed parameters are sent to the sink. The protocol does not suit those applications that require regular transmission of sensed parameters to the sink.

While the working of a few hierarchal routing protocols have been explained to indicate the variations in techniques, topologies, and the parameters optimized in the routing, there exist a huge number of hierarchical routing protocols, of which just a few are listed below:

- Adaptive Threshold-Sensitive Energy-Efficient Sensor Network Protocol (APTEEN) [22],
- Sequential Assignment Routing (SAR),
- Sensor Protocol for Energy-Aware Routing (SPEAR),
- Small Minimum Energy Communication Network (MECN),
- Self-Organizing Protocol (SOP) [23],

Threshold-Sensitive Energy-Efficient Sensor Network Protocol (TEEN)

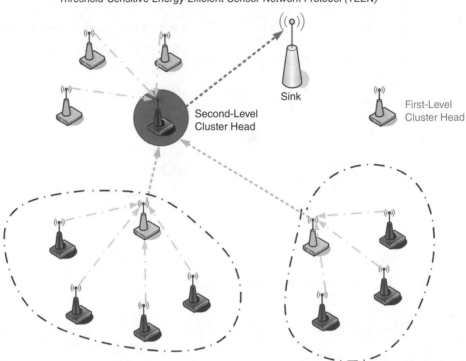

Figure 11.12 A WSN running TEEN protocol.

- Virtual Grid Architecture Routing (VGA),
- Hierarchical PEGASIS [24],
- Energy-Aware Routing [25],
- Hierarchy-Based Anycast Routing (HAR) [26],
- Hierarchy Energy-Aware Routing for Sensor Networks [27],
- Balanced Aggregation Tree Routing (BATR) [28].

11.5 Location-Based Routing Protocols

Location-based routing protocols are highly adaptive to the dynamic nature of sensor networks. The routing overheads are low compared with the other routing protocols as location-based routing protocols use the information about the geographic location of the sensor nodes to forward the packets. The location of a sensor node in a network can either be indicated in terms of its geographic location with a global frame of reference by using a GPS or with a local frame of reference with respect to some local-infrastructure-based positioning system. Apart from these, the location of nodes can also be indicated with reference to each other. Location-based routing can be used for packet delivery with two different kinds of requirement. Firstly, it can be used to move a packet towards the destination node by using the location information of the destination node and make the packet move nearer to the destination with each hop. The protocol can change the route of packet forwarding dynamically if the destination node moves to a different position while the packet is already on the move through the sensor network towards the destination node. Secondly, location-based routing can be used to deliver a data packet to one or more of the sensor nodes that are available in the predesignated location for which the packet has been forwarded. This kind of packet forwarding is also known as geocasting.

Location-based routing has a low memory demand from the sensor nodes. An elaborative routing table is not required to be maintained in the nodes, as only the location information about the neighbors will suffice for routing. Location-aware routing can lead to a single path from the source to the destination or can be used to discover multiple paths, depending on the application scenario. Single-path forwarding results in energy savings, while multiple paths can increase the reliability of packet delivery and can also support load distribution. In the case of multiple paths, avoiding a common node or link among the multiple paths will further enhance the reliability and load-sharing capability.

The most common routing strategy used in location-aware routing is the greedy approach, whereby a node forwards the packet to one or more of its neighbors that are nearer to the destination than the node itself. If only a single path has to be selected, the greedy approach leads to forwarding of the packet to the neighbor that is nearest to the destination in terms of distance. Such a single-path greedy-based approach may lead the packet to a dead end when it reaches a node, which might be very near to the destination but cannot reach the destination in a single hop and there are no intermediate nodes available between the node and the destination to carry the packet forward. In such scenarios, the packet has to be backtracked. Hence, the intermediate nodes should maintain a record of which packet has been forwarded to which node so as to avoid forwarding it to the same neighbor during backtracking after reaching a dead end. Another approach to recover from a dead end in a greedy approach is to use flooding, partial flooding, or directional flooding to forward the packet to neighboring nodes, after which the nodes receiving the packet initiate the greedy approach to forward the packet towards the destination. The greedy approach with location information will also ensure by itself that the route is loop free.

The strategy of a location-based routing changes if the packet is not meant for a particular location but for a specific node, the location of which has to be determined before delivering the packet. The scenario becomes more complex if the destination node is mobile. The awareness of the changing location of the destination has to be passed on to the intermediate node handling the packet to enable the intermediate node to forward the packet in the proper direction. The routing strategy should not only have a mechanism to maintain data about each node and its current location but also be able to make the data available to any node that requires it for routing. The simplest mechanism is that every node keeps transmitting its current location, which propagates throughout the network through intermediate nodes, and all the nodes store this information. If the intermediate node has some other location information stored about the node, the new location information overwrites the previous location information. Time stamping or sequence number may be used to distinguish between two pieces of location information received from a node and to judge which one is recent.

Another approach to get to the destination location can be query based, whereby the source node floods the network with a query seeking the location of the destination node. The intended destination node replies to the flooding by sending its location to the source node. This approach is generally suitable for small, low-density, and low-mobility sensor networks, otherwise the energy and bandwidth consumption will be too high and the destination node might change its location by the time a packet reaches near the destination. An alternative to these approaches is to use some predesignated intermediate nodes as server nodes to which all sensors in their zone send location information. The source node intending to send a packet to the destination node can get the location of the destination node from the server node of the destination zone. The intermediate nodes in the routing path of the packet can also keep checking for any change in the location of the destination node from the server node of the destination zone as the destination node will keep the server node of its zone updated about its changing locations. A zone may have one or more server nodes, and there is a proper handoff mechanism between the server nodes of two separate zones when a node moves from one zone to another zone.

Location-based routing is also used in conjunction with some other static or dynamic routing protocol to create a hybrid routing protocol. The location-based routing is used for long-distance transmission, and the static or dynamic routing is used once the packet reaches the vicinity of the destination. This leads to energy savings and less memory requirement in the sensor nodes, as only the routing information of a small area comprising nodes up to a few hops from the sensor nodes has to be stored in them and not the entire network. Another scenario where a hybrid routing with location-aware routing as one of the components is used is a network in which certain areas do not support location-aware routing. There can be regions of non-availability of location information owing to electromagnetic interference, jamming, the need for electromagnetic silence, or non-availability of GPS information. Such portions of the network may be catered for by static or dynamic routing, while the remaining network can use location-aware routing.

Location-based routing has been of interest to researchers since early 2000, and a good number of such protocols were proposed even during these initial years, some of which are listed below:

- Location Aided Routing (LAR) [29],
- Geographic Adaptive Fidelity (GAF) [30],
- Geographical and Energy-Aware Routing (GEAR) [31],
- Time-Based Positioning Scheme (TPS) [32],
- Two-Tier Data Dissemination (TTDD) [33],
- SPEED [34].

11.6 Multipath Routing Protocols

The multipath routing protocols discover multiple paths from source to destination as well as use these multiple paths to forward the data from the source to the destination. The protocols have three main stages: *path discovery* to locate more paths between any source–destination pair or from any sensor node to the sink node, *distribution of traffic* from the source across multiple paths to the destination, and *path maintenance* to cater for topological changes, scalability, and availability of new paths. All three stages are based on certain criteria, which might be based on the application supported,

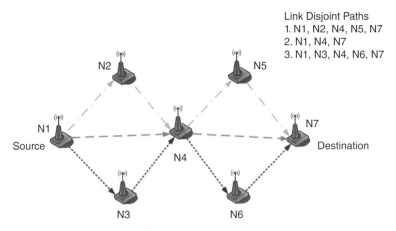

Link Disjoint Paths
1. N1, N2, N4, N5, N7
2. N1, N4, N7
3. N1, N3, N4, N6, N7

Figure 11.13 A WSN with link disjoint paths.

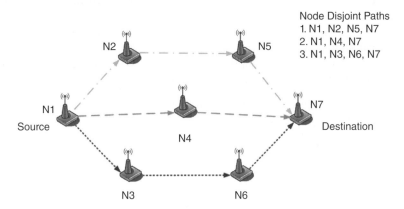

Node Disjoint Paths
1. N1, N2, N5, N7
2. N1, N4, N7
3. N1, N3, N6, N7

Figure 11.14 A WSN with node disjoint paths.

reliability, QoS, energy constraint, or any other parameters. The major advantages of multipath routing are fault tolerance, load balancing, bandwidth aggregation, and faster delivery.

To find the best possible paths from source to destination during path discovery, disjoint paths are discovered so as to avoid common points of failure. The paths can be link disjoint or node disjoint, as indicated in Figures 11.13 and 11.14 respectively. If the set of paths are link disjoint, no two paths can have a common link. This implies that, for a typical WSN topology, in link disjoint paths a few nodes can be common across various paths. However, in node disjoint paths, as no two paths can have a common node except the source and destination nodes, this in itself implies that there can be no two common links either in the disjoint paths. A link failure brings down one path in link disjoint paths, and a node failure brings down one path in node disjoint paths. However, a node failure can bring down multiple paths in the case of link disjoint paths. Thus, node disjoint paths ensure link disjoint paths, but the reverse may not be true. Node disjoint paths are most effective in fault tolerance and bandwidth aggregation as they do not have a common point of failure or congestion.

The traffic distribution over multiple paths can have various strategies. The entire traffic between a source–destination pair may be passed through a single path that is most efficient, reliable, or fault tolerant, and the other paths are kept as backup paths. Alternatively, the traffic may be distributed in the form of different data streams, each moving along a different path. The distribution of traffic over various paths is taken care of by the path selection algorithm, which uses different types of metric to select the paths. A process also has to be in place to reassemble the data streams received from various paths in a single data stream at the destination.

Fault tolerance is achieved by using multiple paths to send not only different data packets but also packets with a certain amount of redundancy along different paths to ensure fault tolerance in the case of path failure. To support load balancing, the multipath routing protocol can divert traffic to alternative paths if the path along which the data is presently being transferred has congestion, is facing latency, or is overutilized. A WSN may have multiple low-bandwidth links, which can be aggregated to achieve higher bandwidth. The data is split into a number of streams, each passed through a different low-bandwidth link to the destination, giving an effective high bandwidth at the destination. This reduces the delay in delivery of data packets from source to destination, as it also saves on the time of alternative route discovery in the case of a failed node or link.

Continuous research in the area of multipath routing has led to the development of a number of multipath routing algorithms, a few of which are named below:

- energy-aware routing for low-energy ad hoc sensor networks [13],
- highly resilient, energy-efficient multipath routing [35],
- an energy-efficient multipath routing protocol [36].

11.7   Query-Based Routing Protocols

Query-based routing protocols are the category of routing protocols where a query is sent to a wireless sensor network to obtain the data values sensed by the sensor node at a particular time or over a time interval. The query may also be sent to detect whether an event of interest is being detected by the sensor network. Query-based routing is generally destination-initiated reactive routing. The query can be generated by the sink node, any sensor node, or a system interfaced to the wireless sensor network. As query-based routing is generally reactive, it adapts even in dynamic network topologies and can discover more than one route between the two nodes, leading to load sharing or recovery from a particular route disruption.

The querying node may be interested in getting information from a specific sensor node for specific data, or from a cluster of sensor nodes within a zone for event detection, or from all the sensor nodes in a wireless sensor network for the purpose of data aggregation and analysis. In a wireless sensor network running query-based routing, when the network initializes or the querying node begins its operation, it may not be aware of the exact node or the specific zone from which the reply is of interest. Hence, initially the operation may be in classical broadcast mode. This fetches data from all the sensors, analysis of which may help in reducing the area of query, leading to identification of the zone of interest and reduction in zone size with time to avoid querying any unwanted node.

A query is generated by the querying node and sent in the wireless sensor network so as to reach the relevant sensor node for reply to the query. This query packet has to pass through a number of intermediate sensor nodes. Any intermediate node may be aware of the route to the intended target nodes for the query and forward the query through that route. Alternatively, the intermediate node may not be aware of the route to the target nodes for the query and hence may use any routing strategy such as flooding, gossip, or rumor routing. To ensure quick and assured discovery, flooding or selective flooding may be preferred to locate the target node for the query [37].

The route query moves from the querying source, which is the intended destination, towards the expected source of data, searching for the path and marking the path traversed so as to use it during return of the data packet from source to destination, subject to links being bidirectional. It should be ensured that the query packet does not enter into infinite loops among the nodes of the wireless sensor network. This will not only lead to delays in determination of the route, it will also lead to traffic congestion. The simplest way to avoid infinite loops in a query packet is to use a time-to-live (TTL) field in the query packet. The value of TTL is generally kept equal to the diameter of the network in the case of a static network, and a few hops extra in the case of a dynamic network. Restricting the value of TTL to these parameters helps in discovering the optimal routes in terms of number of hops, one among which will be the shortest route. However, if no restrictions are imposed on this scheme, all the intermediate nodes forward the route query to all their neighbors except the one from which they have received the query. The scheme can be better regulated in terms of reduction in network traffic if each intermediate node is constrained to forward the query packet only to one of its neighboring nodes. Although this will lead to a reduction in network traffic, it will not guarantee discovery of the shortest path, and the nodes will have to maintain data about all the query packets they have forwarded to avoid forwarding the same packet again if received from some other neighboring node [38].

Once a query reaches the destination node from the querying node, the reply has to go back from the destination node to the querying node. There are several techniques used to preserve the path information for traversal of the reply data. The simplest technique is that, while the query packet moves from the querying node to the destination node, it keeps appending to its data packet the addresses of all the nodes through which it has traversed so that the same can be used to send back the reply packet, and also be used by the nodes in the querying path to store the routing path to the destination node. If the size of the network is big or the optimum path has not been discovered by the query packet, the query packet and the reply packet can become too big in size, as they will contain the entire path information. Alternatively, each intermediate node forwarding the query packet in the sensor network can store the packet identifier along with the address of the previous hop node so that this information can be used to send back the reply data from the destination node to the query source. This reduces the packet size but leads to an overhead of each node maintaining details about every packet forwarded by it. Apart from these schemes, the route from destination node to query source can be discovered independently of the route discovered from delivery of query packet from querying node to the destination. This is essentially one of the techniques that the networks with unidirectional link connectivity have to rely on.

An example of query-based routing protocol is Energy-aware Query-Based Routing Protocol for WSNs (EQR), which also aims to achieve energy balancing and energy saving.

11.8 Negotiation-Based Routing Protocols

Sensor nodes are generally deployed in sufficient quantities to ensure coverage of interesting events with accuracy even in the case of failure of a few boundary or intermediate nodes. There may be critical deployment scenarios where a variety of sensors have been deployed collaboratively to detect an event with high reliability and avoid false alarms. For example, a fire may be detected by a smoke sensor, light sensor, heat sensor, and video sensor simultaneously, each reporting 'Yes' to fire detection in the case of a binary mode of operation for detection of fire in terms of 'Yes' and 'No'. This leads to redundant event detection and duplicate data transmission among the nodes. Each extra or avoidable copy of the data being transmitted through the network consumes power in transmission and processing and occupies memory, both of which are in scarcity in a wireless sensor network. Negotiation-based protocols [39, 40] are designed to address these issues of data implosion and detection of the same event by multiple sensors, which also leads to generation of similar data for transmission across the network. Furthermore, negotiation-based protocols require a very limited view of the sensor network for their operation and can work well with awareness of single-hop neighbors only. The Sensor Protocol for Information via Negotiation (SPIN) family of routing protocols [41], i.e. SPIN-EC, SPIN-RL, SPIN-BC, SPIN-PP, SPIN-1, and SPIN-2, are the most common examples of negotiation-based routing protocols.

Negotiation-based routing protocols conserve energy, reduce the amount of data being communicated, and hence also reduce the computational requirements. In this class of protocol, the nodes, which are generally in the neighborhood, negotiate with each other before transmission of the actual data to ensure that only useful data is transmitted. The two most common negotiations are for the following purposes:

Negotiation ensuring requirement of data. Before sending the data, the transmitting node communicates with the receiving node in its neighborhood to check whether that neighboring node requires the data that is now ready for transmission with the transmitting node or this particular receiving node in the neighborhood is already possessing a copy of the data received from some other node.

Negotiation enquiring of ability to forward data. The transmitting node checks whether the receiving node has sufficient energy and willingness to forward the data packet. This scenario is for those particular cases where the neighboring node has to act as an intermediate node to transfer the data further in the network to some other nodes so as to enable the data to reach the destination node or sink node. The neighboring nodes' agreement to forward it and not drop the data because of energy constraint is negotiated before transmission of the data.

For the negotiation to occur across the nodes, the transmitting node should be able to share with the receiving node the basic information about the characteristics of the data in terms of the time of data creation, the location of generation of the data, intermediate nodes through which the data has traversed, or the type of data. One or more of these data descriptors help the receiver to take a decision as to whether it needs to receive the actual data from the sender. The data descriptor should be sufficiently small compared with the data itself to make the negotiation protocol efficient. Either the complete data descriptor or a predefined field in the data descriptor should be the same for the same data. The format of the data descriptor may be defined as per the application of the sensor network or based on the aggregation of heterogeneous sensors.

Negotiation occurs before the transmission of the actual data. A receiver does not agree to receive the data from the sender if it has already received a copy of it from some other node, and thus negotiation avoids data implosion. The receiver also prevents the transmitter from sending the data to it if it is of the same event or the same location and time that some other sensor in the area has already captured and has reached the receiving node prior to negotiation with the transmitting node. This helps to avoid redundant data transmission from sensors with overlapping sensing areas. The receiving node may not have sufficient energy left with it, and thus may now be interested only in receiving data of direct interest to it, but not willing or capable of participating in data forwarding to the destination node. Its non-willingness to act as an intermediate node in data transmission can also be conveyed to the transmitting node during negotiation to stop the transmitting node from forwarding the data to this energy-deficient node.

The number of steps involved in negotiation in terms of rounds of communication between the transmitting node and the receiving node should be few. The amount of data transmitted between the nodes during negotiation should also be optimally low so as to keep the negotiation overhead at the minimum. Furthermore, the process of negotiation should not be so intense and detailed that it consumes more power, computation, and bandwidth than the actual data transmission. Still, there may be instances when the data might be transmitted between any two nodes without negotiation as the transmitting node may judge it to be more cost effective if transmitted directly to the receiver without negotiation. These cases arise when the size of the data is either too small or the data is very critical and redundancy has to be maintained in the network for certain areas of the network or particular events.

11.9   QoS Routing Protocols

The applications of WSNs are on the rise, with an unimaginable diversity of requirements in the use of WSNs, deployment conditions, and desired output. Quality of Service (QoS) is interpreted as the measurement of the network parameters being delivered to the users as against the requirement of the user. There are certain applications where a real-time output is desired from the WSN [42]. The real-time output can be demanded from the network on a continuous basis or only during detection of an interesting event. For example, the output from a video sensor may not be required on a real-time basis until any motion is detected in its area of visualization, and once the video sensor detects a motion, real-time data from the video sensor is desired at the control station. Latency beyond a certain limit is not acceptable in this scenario. There are applications where latency can be tolerated to a certain extent, but reliability of the data cannot be compromised. An example of such a scenario is a sensor network to study the activity of a volcano when people are already away from it and no human life is at stake even if the data is received a few seconds late. However, reliability is important as the data will help to formulate the actual volcanic model and there is no other means to cross-check the data. Use of sensor networks for tsunami prediction and ballistic missile defense are examples of applications where latency and reliability are of equal importance. Thus, the QoS requirement in a WSN is application dependent.

With advancement in sensor devices supported by technologies such as nanomaterials and MEMSs, the output of a sensor device is no longer limited to plain discrete text and numbers. Continuous data from sensors, including video, voice, and image, has posed new challenges, as these requirements cannot tolerate network latency, lower bandwidth, interference and packet drops, data loss, or out-of-order delivery. The fulfillment of these requirements calls for computational resources, memory, and accelerated operations, which all consume power. Over and above these, a QoS routing protocol is also required to route the packets for effective delivery across the network.

11.9.1 Challenges

QoS routing protocol is also responsible for balancing between energy consumption and assurance in terms of latency, throughput, reliability, or hop count. The QoS may not be dependent on a single metric, but a combination of two or more metrics. This further adds to the complexity of QoS routing, which now has to balance between multiple QoS metrics in a power-saving routing algorithm.

Although energy is the primary constrained resource in a WSN, there are other constraints and challenges in providing QoS [43] in a WSN, some of which are as follows:

- Lower processing capability, limited queuing memory, small transmission range, dependence on other nodes for data forwarding.
- Compared with sensor nodes, there are fewer cluster coordinators, sink nodes, and gateways, and in the case of continuous sensing by sensor nodes, this leads to traffic congestion at these points.
- Multiple sink nodes may also call for variation in the QoS with respect to each sink node.
- There is a great amount of redundancy in detected data as well as data packets flowing in the network, which adds to reliability as well as energy consumption. Data fusion and aggregation techniques are used to reduce the undesirable redundancy, but these further add to energy consumption and processing delays.
- The sensor nodes as well as the sink nodes are mobile. A node may also keep toggling between active and sleep states to save power, adding to the problem of QoS implementation.
- QoS protocol should ensure that the energy consumption is nearly equal across all nodes so as to avoid power drain of a small set of sensor nodes.
- In a heterogeneous network there might be a variety of sensors with a different frequency of traffic generation, different bandwidth requirement, and different QoS requirement. Sensor clouds are a live example of such a network where there can be sensors related to temperature, humidity, wind flow, and video for a particular geographical area.
- The network topology is dynamic, the transmission medium in not reliable, and all the sensor nodes may not be uniquely identifiable.

11.9.2 Approach to QoS Routing

The QoS in a WSN can either be application driven or network driven. As discussed earlier, the field of application where the WSN is in use may dictate the QoS parameters. The application-driven QoS parameters generally are latency, data integrity, reliability of data fusion, complete coverage (overlapping of sensed areas), and power saving from redundant sensors by being in sleep mode. The most common application-specific QoS requirements are latency and reliability. The network-specific QoS ensures minimal usage of sensor resources to deliver application-specific QoS. These network-specific QoS parameters are generally common for all WSNs and are the underlying parameters for data delivery from the sensor to the sink. The data delivery model being used for the network QoS can be distinguished into three categories: event driven, query driven, and data driven. In the event-driven model, the sensor transmits data to the sink on detection of an event or a parameter beyond threshold. In a query-driven model, the sink node enquires about certain parameters from all the sensors or specific sensors in a selected region. In a data-driven model, the sensor nodes keep sending data to the sink either continuously or at prespecified time intervals.

In QoS-based routing protocol, the route from source to destination has to be computed on the basis not only of the various resources available in the network but also the QoS requirements of the application so as to ensure that the QoS service delivered by the WSN matches the QoS requirement of the user. The QoS routing protocol, working on changing topology and unreliable state information about the sensor nodes, has to ensure data flow through the network at a minimum assured bandwidth with less than specified latency. Energy consumption is a parameter that always has to be minimized, along with any other metrics being considered for QoS.

11.9.3 Protocols

Some of the common QoS-based WSN protocols are listed below:

- Sequential Assignment Routing (SAR) [44],
- Multiconstrained QoS Multipath Routing (MCMP) [45],
- Energy Constrained Multipath (ECMP) Routing Protocol [46],
- SPEED [34],
- Multipath and Multispeed Routing (MMSPEED) [47],
- Message-Initiated Constrained-Based Routing (MCBR) [48],
- Energy-Efficient and QoS-Aware Multipath-Based Routing (EQSR) [49],
- QoS-Based Energy-Efficient Sensor Routing (QuESt) Protocol [50],
- Reliable Information Forwarding Using Multipath (ReInForM) [51],
- Directed Alternative Spanning Tree (DAST) [52],
- Multiobjective QoS Routing (MQoSR) [53].

References

1 C. Chong and S.P. Kumar. Sensor networks: evolution, opportunities, and challenges. *IEEE*, **91**(8):1247–1256, 2003.

2 C. Townsend and S. Arms. Wireless sensor networks: principles and applications. *Sensor Technology Handbook, Volume 1*, pp. 439–450, Elsevier, Burlington, 2005.

3 C. J. Leuschner. The design of a simple energy efficient routing protocol to improve wireless sensor network lifetime, University of Pretoria. http://upetd.up.ac.za/thesis/available/etd-01242006-091709/unrestricted/00dissertation.pdf, 2005.

4 Q. Jiang and D. Manivannam. Routing protocols for sensor networks. *1st IEEE Consumer Communications and Networking Conference*, **2**(3):93–98, 2004.

5 J. N. Al-Karaki and A.E Kamal. Routing techniques in wireless sensor networks: a survey. *IEEE Wireless Communications*, **11**(6):6–28, 2004.

6 I. F. Akyildiz, W. Su, Y. Sankarasubramaniam, and E. Cayirci. Wireless sensor networks: a survey. *Computer Networks*, **38**(4):393–422, 2002.

7 K. Akkaya and M. F. Younis. A survey on routing protocols for wireless sensor networks. *Ad Hoc Networks*, **3**(3):325–349, 2005.

8 S. Hedetniemi and A. Liestman. A survey of gossiping and broadcasting in communication networks. *Networks*, **18**(4):319–349, 1988.

9 C. Intanagonwiwat, R. Govindan, and D. Estrin. Directed diffusion: a scalable and robust communication paradigm for sensor networks. *6th Annual International Conference on Mobile Computing and Networking*, pp. 56–67, ACM, 2000.

10 D. Braginsky and D. Estrin. Rumor routing algorthim for sensor networks. *1st ACM International Workshop on Wireless Sensor Networks and Applications*, pp. 22–31, ACM, 2002.

11 C. Schurgers and M.B. Srivastava. Energy efficient routing in wireless sensor networks. *IEEE. Military Communications Conference (MILCOM) – Communications for Network-Centric Operations: Creating the Information Force, Volume 1*, 2001.

12 W. Heinzelman, J. Kulik, and H. Balakrishnan. Adaptive protocols for information dissemination in wireless sensor networks. *5th Annual ACM/IEEE International Conference on Mobile Computing and Networking*, pp. 174–185, ACM, 1999.

13 R. Shah and J. Rabaey. Energy aware routing for low energy ad hoc sensor networks. *IEEE Wireless Communications and Networking Conference (WCNC), Volume 1*, pp. 350–355, 2002.

14 M. Chu, H. Haussecker, and F. Zhao. Scalable information-driven sensor querying and routing for ad hoc heterogeneous sensor networks. *International Journal of High Performance Computing Applications*, **16**(3):293–313, 2002.

15 Y. Yao and J. Gehrke. The cougar approach to in-network query processing in sensor networks. *SIGMOD Record*, **31**(3):9–18, 2002.

16 N. Sadagopan, B. Krishnamachari, and A. Helmy. The ACQUIRE mechanism for efficient querying in sensor networks. *IEEE 1st International Workshop on Sensor Network Protocols and Applications*, pp. 149–155, 2003.

17 F. Ye, A. Chen, S. Lu, and L. Zhang. A scalable solution to minimum cost forwarding in large sensor networks. *IEEE 10th International Conference on Computer Communications and Networks*, pp. 304–309, 2001.

18 V. Rodoplu and T. H. Meng. Minimum energy mobile wireless networks. *IEEE Journal on Selected Areas in Communications*, **17**(8):1333–1344, 1999.

19 W. Heinzelman, A. Chandrakasan, and H. Balakrishnan. Energy-efficient communication protocol for wireless microsensor networks. *IEEE 33rd Annual Hawaii International Conference on System Sciences, Volume 2*, p. 10, 2000.

20 S. Lindsey and C. S. Raghavendra. PEGASIS: power-efficient gathering in sensor information systems. *IEEE Aerospace Conference, Volume 3*, pp. 3-1125–3-1130, 2002.

21 A. Manjeshwar and D. P. Agrawal. TEEN: a routing protocol for enhanced efficiency in wireless sensor networks. *International Parallel and Distributed Processing Symposium, Volume 3*, p. 30189a, 2001.

22 A. Manjeshwar and D. P. Agrawal. APTEEN: a hybrid protocol for efficient routing and comprehensive information retrieval in wireless sensor networks. *IEEE Computer Society, International Parallel and Distributed Processing Symposium, Volume 2,* pp. 195–202, 2002.

23 L. Subramanian and R. H. Katz. An architecture for building self-configurable systems. *1st Annual Workshop on Mobile and Ad Hoc Networking and Computing (MobiHOC),* pp. 63–73, 2000.

24 S. Lindsey, C. S. Raghavendra, and K. Sivalingam. Data gathering in sensor networks using the energy*delay metric. *IEEE Computer Society, International Parallel and Distributed Processing Symposium, Volume 3,* San Francisco, CA, 2001.

25 M. Younisa, M. Youssef, and K. Arisha. Energy-aware routing in cluster-based sensor networks. *10th IEEE/ACM International Symposium on Modeling, Analysis and Simulation of Computer and Telecommunication Systems (MASCOTS 2002),* pp. 129–136, 2002.

26 N. Thepvilojanapong, Y. Tobe, and K. Sezaki. HAR: hierarchy-based anycast routing protocol for wireless sensor networks. *2005 Symposium on Applications and the Internet,* pp. 204–212, 2005.

27 M. Hempel, H.Sharif, and P. Raviraj. HEAR-SN: a new hierarchical energy-aware routing protocol for sensor networks. *38th Annual Hawaii International Conference on System Sciences,* Hawaii, USA, 2005.

28 H. S. Kim and K. J. Han. A power efficient routing protocol based on balanced tree in wireless sensor networks. *1st International Conference on Distributed Frameworks for Multimedia Applications,* pp. 138–143, 2005.

29 Y. B. Ko and N. H. Vaidya. Location-Aided Routing (LAR) in mobile ad hoc networks. *Wireless Networks,* **6**(4):307–321, 2000.

30 X. Ya, J. Heidemann, and D. Estrin. Geography-informed energy conservation for ad hoc routing. *7th Annual International Conference on Mobile Computing and Networking, MobiCom '01,* pp. 70–84, New York, NY, USA, ACM, 2001.

31 Y. Yu, R. Govindan, and D. Estrin. Geographical and energy aware routing: a recursive data dissemination protocol for wireless sensor networks. UCLA Computer Science Department Technical Report ucla/csd-tr-01-0023, 2001.

32 X. Cheng, A. Thaeler, G. Xue, and D. Chen. TPS: a time-based positioning scheme for outdoor wireless sensor networks. *INFOCOM 2004, 23rd Annual Joint Conference of the IEEE Computer and Communications Societies, Volume 4,* pp. 2685–2696, 2004.

33 F. Ye, H. Luo, J. Cheng, S. Lu, and L. Zhang. A two-tier data dissemination model for large-scale wireless sensor networks. *8th Annual International Conference on Mobile Computing and Networking,* pp. 148–159, ACM, 2002.

34 T. He, J. Stankovic, C. Lu, and T. Abdelzaher. SPEED: a stateless protocol for real-time communication in sensor networks. *IEEE 23rd International Conference on Distributed Computing Systems,* pp. 46–55, 2003.

35 D. Ganesan, R. Govindan, S. Shenker, and D. Estrin. Highly-resilient, energy-efficient multipath routing in wireless sensor networks, ACM SIGMOBILE. *Mobile Computing and Communications Review,* **5**(4):11–25, 2001.

36 Y. M. Lu and V. W. S. Wong. An energy-efficient multipath routing protocol for wireless sensor networks. *International Journal of Communication Systems,* **20**(7): 747–766, 2007.

37 E. Ahvar, R. Serral-Gracià, E. Marín-Tordera, X. Masip-Bruin, and M. Yannuzzi. EQR: a new energy-aware query-based routing protocol for wireless sensor networks. *Wired/Wireless Internet Communication*, pp. 102–113, Springer, 2012.

38 M. R. Pearlman and Z. J. Haas. Improving the performance of query-based routing protocols through diversity injection. *Wireless Communications and Networking Conference (WCNC 1999), Volume 3*, pp. 1548–1552, 1999.

39 J. Kulik, W. Heinzelman, and H. Balakrishnan. Negotiation-based protocols for disseminating information in wireless sensor networks. *Wireless Networks*, **8**(2/3): 169–185, 2002.

40 A. Okunoye and O. B. Longe. Enhanced negotiation-based routing in wireless sensor networks. *Computing, Information Systems & Development Informatics*, **3**(4):66–78, 2012.

41 L. J. García Villalba, A. L. Sandoval Orozco, A. T. Cabrera, and C. J. Barenco Abbas. Routing protocols in wireless sensor networks. *Sensors*, **9**(11):8399–8421, 2009.

42 R. Sumathi and M. G. Srinivas. A survey of QoS based routing protocols for wireless sensor networks. *Journal of Information Processing Systems*, **8**(4): 589–602, 2012.

43 B. Bhuyan, H. K. Deva Sarma, N. Sarma, A. Kar, and R. Mall. Quality of service (QoS) provisions in wireless sensor networks and related challenges. *Wireless Sensor Network*, **2**(11):861–868, 2010.

44 K. Sohrab, J. Gao, V. Ailawadh, and G. J. Pottie. Protocols for self-organization of a wireless sensor network. *IEEE Personal Communications*, **7**(5):16–27, 2000.

45 X. Huang and Y. Fang. Multiconstrained QoS multipath routing in wireless sensor networks. *Wireless Networks*, **14**(4):465–478, 2008.

46 A. B. Bagula and K. G. Mazandu. Energy constrained multipath routing in wireless sensor networks. *Ubiquitous Intelligence and Computing*, pp. 453–467, Springer, 2008.

47 E. Felemban, C. G. Lee, and E. Ekici. MMSPEED: Multipath Multi-SPEED Protocol for QoS guarantee of reliability and timeliness in wireless sensor networks. *IEEE Transactions on Mobile Computing*, **5**(6):738–754, 2006.

48 Y. Zhang and M. Fromherz. Message-initiated constraint-based routing for wireless ad-hoc sensor networks. *1st IEEE Consumer Communications and Networking Conference*, pp. 648–650, 2004.

49 J. Ben-Othman and B. Yahya. Energy efficient and QoS based routing protocol for wireless sensor networks. *Journal of Parallel and Distributed Computing*, **70**(8): 849–857, 2010.

50 N. Saxena, A. Roy, and J. Shin. QuESt: a QoS-based energy efficient sensor routing protocol. *Wireless Communications and Mobile Computing*, **9**(3): 417–426, 2009.

51 B. Deb, S. Bhatnagar, and B. Nath. ReInForM: reliable information forwarding using multiple paths in sensor networks. *28th Annual IEEE International Conference on Local Computer Networks LCN'03*, pp. 406–415, 2003.

52 P. Ji, C. Wu, Y. Zhang, and Z. Jia. DAST: a QoS-aware routing protocol for wireless sensor networks. *IEEE International Conference on Embedded Software and Systems Symposia ICESS Symposia'08*, pp. 259–264, 2008.

53 H. Alwan and A. Agarwal. MQoSR: a multiobjective QoS routing protocol for wireless sensor networks. *ISRN Sensor Networks*, 2013:**1–12**, 2013.

Abbreviations/Terminologies

AC   Authentication Confidentiality
ACQUIRE   Active Query Forwarding in Sensor Networks
APTEEN   Adaptive Threshold-Sensitive Energy-Efficient Sensor Network Protocol
ASIC   Application-Specific Integrated Circuit
BATR   Balanced Aggregation Tree Routing
CADR   Constrained Anisotropic Diffusion Routing
CDMA   Code Division Multiple Access
DARPA   Defense Advanced Research Projects Agency
DAST   Directed Alternative Spanning Tree
DoS   Denial of Service
ECMP   Energy Constrained Multipath Routing Protocol
EQR   Energy-Aware Query-Based Routing
EQSR   Energy-Efficient and QoS-Aware Multipath-Based Routing
GAF   Geographic Adaptive Fidelity
GEAR   Geographical and Energy-Aware Routing
GPS   Global Positioning System
HART   Highway Addressable Remote Transducer Protocol
HAR   Hierarchy-Based Anycast Routing
ISA   International Society of Automation
LAN   Local Area Network
LAR   Location-Aided Routing
LEACH   Low-Energy Adaptive Clustering Hierarchy
MCBR   Message-Initiated Constrained-Based Routing
MCFA   Minimum-Cost Forwarding Algorithm
MCMP   Multiconstrained QoS Multipath Routing
MECN   Minimum-Energy Communication Network
MEMS   Microelectromechanical System
MIT   Massachusetts Institute of Technology
MMSPEED   Multipath and Multispeed Routing
MQoSR   Multiobjective QoS Routing
PEGASIS   Power-Efficient Gathering in Sensor Information Systems
QoS   Quality of Service
QuESt   QoS-Based Energy-Efficient Sensor routing
ReInForM   Reliable Information Forwarding Using Multipath
SAR   Sequential Assignment Routing
SoC   System on Chip
SOP   Self-Organizing Protocol
SPEAR   Sensor Protocol for Energy-Aware Routing
SPIN   Sensor Protocol for Information via Negotiation
SPINS   Security Protocol for Sensor Network
TDMA   Time Division Multiple Access
TEEN   Threshold-Sensitive Energy-Efficient Sensor Network Protocol
TPS   Time-Based Positioning Scheme
TTDD   Two-Tier Data Dissemination
TTL   Time-to-Live

UAV   Unmanned Aerial Vehicle
VGA   Virtual Grid Architecture Routing
WSN   Wireless Sensor Network

Questions

1   Mention the typical characteristics of a WSN.

2   Make a list of at least 20 applications of wireless sensor networks. Thereafter, group these applications into various fields such as medical, industrial, defense, vehicular, home, environmental, and any other application.

3   Draw a block diagram of a wireless sensor node, clearly showing all the components. Explain the role of each component in the sensor node.

4   Explain the various networking topologies in a WSN.

5   Describe the various classes of protocols in a WSN.

6   State the requirements and the characteristics of a self-healing sensor network.

7   Differentiate between flat routing and hierarchical routing, clearly mentioning their advantages and disadvantages.

8   Differentiate between rumor routing and gossiping.

9   Why is location-based routing required? Mention its advantages.

10   Explain the main stages in a multipath routing protocol.

11   State the requirements of a negotiation-based routing.

12   Mention the names of any two routing protocols from each of the following classes of routing:
    A flat routing,
    B hierarchical routing,
    C location-based routing,
    D multipath routing,
    E query-based routing,
    F QoS routing.

13   Explain in detail, with an example, the working of the following protocols:
    A SPIN,
    B LEACH,
    C PEGASIS,
    D flooding,
    E direct diffusion.

14 State whether the following statements are true or false and give reasons for the answer:

A The size of a sensor node should be less than 1 cm.

B A sensor node should essentially have a receiver and a transmitter.

C Event-driven routing in a WSN is a source-based routing.

D The parameters for QoS routing in a WSN are hop count and bandwidth.

E Even in negotiation-based routing, some data may be sent directly to the neighbor without any negotiation.

F Query-based routing can discover multiple paths between source and destination.

G A node disjoint path may not be link disjoint.

H Location-based routing is completely dependent on a GPS to know the location of the nodes.

I TEEN is a hierarchical routing protocol.

J If a node has battery power but still is not receiving or transmitting data, this indicates that the node has either gone faulty or been captured by an adversary.

15 For the following, mark all options which are true:

A SPIN does not use this type of message:
- advertisement,
- request,
- ack,
- data.

B In a typical hierarchical routing, a group of nodes at the lowest hierarchical level send their data to:
- cluster head,
- sink,
- gateway.
- neighboring nodes.

C Location-based routing can use any of the following for location awareness:
- cluster head,
- GPS,
- local frame of reference,
- position with respect to each other.

D Which is not a stage in multipath routing:
- path discovery,
- leader election,
- distribution of traffic,
- path maintenance.

E Which is not a transmission standard in WSN:
- Bluetooth,
- ZigBee,
- Wireless HART,
- IEEE 802.11.x.

Exercises

1 Consider a wireless sensor network as indicated in the figure below where the links indicate connectivity between the nodes. The routing in the WSN is based on flooding and a packet has to be transmitted (a) from node 11 to node 16, (b) from node 2 to node 1, (c) from node 8 to node 10, and (d) from node 4 to node 16. How many copies of the packet will be generated in the network for each of the above?

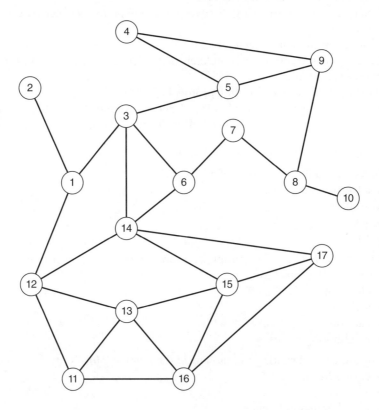

2 Sensor nodes have been fitted on deers in a forest. Each sensor node has a variety of sensors available in it that can detect position (GPS), acceleration/speed, humidity, body temperature, and atmospheric temperature. The sensor nodes form a part of a WSN that has rumor routing implemented in it. It is planned to locate the deer in the forest that have high body temperature owing to heavy rainfall. Make necessary assumptions and make a representative diagram of this WSN. Indicate the contents of any agent at three different nodes. Indicate the flow of packets, the contents of the event table, and the contents of the agent for the process to gain locations of the deers with high body temperature due to heavy rainfall.

3 A drawback of SPIN is that it 'does not guarantee delivery of data between any two nodes because, if the intermediate nodes are not interested in the data, the data cannot cross over the region of uninterested nodes and reach the destination that is

interested in the data'. Draw an indicative WSN and mark a sink node and a node generating data of interest. Place the other sensor nodes accordingly to prove the correctness of the statement above and explain the process with 'advertisement', 'request', and 'data' packets moving in the WSN.

4 A WSN has to implement hierarchical routing. There are 10 000 sensor nodes in the WSN. Each cluster head has resources to communicate with a maximum of ten nodes. Why does a multilevel hierarchy have to be implemented and why will it not be possible to establish a two-level hierarchy? How many levels of hierarchy will have to be formed, and what will be the minimum number of cluster heads that have to be formed at each level? What will be the maximum number of hops required for transmission of a packet between any two sensor nodes?

5 Formulate an equation to determine the threshold value for election as cluster head in LEACH. Necessary parameters may be taken as required. Prove that the value equated through this equation is not skewed.

6 Assume that the network indicated in exercise 2 has TEEN implemented over it. The hard threshold for temperature has been set as 103 °F, and the soft threshold is 2 °F. The following temperatures in °F are being detected by the temperature sensor in this sequence – 99, 102, 100, 103, 100, 104, 105, 106, 103, 105, 106, 107, 108, 104, 102, 101. Please indicate the temperatures that will be transmitted by the WSN to the sink and in the same sequence in which they will be transmitted.

7 Consider the WSN indicated in exercise 1. Mark two node disjoint paths between node 7 and node 12. Can two paths between any two nodes be detected that are node disjoint but not link disjoint?

8 Assume a WSN as indicated in exercise 1 that uses TTL to avoid routing loops. This particular WSN is extremely dynamic. Bearing in mind the present arrangements of the nodes and the fact that the sensors have high mobility, what should be the minimum value of TTL?

12

Routing in 6LoWPAN

12.1 Introduction

IPv6 over Low-Power Wireless Personal Area Networks (6LoWPAN) is a protocol defining flow of Internet Protocol Version 6 (IPv6) packets over a resource-constrained, low-power, and short-range wireless network. Low-power wireless personal area networks are based on the IEEE 802.15.4 standard. IEEE 802.15.4-compliant devices have short communication range and limited bandwidth, memory, processing power, and energy requirements. Based on these restrictions, it is generally assumed that 6LoWPAN transmits a small amount of data, but the protocol has no restriction on the amount of data that it can transmit [1]. These networks are also referred to as low-power and lossy networks (LLNs), as the nodes are highly constrained and the characteristics of the network, radio links, and nodes lead to unstable connectivity [2]. 6LoWPAN aims at extreme network scalability and providing IP-based connectivity even to the smallest of sensors and actuators that have a requirement to communicate among each other or with the external world over the Internet.

Network Routing: Fundamentals, Applications, and Emerging Technologies, First Edition.
Sudip Misra and Sumit Goswami.
© 2017 John Wiley & Sons Ltd. Published 2017 by John Wiley & Sons Ltd.
Companion website: www.wiley.com/go/misra2204

12.1.1 IP for Smart Objects

Smart objects are devices fitted with sensors or actuators [2] and have a communication capability. Smart objects have many emerging applications in the field of home automation, hospital and patient management, vehicular tracking, industrial monitoring, data collection, and warning systems for hazardous areas and related infrastructure and defense usages. Depending on the application, the sensors may provide real-time data or transmit data on availability of a particular threshold or after a specified interval of time. The time interval between data transmission may span from a few minutes to a few days, as the smart objects have to run on the same battery for a few years depending on their application and communication pattern and protocols. The communication with the smart objects can be unidirectional, as in case of RFID tags, or bidirectional, as in the case of sensors communicating using Bluetooth, ZigBee, or 6LoWPAN.

Most of the sensor networks are based on proprietary protocols, which hampers internetwork end-to-end communication. In order to enable such proprietary networks to be a part of other networks, a complex protocol translation gateway model is required. Such gateways lead to communication delays, inefficient routing, and poor quality of service (QoS). For sensor networks comprising thousands of sensor nodes, an IP-based network architecture [3] can provide an efficient, interoperable, and scalable communication solution with a greater commercial adoption than the proprietary counterparts.

The benefits of using IP over sensors can be established on the basis of the widespread use and popularity of the Ethernet. The amount of research going into it has enhanced its speed from 10 Mbps during inception to the present-day few hundred Gbps, and that too in such a short span of time of its development. Furthermore, the wireless link based on IEEE 802.11 is also IP based [4]. IP-based communication is highly scalable and flexible, ensures QoS, and has high availability and reliability. It supports many applications, protocols, communication technologies, and devices. IP is extensively interoperable and has established security mechanisms (using access control, firewall, authentication, and security models) and naming, addressing, translation, lookup, and discovery protocols. It has an established proxy mechanism with network address translation (NAT), load balancing, and caching for higher-level services [5].

Until recently it was thought that IP-based applications are memory and bandwidth intensive [6] and cannot be scaled down in terms of memory, computation requirement, and power consumption to make them appropriate for resource-constrained microcontrollers, embedded systems, sensors, and low-power links, as in the case of IEEE 802.15.4, which has a smaller packet size. However, lightweight and compressed versions of IP-based transmission were suggested in the IETF 6LoWPAN draft standard. This makes IPv6-based data transmission possible among smart devices that have low memory and processing power and support a smaller frame size. IP-based communication can enable seamless and secured data transmission between sensors and PCs, laptops, and PDAs across various networks or the Internet, which leads to ease of integration and a wider network interoperability.

There are a number of advantages of using IP in smart objects. A few of these advantages are that IP is open, lightweight, versatile, ubiquitous, scalable, manageable, stable, and end-to-end [3]. IP masks the lower-layer details such as packet size and format and sends information to the destination after locating it and routing the information through intermediate accessible hosts transparently to the user [7]. A few other benefits of using IP technology are its pervasive nature and its integration with existing IP-based applications using the existing network infrastructure without the requirement of any translation gateways [1]. It has been in use for the past many years [8], with fine-tuning of the protocols to make it secure and flawless. IP-based diagnostic tools and management and monitoring tools such as Ping, traceroute, SNMP, openview, and netmanager are also available [5]. However, these tools should be robust enough to manage a dense deployment of nodes, but should not have much overhead. Another important factor for the success and widespread usage of IP is that the standard is open [1] and the specifications are easily available and free, enabling the users to have an insight into it and to develop the supporting applications over it.

IPv6 over IEEE 802.15.4

The reasons for suitability [1] of IPv6 over IEEE 802.15.4 for usage in 6LoWPAN are as follows:

- IPv6 has a huge address space available with it, which can cater for the higher-density addressing requirements of the sensors and embedded systems.
- IPv6 address format can support the IEEE 802.15.4 addressing scheme and its size limitation with due compression.
- It supports autoconfiguration of the network.
- Owing to its huge addressing range, NAT is not required, saving on the overhead of NAT.
- Using IPv6, the sensors can connect to the Internet through a gateway and are open to new techniques [9] such as location aware addressing.

12.1.2   6LoWPAN

IEEE 802.15.4 has the capability to network intelligent devices fitted with sensors that can communicate with each other. This network of sensors can have a wider range of applications if it has the capability to be connected to the existing IP-based local area networks (LANs), wireless networks, wide-area networks (WANs), or the Internet. It will connect homes, factories, warehouses, bridges, malls, and other infrastructures with the real-world applications available on the network. Low-range wireless personal area networks (LR-PANs) have very low battery requirements, leading to a battery life of a few years, but at the same time they have relatively less throughput and a short range of communication. In addition to this, the sensors have constrained memory space and computational power.

ZigBee technology fits best for those applications that can do with low bandwidth availability, but it requires support for mobility, short-range wireless communication, and self-organization. The power consumption to run ZigBee protocol on a sensor is low, which leads to longer battery life. It was mainly used to support distributed automation, remote control, and monitoring applications over a range of less than a 100 m. As ZigBee works on a low data rate, it led to the formation of the IEEE 802.15.4 committee to work on a standard for such data rates. An alliance of various companies either working or interested in the area of ZigBee were also working under the 'Zigbee Alliance'. The ZigBee Alliance and the IEEE joined together for further development of the standard, and the technology was commercially named 'ZigBee' [10].

The IEEE 802.15.4 standard is very less complex than Bluetooth owing to the reduced quality of service (QoS) [11]. IEEE 802.15.4 defines only 45 MAC primitives and 14 PHY primitives [12, 13]. IEEE 802.15.4 is restricted only to the PHY and MAC layers and does not mention any criteria for the upper layers such as network or application layers [14]. Some manufacturers have tried building their proprietary systems for sensors and control devices. However, owing to variation in the standard, interoperability is always a major problem [15]. A common network cannot be created without interoperability between all kinds of sensor and control device across network layer.

IETF established the 6LoWPAN Working Group in 2004 to use IPv6 in formulating a standard for LR-WPAN. The motive of this group was to use IEEE 802.15.4 at the bottom layers and introduce IPv6 at the network layer of LR-WPAN. This relies on the compression of the IPv6 packets to make them suitable for battery-, memory-, and throughput-restrained LR-WPAN-compliant devices [16]. 6LoWPAN uses the AES 128 bit encryption supported by IEEE 802.15.4 for authentication and security. It can also implement any other IP-based advanced network security schemes and can directly integrate with other networks using IP routers. 6LoWPAN may be based on IEEE 802.15.4-2003, or it may be compatible with IEEE 802.15.4-2006 or TG4e [17].

12.1.3 ZigBee

The IEEE 802.15.4 standard mentions the physical and MAC layer issues and protocols in a resource-constrained wireless personal area network (PAN.) The PAN can be based on a star, mesh, or cluster tree topology [10]. ZigBee addresses the upper-layer protocols and application profiles such as routing, data security, monitoring applications, and interoperable application profiles. In the ZigBee network, a full function device (FFD) can play the role of a PAN coordinator, ZigBee router, or end device. A reduced function device (RFD) can take on the functionality of only an end device. The FFDs can form a peer-to-peer network in a mesh topology. The RFDs form a star network with an FFD in a master–slave configuration [18]. There is only one PAN coordinator in a ZigBee network, which allows the devices in the network to communicate with each other using multihop transmissions. The other roles performed by the PAN coordinator are setting up the network, transmitting beacon messages, routing messages, and storing node information. ZigBee supports a self-healing, self-forming network based on CSMA-CA for data transmission. A ZigBee node spends most of its time snoozing, and the bootstrapping time for a hibernating node is 15 ms [12].

12.1.4 ZigBee vs 6LoWPAN

A few major differences between Zigbee and 6LoWPAN are listed below:

- 6LoWPAN is an alternative to ZigBee and is an open network based on Internet Protocol. 6LoWPAN already has an open-source version and provides easily implementable models for Internet connectivity [19].
- ZigBee standardizes application profiles but still has a few drawbacks in networking, while 6LoWPAN has better networking with IPv6, but lacks application standard interoperability [8].
- The network header in 6LoWPAN comprises dispatch and header compression HC1 and HC2, followed by hop count and UDP header, while the ZigBee APDU frame

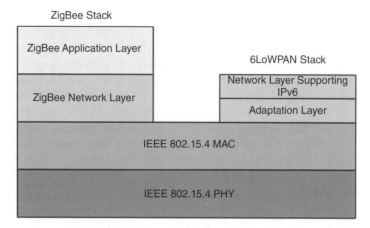

Figure 12.1 A schematic representation of ZigBee and 6LoWPAN stacks.

format [5] comprises frame control bit fields, destination endpoint, cluster identifier, profile identifier, source endpoint, and APS counter in its network header.

• As reported in a survey [8], among the shipped IEEE 802.15.4 chips, about 55% support 6LoWPAN, while only 30% support ZigBee.

Figure 12.1 shows the difference between the ZigBee and 6LoWPAN stacks.

12.2   6LoWPAN Fundamentals

6LoWPAN has attractive features of autoconfiguration, IPv6 fragmentation, mesh forwarding, mobility support, and up to 80% header compression. It can support various types of radio link, such as IEEE 802.15.4, IEEE 802.15.4a, and even a low-bandwidth powerline [8]. The major characteristics of LoWPAN are its small packet size, provision for 16 bit as well as 64 bit MAC address, low bandwidth as defined in ZigBee radio, support of star and mesh topology, low power consumption, low cost, tremendous scalability, mobile and random deployment, and secured network connectivity even with non-cooperative and compromised devices. As the physical layer can support a maximum packet size of 127 bytes, this leads to a maximum frame size of 102 bytes in the MAC layer. The security protocols available in IEEE 802.15.4 are AES-CCM-32, AES-CCM-64, and AES-CCM-128, which have headers of 9, 13, and 21 bytes respectively. This finally leaves 81 bytes for the data [1]. IEEE 802.15.4 supports two types of addressing – 64 bit and 16 bit addressing. The 16 bit address is allocated to a device only after its association with a PAN coordinator [20] or other nodes in the PAN, and the address is unique only within that particular PAN.

The design space of 6LoWPAN is varied [2]. The nodes can be deployed in an organized manner or scattered, and they can be deployed all at one time or incrementally, making it highly scalable. The network may comprise a few nodes to several thousand nodes. This leads to varying density of nodes and change in their locations. The nodes in the 6LoWPAN are 'always connected', but certain nodes may be under hibernation while others may come in the transmission range or leave the signal availability region owing to their mobility. This leads to difficulty in determining the active nodes in the network. It supports star as well as mesh topology. A single hop may be sufficient for communication in a star topology, while multihop data transmission with routing capabilities is required for a mesh or hierarchical topology. Researchers have reported various routing mechanisms suitable for LoWPAN, some of which are localization based, reputation based, topographical, data centric, address centric or event driven. The data transmission pattern may be point to point (P2P), point to multipoint (P2MP), or multipoint to point (MP2P), depending on the usage, architecture, and routing pattern. If the LoWPAN is assigned for a strategic application, security considerations and QoS have to be ensured in the resource-constrained lossy network.

12.2.1   Architecture

IEEE 802.15.4 defines two types of node – FFD and the RFD. However, in the case of LoWPAN the nodes are distinguished as LoWPAN hosts, LoWPAN routers, and LoWPAN mesh nodes [2]. The LoWPAN hosts are the end nodes, which are the source or sink nodes for IPv6 datagrams and can comprise FFDs or RFDs. The LoWPAN routers and the LoWPAN mesh nodes are FFDs and help in data transmission between the source and destination nodes by appropriate forwarding of data in multihop mode. The LoWPAN routers operate in the IP layer and perform routing. The edge routers connect the LoWPAN to other IP-based networks or other LoWPAN networks. The mesh nodes operate on top of the link layer for the forwarding of data using link addresses. The 6LoWPAN stack [21] has the 6LoWPAN-specific applications at the top, using the socket interfaces to connect to the TCP/UDP at the next layer, followed by IPv6 and the adaptation layer just below it, which makes 6LoWPAN utilize the IP over IEEE 802.15.4. The last two layers are those defined in IEEE 802.15.4 and are its link layer and physical layer.

Gateways are devices that have the IPv6 network and the IEEE 802.15.4 network at their two ends. Each PAN can have its own gateway or multiple PANs can have a common gateway [20]. The gateway implements the adaptation layer function [22]. The gateways also perform address translation between IPv6 and the 16 bit/64 bit address to make the addresses compliant with the IPv6 network and the IEEE 802.15.4 network. A typical network diagram showing the network connectivity of a 6LoWPAN with the Internet through a gateway is shown in Figure 12.2.

12.2.2   Header Format and Compression

Although 6LoWPAN has no limitation on the packet size that a sensor can transmit, given that it is bandwidth, energy, and processing power constrained, the LoWPAN applications generally generate small packets. By adding headers across all layers to establish IP connectivity, it should confine itself in a single frame [1]. The control packet should also fit within a single IEEE 802.15.4 frame.

IPv6 packets are transmitted over the data frames of IEEE 802.15.4 with a request for acknowledgement to help link layer recovery [23]. The other frames supported by IEEE 802.15.4 are the beacon frames, the MAC command frames, and the acknowledgement frame. The source and destination addresses should be included in the IEEE 802.15.4 frame header, with the optional inclusion of the source and destination PAN ID. The 16 bit address, which is unique within the PAN, is used for addressing within the

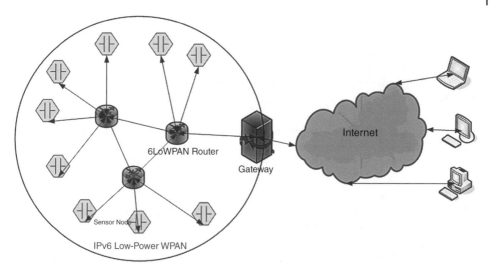

Figure 12.2 Interoperability of 6LoWPAN over the Internet.

6LoWPAN and is valid until its association with the PAN coordinator. It is assumed that each PAN maps to a particular IPv6 link, and the PAN ID of this link should match the destination PAN ID included in the frame [22]. The 16 bit destination address is included in the frame to support broadcast.

An IPv6 packet cannot fit in a IEEE 802.15.4 frame because the MTU of an IPv6 packet is 1280 bytes. IEEE 802.15.4 can support a maximum frame size of 127 bytes in the physical layer with a frame overhead of 25 bytes and 102 bytes for the MAC layer. The link layer security has an overhead of 21, 13, or 9 bytes in the case of AES-CCM-128, AES-CCM-32, and AES-CCM-16 respectively. The IPv6 header uses 40 bytes [5], thus leaving 41 bytes for the upper layer if AES-CCM-128 is used. Of these 41 bytes, some bytes are used by the upper-layer protocol, leaving even fewer bytes for the application data. The minimum packet size in IPv6 is 1280 bytes, and thus an intermediate adaptation layer [21] is required to support this. The primary functions of the adaptation layer [9] are TCP/IP header compression, handle packet fragmentation and reassembly, support routing in edge node, neighbor discovery, and multicast support.

The datagrams of 6LoWPAN are transmitted over IEEE 802.15.4 radio. These encapsulated datagrams are prefixed with an encapsulation header. An IPv6 header contains source and destination addressing, hop-by-hop options, routing information, fragmentation, destination options, and payload [24]. The LoWPAN header comprises mesh addressing, hop-by-hop options, fragmentation, and payload. RFC 4944 [22] describes the frame format of LoWPAN-encapsulated IPv6 datagrams, LoWPAN-encapsulated LoWPAN_HC1-compressed IPv6 datagrams, LoWPAN-encapsulated LoWPAN_HC1-compressed IPv6 datagrams for mesh addressing, LoWPAN-encapsulated LoWPAN_HC1-compressed IPv6 datagrams for fragmentation, LoWPAN-encapsulated LoWPAN_HC1-compressed IPv6 datagrams for mesh addressing as well as fragmentation, and LoWPAN-encapsulated LoWPAN_HC1-compressed IPv6 datagrams for mesh addressing, and the broadcast header to support mesh broadcast/multicast. When multiple LoWPAN headers are present in the same packet, the mesh addressing header should appear first, followed by the broadcast header and then the fragmentation header. In addition to these, the header contains dispatch value, definition of header fields, and their ordering constraints.

In the dispatch header, the first two bits are set to 00 or 01, which identifies the dispatch type. The dispatch header is 6 bit and indicates the type of header that follows it. The header types can be NALP (Not a LoWPAN frame), IPv6, LoWPAN_HC1, LoWPAN_BC0, and ESC. Thus, the type of header is specified by the dispatch header [20]. In the mesh addressing header, the first two bits are 10, followed by a bit each for V (Very First/Source) and F (Final Destination). V is 0 if the source address is a 64 bit address, and 1 if it is a 16 bit address. F is 0 if the destination address is a 64 bit address, and 1 if it is a 16 bit address. Thereafter it has four bits for 'hops left', followed by the originator address and the final destination address.

The entire IPv6 payload may fit in a single IEEE 802.15.4 frame, and in this case it does not require any fragmentation. But if the payload does not fit in a single frame, it is split into link fragments [5] of multiples of eight bytes each, except for the last fragment. The first link fragment comprises an 11 bit datagram size and a 16 bit datagram tag, and the remaining fragments have an 8 bit datagram offset in addition to the datagram size and datagram tag. The datagram size specifies the size of the IP packet before fragmentation, and its value remains the same across all the frame fragments. The datagram tag has an incremental value for successive fragments and is set by the sender. Although the initial value is not defined, it wraps to 0 when the value reaches 65 535.

The recipient of the link fragments uses the source address, destination address, datagram size, and datagram tag to identify the fragments that are addressed to it and belong to a particular datagram. Thereafter it reconstructs the entire datagram from the fragments by getting the size from the datagram size tag and datagram offset to determine the location of each fragment in the datagram [22]. The datagram tag supports reassembly in the case of node disassociation.

A number of IPv6 header values are common in the 6LoWPAN networks, and these can be used efficiently to compress them and construct the HC1 header. In the IPv6 header in 6LoWPAN, IPv6 is the version, the source and destination address are link local, the interface identifiers for source and destination can be obtained from the link layer source and destination address, the packet length can be obtained from the datagram size field in the fragment header or from the frame length field in the link layer, the traffic class and the flow label are 0, and the next header is UDP, ICMP, or TCP. Hop limit is a field that cannot be compressed. The common IPv6 header can be compressed into 16 bits instead of the 40 bytes used in IPv6 [22]. Out of the 16 bits, eight bits are used for HC1 encoding, and the remaining eight bits for the hop count. In HC1 encoding [5, 17], the first two bits encode the source address, the next two bits encode the destination address, the fifth bit specifies the traffic class and flow label, the sixth and seventh bits specify the next header, and the last bit is set to 1 if HC2 encoding follows the HC1 encoding and is set to 0 if there is no further header compression after HC1. The two bits used to identify the next header point to no compression, UDP, ICMP, and TCP with a value of 00, 01, 10, and 11 respectively. The traffic class and flow label is 0 if the bit is set to 1, and there is no compression if it is 0. The two bits help in determining the source and destination address as each bit specifies whether it is prefix compressed or interface identifier compressed.

12.2.3 Network Topology

The IEEE 802.15.4 supports two types of device – FFD and RFD. An FFD can send as well as receive data from an FFD as well as an RFD. However, an RFD can communicate only with an FFD. These limitations lead to two types of topology in LoWPAN – mesh and star. FFDs also forward the packets at the link layer. The routing protocols for 6LoWPAN should have less overhead on the packet size during routing, preferably independently of the number of hops and topological changes [1]. The sensor nodes are generally deployed in excessive numbers with restrained input capability and are ad hoc in nature, leading to location ambiguity. The device should be capable of bootstrapping easily, and the network should be self-healing in nature to address such uncertainty of the deployed devices. The routing algorithms should be optimized for minimal power consumption, computational requirement, and memory usage. The constrained memory availability limits the maintenance of routing tables in the FFDs. As the RFDs hibernate frequently to save on battery power, the routing protocol should have techniques incorporated to support hibernating nodes.

A LoWPAN can be formed as a 'mesh under' or 'route over' configuration [2]. In the 'mesh under' configuration the data transmission is through the intermediate mesh nodes, which operate at the link layer or the adaptation layer and use the link address as the identifier for forwarding the data. In the 'route over' configuration, the packet is transmitted in multiple hops using IP through the intermediate FFDs, which act as the routers.

12.2.4 Neighbor Discovery

The 6LoWPAN has lossy, short-range, low-bandwidth, and low-power links with sleeping nodes, and thus multicasting cannot be used for neighbor discovery as in the case of IPv6. The process of neighbor discovery in IPv6 has been described in RFC 4861 [25], along with the process of router advertisement and discovery, neighbor advertisement and solicitation, redirection, address resolution, and duplicate address detection. However, this cannot be used for 6LoWPAN as it has two types of addressing scheme – 16 bit addressing and 64 bit addressing, the network does not have multicast capability at the link layer, the sleeping nodes do not respond to signaling messages, a node may periodically change its parent for routing, and the datagram requires fragmentation and header compression.

Various optimizations have to be made to IPv6 neighbor discovery to make it suitable for 6LoWPAN [26]. It avoids multicast flooding, minimizes link-local multicast frequency, uses whiteboard in edge routers for duplicate address detection, duplicate address detection is across the entire 6LoWPAN, and next-hop determination is simplified, but neighbor unreachability detection is not performed, address resolution is not performed, and node registration and whiteboard resolution have to be introduced afresh. Contextual-information-transformation-based service discovery architecture for a 6LoWPAN can be used to locate a service in the proximity of the node [27]. The mechanism works even in a 6LoWPAN Internetworked with an external IP network and can discover user-defined services from within the 6LoWPAN as well as outside it.

When a node comes in an LoWPAN, it listens to the router advertisements from the routers in the network or broadcasts a router solicitation expecting a response from

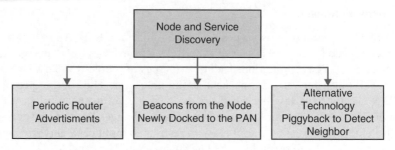

Figure 12.3 Different approaches to node and service discovery in 6LoWPAN.

any local router. On receiving the router advertisement, the node gets an address by stateless address autoconfiguration and selects its default router. Thereafter the node sends a unicast node registration message to the edge router directly or through intermediate routers to register itself with the edge router, to which the edge router replies with a node confirmation either directly or through the intermediate routers and makes an entry of it in its whiteboard. This makes the node a part of the 6LoWPAN and ready for communication [17, 26] to any IPv6 node either within the LoWPAN or outside it. However, intermediate routers are not available in 'mesh under' configuration.

In addition to the process of the router transmitting periodic advertisements of node discovery, there are two other modes of node and service discovery supported by 6LoWPAN. In the second mode of node discovery, a sensor, after coming into the PAN, sends beacons to which a sink node may listen and perform the registration of the node, and the docking node receives a TTL from the sink node and becomes a part of the network. In this case the sink node has information about all the nodes in the network and does not require a method for node removal as it uses TTL. However, it is difficult to calculate an optimal TTL. Alternatively to these two modes, RFID may be used for neighbor discovery as it saves all broadcast messages, but requires integration of RFID with 6LoWPAN. These three types of node and service discovery are described as the YouCatchMe, ICatchYou, and Some1CatchMe approaches respectively [28] and are depicted in Figure 12.3.

12.2.5 Routing

The 6LoWPAN or IEEE 802.15.4 does not specify how the mesh network can be formed with multihop routing. Thus, the multihop routing should either be supported at the IP layer or at the link layer. The design space and 6LoWPAN routing requirements can be discussed with respect to the device properties, link properties, network characteristics, and 'mesh under' forwarding [24]. The existing multihop routing protocols cannot be used in 6LoWPAN as there are various types of node in the 6LoWPAN performing various roles such as routers, gateways, hosts, or edge nodes, and PAN coordinators. Existing routing protocols do not support these 6LoWPAN device types. Moreover, the power, memory, and computational requirements are more stringent in the 6LoWPAN nodes, the existing protocols cannot handle the transmission pattern of the hibernating nodes [29, 30] and lossy links, and a 6LoWPAN network may either be a transit network or a stub network.

6LoWPAN uses a 'route over' approach by IP routing during routing at the IP layer and a 'mesh under' approach by multihop communication using the MAC address during routing below the link layer. In a 'route over' configuration, the edge routers as well as the intermediate nodes perform layer 3 routing for multihop communication. In a 'mesh under' configuration, only an edge router is a IPv6 router and the other nodes are the mesh nodes or the LoWPAN host, which operate at layer 2. This leads to a single IP hop with multiple radio hops [5] if a node wants to transmit a packet from a node to the edge router. The routing packets should fit into a single IEEE 802.15.4 frame to prevent fragmentation and reassembly [22]. As there is heavy resource constraint, the routing protocols should be designed so that the packets are delivered with a certain threshold probability depending on the criticality of the application [23, 31, 32]. The routing protocol should consider the latency requirement of the application and the 6LoWPAN link latency [29]. The routing protocol should support scalability from a few nodes to a few thousand nodes [33] and should have a procedure for route reallocation with minimum energy consumption in control messages for route discovery [34]. The routing protocol should also support the mobile nodes as well as the movement of the entire 6LoWPAN. It should also support point-to-point, point-to-multipoint, and multipoint-to-point transmission patterns with minimal multicast traffic [35]. The control messages should be securely delivered [36]. The protocol should support 16 bit addressing as well as 64 bit addressing.

Owing to memory constraints in 6LoWPAN nodes, a very limited entry routing table can be created with the interface identifiers. So, a hierarchical routing based on the 16 bit address can be designed, which uses the 6LoWPAN message format in the adaptation layer [37]. However, this architecture does not incorporate the feature of path recovery in the case of a node failure. This can be improved upon by an extended hierarchical routing [38], which can have a feature of recovering the routing path. This mechanism adds two new entries to the routing table. The first new entry is called the neighbor replace parent (NRP), which points to an upstream node to which a node can transmit data in the case of failure of the parent. The second new entry is called the neighbor added child (NAC), which the parent node added in the NRP makes for this newly added child.

12.3 Interoperability of 6LoWPAN

With increase in the applications of 6LoWPAN, the requirement for mobility of nodes among various PANs and the interoperability of 6LoWPAN is increasing. A sensor mode may move from one 6LoWPAN network [39] to another, but as the related devices are energy constrained, it is challenging to support such mobility as it involves handoff techniques, authentication, and information exchange and storage. The mobility-related tasks should be delegated to the PAN coordinator or some other FFD in the network, depending on the resources available to the device. The mechanism for supporting mobility should be incorporated in the adaptation layer and not the IP layer, as it reduces the number of bits transmitted over the radio channel by the 6LoWPAN device. The inter-PAN handover time should be minimum, and it should have awareness about the surrounding PANs to enhance the reliability of the visiting node.

A few major research studies reported in the field of 6LoWPAN interoperability are as follows:

- A model for interoperability between NEMO [40] and 6LoWPAN has been suggested with a routing scheme to support the network mobility [9, 41].
- Internetworking between ZigBee and 6LoWPAN [42] using IPv6 prefix delegation, IPv6 multicast, and extended IP switching.
- Use of 6LoWPAN in IPv4 networks [43] by adding two more values in the prototype tag in the adaptation layer for IPv4 and compressed IPv4 and adding for IPv4 and IPv6 a dual stack in the gateway. NAT is incorporated to support IPv4 for the huge number of sensors.

An inter-PAN mobility support mechanism has been suggested [20] by providing a distributed version of the inter-PAN mobility scheme [44]. It assumes that each PAN has a unique PAN ID, and has different communication frequency ranges and a separate gateway that connects it to the external IP network. Gateways are responsible for fragmentation and reassembly of IP packets to make them IEEE 802.15.4 frame size compliant. When a mobile node enters a 6LoWPAN, it detects the frequency range of the PAN and associates itself with any node (parent) in the network, and this information is sent by the parent to the gateway with the 16 bit address of its parent and the fixed ID of the mobile node. The gateway maintains binding information of all the mobile nodes and their parents and updates this information to the other gateways in the domain, including the parent gateway of the mobile node.

The mobile node may disassociate from its parent and may associate with some other node (parent) within the same PAN, thus changing the binding information in its gateway, but such types of update occurring within a PAN are not updated with the other gateways. The mechanism has provision for inter-PAN communication where two mobile nodes in different PANs communicate with each other, as well as interdomain communication, which is used to forward a data packet received at the home network of the PAN to the mobile node where it is presently available. HMIPv6 [45], FMIPv6, and MIPv6 [46] are a few tunnel-based mobility protocols that prevent packet losses during mobility and help in continuity of communication with the outer world during mobility [20]. These are host-based protocols in which the IP layer carries the mobility-related packets [47].

The gateway architecture of 6LoWPAN routing [35] supports interoperability between 6LoWPAN and external IPv6 networks. This is achieved through compressing or decompressing the IPv6 packets and mapping the IEEE 802.15.4 16 bit addresses with the IPv6 address [9].

12.4 Applications

There is an increasing market demand for IEEE 802.15.4 chips. Millions of IEEE 802.15.4 chips are being manufactured and shipped annually, and this value is expected to rise with a growth rate of 20–30% [8]. The 6LoWPAN Working Group has set out in detail in its Internet Draft on Design and Application Spaces for 6LoWPAN its various application domains in the field of industrial monitoring [19], structural monitoring, healthcare, networked home [21, 48], childcare [9], vehicular telematics, agricultural

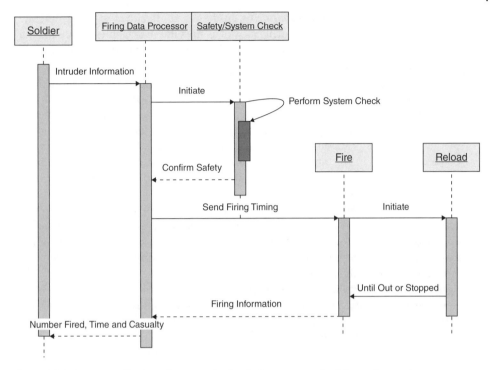

Figure 12.4 Sequence diagram of communication between a smart soldier and a smart weapon system.

monitoring [2], energy metering [8, 49], situational awareness and precision asset tracking for defense or firefighting [50], and environmental monitoring and sensing [20]. A sequence diagram depicting the data communication between a smart soldier and a smart weapon system fitted with a variety of sensors is depicted in Figure 12.4. These communications are in very short range and require much less data transmission. The actual data is not transmitted over the network, rather trigger signals are transmitted on threshold values, and hence it can be supported well by a 6LowPAN. Some of the other applications of 6LoWPAN are discussed below.

a) *Industrial automation.* There are a number of application fields in industrial monitoring to reduce the downtime of machinery and automate the process of equipment monitoring and data transfer and analysis. Industrial monitoring involves various requirements such as process control and monitoring, equipment surveillance, asset tracking, and monitoring of the storage environment for temperature, humidity, and pressure critical stocking requirement.

b) *Strength and security of structures.* Structural monitoring using 6LoWPAN provides real-time monitoring of critical infrastructure or of structures under extreme temperature, pressure, or radiation effects leading to deterioration of the strength of the material and fatigue. Such networks provide efficient, routine, and reliable monitoring of structures. The network can also be used for security monitoring of structures.

c) *Wireless patient and hospital monitoring.* In the case of health care management, it can be effective in patient monitoring as well as hospital management. Patients may be fitted with various sensors forming a wireless body sensor network (WBSN) [20] to track their health status in a non-invasive manner (to track motion, temperature, heart beat, or pulse rate) or in an invasive manner (to track blood content or nervous, digestive, or respiratory system) [51]. This leads to a wireless patient monitoring environment in an intensive care unit of the hospital. In hospital management, it can be used to maintain the temperature and humidity of the storage for blood and medicine. It can alert about depleting pressure in oxygen cylinders, increase in suspended material in the hospital environment as a result of pollution, which can effect patient health, tracking of hospital equipment, and timely consumption of medicine by patients.

d) *Convenient homes.* Smart and convenient homes [9] deploy the technology to integrate smart devices in the home and connect them to the Internet for information exchange. It is used for the surveillance and safety of houses, maintenance of gardens, health monitoring of occupants, and control of home appliances, light and temperature, and entertainment systems.

e) *Agricultural support.* Sensors can be fitted on farms to monitor the temperature, water level, humidity, and soil condition of farmland. Humidity sensors can activate sprinkler actuators to manage the sprinklers to maintain the water level on the farm. The soil sensors can generate an alarm on depletion of minerals in the soil that are specifically required for a particular crop, thus making it easier to decide on the type of fertilizer and its area of spread. The crop condition can also be forwarded to a centralized data processing center, which can analyze the data to predict the crop yield and arrange for its market accordingly.

f) *Management of transportation system.* The sensors can be incorporated in vehicles to track the health of the vehicle. They can be fitted in traffic lights, roadside base stations, highway patrol vehicles, emergency service vehicles such as tow vehicles or ambulances, and approaching petrol pumps. The network provides a variety of safety and entertainment services along with traffic flow optimization and path guidance for traffic and accident avoidance.

g) *Real-time meter reading.* Automated meter reading (AMR) provides real-time meter reading to the consumer and can be utilized to device intelligent applications for real-time consumer information, demand response, and building automation. 6LoWPAN can be used to provide an end-to-end IP-based communication infrastructure between the meter and the power station and the automation system. 6LoWPAN is made available at the meters with gateway routers connecting it to a backbone Ethernet or cellular network [49]. The technology can be used for electric, water, or gas meters for domestic usage or in vehicle meters for vehicular telematics applications.

The home automation network may not have a very critical data transmission requirement demanding a QoS, but the security and surveillance system should ensure authentication, availability, and secrecy. Moreover, when the home grid gets connected to the Internet for data exchange through a 6LoWPAN edge router, which acts as a gateway to the Internet, security considerations become critical, and security protocols should be used at various levels, such as PKI and IPSec, in addition to the AES provided by IEEE 802.15.4. The 6LoWPAN also has a promising role to play in vehicular networks.

12.5   Security Considerations and Research Areas

The applications running over 6LoWPAN should ensure integrity, confidentiality, and authentication. The network should be free from malicious nodes and from active as well as passive attacks, such as denial of service, spoofing, and man-in-the-middle attacks. These network security features can be provided at various layers of the network, starting from the application layer to the physical layer. 6LoWPAN utilizes the AES-based link layer security of IEEE 802.15.4. However, the protocol does not specify any details about key management or security in higher layers. Although IEEE 802.15.4 provides link layer security to 6LoWPAN, it should also address other security aspects such as secured bootstrapping of a node in the network, key establishment, key exchange, and key management [1]. It should also adopt scaled-down versions of TLS, IPSec, SSL, and WEP to make it IP-based sensor network compliant [8]. Various crypto suites should also be scaled down and tested for their suitability in 6LoWPAN.

The security requirement differs with the type of 6LoWPAN application [2]. In safety-critical use, such as structural monitoring, QoS has to be ensured along with secured transmission. The integrity of the data is critical in such cases, as a change in the data can lead to a wrong diagnostic of the strength of the building. In the case of secured home or hospital networks connected to the Internet, data exchange about depleting products and online ordering on the Internet require authentication and non-repudiation. A lighter version of PKI can be used in such cases. Patient monitoring systems should ensure privacy of the patient data and role-based access to the data.

The interface identifier obtained from the MAC address is used to provide a unique identification, but it has no mechanism to prevent its duplication by an attacker. The LoWPAN hosts used in sensing and security applications have more RFDs than FFDs, and these RFDs are resource constrained, implement the minimum security features, and rely on AES encryption and authentication. The issue of secure configuration and management [22] during field installations should be addressed, along with end-to-end security in the case of communication with IPv6 peers.

As 6LoWPAN transmits through radio links, it is vulnerable to spoofing, altered routing, and other attacks faced by a wireless network [35, 52]. The nodes are deployed in the field and are susceptible to tampering, capture, or cloning. As the 6LoWPAN forms a multihop network with lossy links and slow transmission, this leads to delay in detection of attacks. In order to ensure confidentiality, authentication, integrity, and time stamping, time synchronization, authenticated broadcasts, link verification, and secure self-organization of the multihop routing are required, in addition to the MAC layer AES cryptography.

A list of goals that should be accomplished to incorporate the best practice for transmitting IP packets and the related protocols of the upper layers has been documented by IETF in its RFC 4919 [1]. Although advances have been made with respect to some of the goals, the majority of them still require further research:

- A fragmentation and reassembly layer below the IP layer can be developed to synchronize the mismatch in the packet size of IEEE 802.15.4 and IPv6 [53]. The existing protocol leaves far fewer bytes for data transmission and calls for excessive fragmentation and reassembly even in the case of a small data packet.

- The header compression techniques can be more standardized with respect to 6LoWPAN, and if the existing header compression techniques do not meet the requirements, fresh specifications can be framed.
- An interface identifier can be generated for each IEEE 802.15.4 device so as to make stateless address autoconfiguration simpler.
- Routing protocols should be tailor made for 6LoWPAN so as to enable the routing packet to be of an appropriate size to fit in a IEEE 802.15.4 frame [52].
- Network management protocols should be developed that have less overhead but can support a densely populated network. SNMPv3 or a mutation of it should be rigorously tested for its effectiveness of usage in 6LoWPAN.
- The heavyweight application protocols, such as SOAP, HTML, HTTP, XML, and REST, can be compactly encoded [54] to make them suitable for transmission over 6LoWPAN.

References

1 N. Kushalnagar, G. Montenegro, and C. Schumacher. IPv6 over low-power Wireless Personal Area Networks (6LoWPANs): overview, assumptions, problem statement, and goals, Request for Comments: 4919.

2 E. Kim, N. Chevrollier, D. Kaspar, and J. P. Vasseur. 6LoWPAN Working Group Internet-draft, Design and Application Spaces for 6LoWPANs, Version 4, 1 October 2012.

3 A. Dunkels and J. P. Vasseur. White Paper No. 1: IP for smart objects, Internet Protocol for Smart Objects (IPSO) Alliance, September 2008. www.ipso-alliance.org.

4 D. Culler. Secure, low-power, IP-based connectivity with IEEE 802.15.4 wireless networks, industrial embedded systems, 2007. http://www.archrock.com/downloads/resources/ArchRock.Sum07.pdf

5 D. E. Culler and J. Hui. IP on IEEE 802.15.4 low-power wireless networks. http://www.cs.berkeley.edu/~jwhui/6lowpan/6LoWPAN-tutorial.pdf.

6 J. W. Hui and D. E. Culler. Extending IP to low-power, wireless personal area networks. *IEEE Internet Computing*, **12**(4):37–45, 2008.

7 R. Braden. Requirements for Internet Hosts – Communication Layers, 1989, *RFC 1122*.

8 Z. Shelby. What future for 6LoWPAN?, October 2008, ftp://ftp.cordis.europa.eu/pub/fp7/ict/docs/necs/20081022-wsnnco-04-zshelby_en.pdf.

9 C. S. Hong. Interworking between sensor networks and IPv6 network. http://ngn.cnu.ac.kr/core2006/core_2006_ppt/ChoongseonHong.pdf.

10 S. C. Ergen. ZigBee/IEEE 802.15.4 summary, 2004. http://www.sinemergen.com/zigbee.pdf.

11 ZigBee Alliance. www.ZigBee.org.

12 M. Othman. *Principles of Mobile Computing and Communication*, pp. 83–106, Auerbach Publications, New York, 2007.

13 LAN-MAN Standards Committee of the IEEE Computer Society, Wireless medium access control (MAC) and physical layer (PHY) specifications for low-rate wireless personal area networks (LR-WPANs), IEEE, 2003.

14 R. Poor. Information on the IEEE 802.15.4-2006 revised standard. http://grouper.ieee.org/groups/802/15/pub/TG4b.html.

15 D. E. Culler. Embedded web services and industrial instrumentation standards. http://www.eecs.berkeley.edu/~culler/WEI/lectures/L11-embedded-web.ppt.

16 X. Ma and W. Luo. The analysis of 6LoWPAN technology. *IEEE Pacific-Asia Workshop on Computational Intelligence and Industrial Application, PACIIA, volume 1*, pp. 963–966, 2008.

17 G. Mulligan and C. Bormann. IPv6 over low power WPAN WG (6LoWPAN), July 2009. http://tools.ietf.org/agenda/75/slides/6lowpan-0.pdf.

18 N. Krichene and N. Boudriga. Mesh networking in wireless PANs, LANs, MANs, and WANs, in Y. Zhang, J. Zheng, and H. Hu (eds). *Security in Wireless Mesh Networks*, Auerbach, 2007.

19 W. Webb. Move over ZigBee, make room for 6LoWPAN, 2 August 2007. http://www.edn.com/blog/1710000171/post/640012664.html.

20 G. Bag, H. Mukhtar, S.M. Saif Shams, K.H. Kim, and S. Yoo. Inter-PAN mobility support for 6LoWPAN. *Third International Conference on Convergence and Hybrid Information Technology, ICCIT 2008, volume 1*, pp. 787–792, 2008.

21 6LoWPAN technical overview. http://www.mindteck.com/images/6LoWPAN_overview.pdf.

22 G. Montenegro, N. Kushalnagar, J. Hui, and D. Culler. Transmission of IPv6 packets over IEEE 802.15.4 networks, *RFC 4944.*

23 P. Karn, C. Bormann, G. Fairhurst, D. Grossman, R. Ludwig, J. Mahdavi, G. Montenegro, J. Touch, and L. Wood. Advice for Internet subnetwork designers, *RFC 3819*, 2004.

24 S. Deering and R. Hinden. Internet Protocol, Version 6 (IPv6) specification, *RFC 2460*, 1998.

25 T. Narten, E. Nordmark, W. Simpson, and H. Soliman. Neighbor Discovery for IP Version 6 (IPv6), *RFC 4861*, 2007.

26 Z. Shelby (ed.), P. Thubert, J. Hui, S. Chakrabarti, C. Bormann, and E. Nordmark. Internet-draft, 6LoWPAN neighbor discovery, draft-ietf-6LoWPAN-nd-06, 6LoWPAN Working Group, 22 September 2009.

27 S. H. Chauhdary, M. Cui, J. H. Kim, A. K. Bashir, and M. Park. A context-aware service discovery consideration in 6LoWPAN. *Proceedings of the 2008 Third International Conference on Convergence and Hybrid Information Technology, Volume 1, 11–13 November 2008, ICCIT*, pp. 21–26, IEEE Computer Society, Washington, DC, 2008.

28 R. Mendão Silva. Wireless sensor networks – node discovery and mobility. http://rtcm.inescn.pt/fileadmin/rtcm/Workshop_23_Jun_09/s2p2.pdf.

29 M. Dohler, T. Watteyne, T. Winter, and D. Barthel. Routing requirements for urban low-power and lossy networks, *RFC 5548*, 2009.

30 B. Chen, K. Muniswamy-Reddy, and M. Welsh. Ad-hoc multicast routing on resource-limited sensor nodes. *Proceedings of the 2nd International Workshop on Multi-Hop Ad Hoc Networks: from Theory To Reality, REALMAN '06, Florence, Italy, 26 May 2006*, pp. 87–94, ACM, New York, NY.

31 J. Martocci, N. Riou, P. Mil, and W. Vermeylen. Building automation routing requirements in low power and lossy networks. http://tools.ietf.org/html/draft-ietf-roll-building-routing-reqs-07.

32 D. Networks, P. Thubert, S. Dwars, and T. Phinney. Industrial routing requirements in low power and lossy networks. http://tools.ietf.org/html/draft-ietf-roll-indus-routing-reqs-06.

33 G. Porcu. Home automation routing requirements in low power and lossy networks, draft-ietf-roll-home-routing-reqs-06. http://www.ietf.org/id/draft-ietf-roll-home-routing-reqs-08.txt.

34 S. Lee, E. Belding-Royer, and C. Perkins. Scalability study of the Ad Hoc On-Demand Distance-Vector Routing Protocol. *ACM/Wiley International Journal of Network Management*, March 2003.

35 E. Kim, D. Kaspar, C. Gomez, and C. Bormann. Internet-draft: Problem statement and requirements for 6LoWPAN routing, 6LoWPAN Working Group, 28 July 2009. http://tools.ietf.org/html/draft-ietf-6lowpan-routing-requirements-04.

36 P. Nikander, J. Kempf, and E. Nordmark. IPv6 neighbor discovery (ND) trust models and threats, *RFC 3756*, 2004.

37 K. Kim, S. Daniel Park, and J. Lee. Hierarchical routing over 6LoWPAN (HiLow). http://tools.ietf.org/html/draft-daniel-6lowpan-hilow-hierarchical-routing-01.

38 C. Nam, H. Jeong, and D. Shin. Extended hierarchical routing over 6LoWPAN. *Proceedings of the 2008 Fourth International Conference on Networked Computing and Advanced information Management – NCM 2008, Volume 1*, Washington, DC, pp. 403–405, 2008.

39 M. Durvy, J. Abeillé, P. Wetterwald, C. O'Flynn, B. Leverett, E. Gnoske, M. Vidales, G. Mulligan, N. Tsiftes, N. Finne, and A. Dunkels. Making sensor networks IPv6 ready. *Proceedings of the 6th ACM Conference on Embedded Network Sensor Systems*, Raleigh, NC, USA, 2008.

40 V. Devarapalli, R. Wakikawa, A. Petrescu, and P. Thubert. Network Mobility (NEMO) LoWMob Support Protocol, *RFC 3936*, January 2005.

41 J. H. Kim, C. S. Hong, and K. Okamura. A routing scheme for supporting network mobility of sensor network based on 6LoWPAN. *Proceedings of the 10th Asia-Pacific Network Operations and Management Symposium on Managing Next Generation Networks and Services*, Sapporo, Japan, 10–12 October 2007, in S. Ata and C. S. Hong (eds), *Lecture Notes in Computer Science, volume 4773*, pp. 155–164, Springer-Verlag, Berlin/Heidelberg, 2007.

42 R. Wang, R. Chang, and H. Chao. Internetworking between ZigBee/802.15.4 and IPv6/802.3 network. *Proceedings of ACM SIGCOMM 2007 Workshops*, Kyoto, Japan, 27–31 August, pp. 362–367, 2007.

43 C. Y. Yum, Y. S. Beun, S. Kang, Y. R. Lee, and J. S. Song. Methods to use 6LoWPAN in IPv4 network, *The 9th International Conference on Advanced Communication Technology, Volume 2*, Gangwon-Do, pp. 969–972, 2007.

44 G. Bag, M. T. Raza, K. H. Kim, and S. W. Yoo. LoWMob: intra-PAN mobility support schemes for 6LoWPAN. *Sensors* **9**(7):5844–5877, 2009.

45 H. Soliman, C. Castelluccia, K. E. Malki, and L. Bellier. Hierarchical MIPv6 mobility management, *RFC 4140*, August 2005.

46 D. Saha, A. Mukherjee, I. S. Misra, and M. Chakraborty. Mobility support in IP: a survey of related protocols. *IEEE Network* **18**:34–40, 2004.

47 I. F. Akyildiz, J. Xie, and S. Mohanty. A survey of mobility management in next-generation all-IP based wireless systems. *IEEE Wireless Communications*, **11**:16–28, 2004.

48 M. Alam, S. Dixit, and R. Prasad. *Introduction to Networked Home, Technologies for Home Networking*, ed. by Sudhir Dixit and Ramjee Prasad, pp. 1–25, John Wiley & Sons, 2008.

49 Z. Shelby. IP-based 6LoWPAN networks for AMI. http://whitepapers.techrepublic.com. com/abstract.aspx?docid=951609.

50 J. Zheng and M. Lee. Will IEEE 802.15.4 make ubiquitous networking a reality? A discussion on a potential low power, low bit rate standard. *IEEE Communications Magazine*, **42**(6):140–146, 2004.

51 R. Istepanian, E. Jovanov, and Y. Zhang. Beyond seamless mobility and global wireless health-care connectivity. *IEEE Transactions on Information Technology in Biomedicine*, **8**:405–414, 2004.

52 N. Kushalnagar and G. Montenegro, 6LoWPAN. Overview, assumptions, problem statement and goals. http://www.ietf.org/old/2009/proceedings/05mar/slides/6lowpan-3/6lowpan-4.ppt.

53 S. Deering and R. Hinden. Internet Protocol, Version 6 (IPv6) specification, *RFC 2460*.

54 M. T. Raza, R. Jeatek, S. Yoo, K. Kim, S. Joo, and W. Jeong, An architectural framework for Web portal in ubiquitous pervasive environment. *Seventh Annual Communication Networks and Services Research Conference, CNSR*, pp. 102–109, 2009.

Abbreviations/Terminologies

AES	Advanced Encryption Scheme
AMR	Automated Meter Reading
APDU	Application Protocol Data Units
CBC-MAC	Cipher Block Chaining – Message Authentication Code
CCM	CBC-MAC mode
CSMA-CA	Carrier Sense Multiple Access with Collision Avoidance
FFD	Full Function Device
FMIPv6	Fast Mobile IPv6
HC	Header Compression
HMIPv6	Hierarchical Mobile IPv6
HTML	HyperText Markup Language
HTTP	Hypertext Transfer Protocol
ICMP	Internet Control Message Protocol
IEEE	Institute of Electrical and Electronics Engineers
IETF	Internet Engineering Task Force
IP	Internet Protocol
IPSec	IP Security
IPSO	Internet Protocol for Smart Objects
IPv4	Internet Protocol Version 4
IPv6	Internet Protocol Version 6
LAN	Local Area Network
LLN	Low-Power and Lossy Network
LR-PAN	Low-Range Wireless Personal Area Network
LR-WPAN	Low-Rate Wireless Personal Area Network
MAC	Media Access Control
MIPv6	Mobile IPv6
MP2P	Multipoint to Point

MTU	Maximum Transmission Unit
NALP	Not a LoWPAN Frame
NAC	Neighbor Added Child
NAT	Network Address Translation
NEMO	Network Mobility
NRP	Neighbor Replace Parent
PAN	Personal Area Network
PC	Personal Computer
PDA	Personal Digital Assistant
PHY	Physical Layer
PKI	Public Key Infrastructure
P2MP	Point to Multipoint
P2P	Point to Point
QoS	Quality of Service
REST	Representational State Transfer
RFC	Request For Comment
RFD	Reduced Function Device
RFID	Radio-Frequency Identification
SNMP	Simple Network Management Protocol
SOAP	Simple Object Access Protocol
SSL	Secure Sockets Layer
TCP	Transmission Control Protocol
TLS	Transport Layer Security
TTL	Time to Live
UDP	User Datagram Protocol
WAN	Wide Area Network
WBSN	Wireless Body Sensor Network
WEP	Wired Equivalent Privacy
WPAN	Wireless Personal Area Network
XML	Extensible Markup Language
6LoWPAN	IPv6 over Low-Power Wireless Personal Area Networks

Questions

1 Identify the benefits of using IP in sensor networks.

2 What are the advantages of 6LoWPAN over wired networks as well as other wireless networks?

3 State five applications of 6LoWPAN.

4 Draw a 6LoWPAN stack and explain its various layers.

5 Differentiate between ZigBee and 6LoWPAN.

6 Write the names of different types of node in a LoWPAN and their characteristics.

7 Explain with a diagram the network topology of IEEE 802.15.4.

8 How is a 6LoWPAN datagram transmitted over IEEE 802.16.4 radio?

9 Describe the process of neighbor discovery in 6LoWPAN.

10 Explain the 'route over' approach and the 'mesh under' approach of routing.

11 Explain with a diagram how 6LoWPAN can be used for an application in your surroundings.

12 State whether the following statements are true or false and give reasons for the answer:
A 6LoWPAN networks are known as low-power and lossy networks.
B Smart objects are those that are small in size and accurate in operations.
C Communication in a sensor network can only be unidirectional.
D IP-based applications are memory and bandwidth intensive and cannot be used for sensor networks.
E 6LoWPAN can connect to the Internet through a gateway.
F If an FFD fails, an RFD can take over the role of PAN coordinator.
G IEEE 802.15.4 provides confidentiality to 6LoWPAN transmissions.
H 6LoWPAN provides the network backbone for the Internet of Things (IoT).

13 For the following, mark all options that are true:

A IEEE 802.15.4 defines the following types of node:
- full function device,
- partial function device,
- reduced function device,
- remote function device.

B Which is not a type of node and service discovery in 6LoWPAN:
- Some1CatchMe,
- Every1CatchMe,
- ICatchYou,
- YouCatchMe.

C The frames supported by IEEE 802.15.4 are:
- data frame,
- beacon frame,
- MAC command frame,
- acknowledgement frame.

  D IEEE 802.15.4 provides security-related functionality to:
  - 6LoWPAN,
  - secured bootstrapping,
  - key management,
  - link layer security,
  - digital certificates.

  E IEEE 802.15.4 devices have the following characteristics:
  - short communication range,
  - low bandwidth requirement,
  - low memory requirement,
  - low battery life.

Exercises

1   2 MB of restricted data has to be transferred from one node to other in a 6LoWPAN network using AES-CCM-128 security. How many MAC layer frames have to be formed?

2   Draw a state diagram of a 6LoWPAN node.

3   A 6LoWPAN network has a very high number of nodes 'N' in it. The spread of the network is also very wide, mobile, and random. It has been stated that, to save on energy and time, the maximum permissible hops for data communication between any two nodes even in the worst-case deployment should not be more than $\log N$. What is the worst-case deployment? What topology should be used in this 6LoWPAN network to ensure maximum $\log N$ hops even in the worst case?

4   Identify in Figure 12.2 the sensor nodes that have to be RFDs or FFDs, and those sensor nodes that can be either RFDs or FFDs in terms of IEEE 802.15.4.

5   Draw a sequence diagram to depict the communication between nodes in a 6LoWPAN in terms of exchange of frames.

Part V

Advanced Concepts

13

Security in Routing

13.1 Introduction

Security of a network ensures that the network is available along with its required components and services at the time of requirement. The requirement can extend up to 24 hours per day and 7 days a week (24×7) throughout the year. The network security also has to ensure the integrity and security of the data that is being carried through it [1]. The security of a network is not only ensured by technical solutions and foolproof security products, it also involves policies, operational procedures, operating environment, physical security, social engineering, and many other non-technical aspects. The network loses its significance if it cannot ensure availability and integrity of the data in it, as this is what the networks are established for. The region of presence and severity of a network failure or compromise can vary from an intraorganizational localization to

Network Routing: Fundamentals, Applications, and Emerging Technologies, First Edition.
Sudip Misra and Sumit Goswami.
© 2017 John Wiley & Sons Ltd. Published 2017 by John Wiley & Sons Ltd.
Companion website: www.wiley.com/go/misra2204

a national catastrophe or international influence. An example of organizational effect can be the attack on a local area network, while the failure of a supervisory control and data acquisition (SCADA) system related to electricity distribution can have ripples across the country not only in terms of electrical failure but also in terms of failure of all the systems running on electricity.

Most of the systems relying on a network for their operation are moving towards round-the-clock service availability. This has increased the requirement of network availability and reduced the time for those maintenance operations in the network that require the network to be brought down. Furthermore, the network designs have been optimized to utilize the available bandwidth to an optimum level, reducing the idle time in a network. Network updates and upgrades have to be planned so as not to flood the network, and at the same time they should be performed at those time intervals when the network utilization is comparatively low. The list of services in which the network has one of the highest stakes in customer satisfaction is endless and includes banking, online ticketing/reservations, payment gateways, communication networks, defence networks, and all applications related to e-commerce, e-business, e-education, and e-governance.

The network provides global access to the data even of a small local department store. On the other hand, the users seeking information and data are of global distribution and thus the data can be sought over the network at any time of day whether the department store is open or closed. When data access is universal, so is the attack profile on the data flowing through the network. Network failure may not always be related to an attack, but may occur because of a simple failure of the network equipment. The network equipment may fail owing to malfunctioning, power failure, or loss of connection. A network administrator would generally say that network failure is rarely due to a deliberate attempt, but mainly due to malfunctions or misconfigurations. But whatever the underlying reason may be, the unavailability will reflect badly and affect a much wider geographical spread than the geographical presence of the organization and its products and services.

Other factors that are equally strong in affecting network availability are natural phenomena and external effects. As the networks have gone global, so have the points of influence and the areas of breakdown. A natural disaster at a location may disrupt operations in some other place of the globe, as the data center, network operations center, or data storage of the service might have been in the area hit by the disaster. Internet outage over a large geographical area owing to cutting of submarine cable somewhere deep in the ocean is not unheard of. Redundancies, disaster recovery sites, alternative network routes, scaling-up of operations, and incidence handling policies are common solutions to this category of problem.

The security risks to a network can be classified into three categories:

1) *Bugs and misconfiguration.* Misconfiguration of the intermediate network devices may leave a foothold for any user to have unauthorized access to the network devices. Once the user gains access to an intermediate network device, it can be used to reconfigure the device, listen to the data flowing through it, create new entry points for future exploitation, or simply bring down the device or make it inaccessible. Similar damage can be caused by any bug/flaw in the software running in the network devices. The bug may be activated all by itself and affect the network operation when that portion of the code that contains the bug is run. Alternatively, the bug/

flaw may be discovered by someone and used to affect the network in the form of zero-day attack or, keeping silent about its detected presence, and use to exploit the network until it is discovered by the software developers or the network administrator and patched.

2) *Interception.* The data being sent from one device to another may be eavesdropped at the device level or while it is travelling from one device to the other through a guided medium or in wireless mode. Thus, the entire network path is open to eavesdropping. Confidentiality of data through encryption is a security measure used to ensure secrecy of the data even if it is snooped. It is important in a network to detect whether it is being snooped and to prevent eavesdropping.

3) *Active attacks.* An active attack brings down the network by affecting the network devices, makes the network unavailable, decrypts the cryptic data to rip open the confidentiality and privacy, or simply makes the network unusable by flooding it with unwanted data. Social engineering is commonly used by unscrupulous network users to fool the administrator into giving away critical details, passwords, and configurations, who then use the information to attack the network.

A network can be secured by ensuring not only point-to-point link security but also security of all the devices at the periphery. Proper security mechanisms should be in place at the receiver as well as at the transmitter ends. All the intermediate devices and links through which the data passes should be secured. In the present-day scenario, even information such as the amount of data, the rate of data transfer, the time of data transfer, the originator, and the destination can lead to sufficient information for a successful network attack. The side channel attacks in the network work on features such as power usage, CPU utilization, and memory usage of the devices.

13.1.1 Network Sniffer

A sniffer is a system (hardware or software) that listens to the data flowing through a network and that may capture and store the sniffed data for analysis or use. Network sniffers do not disrupt network traffic or alter the data flowing through the network. Sniffers are not necessarily a hazard to network security because there are many network management and network security tools that have to run sniffers to keep a check on the health of the network. Network sniffers are generally a part of the network management software and the intrusion detection systems, and here a packet flowing through the network is analyzed for its compliance with the network policies.

In the case of networks that do not use switching, the packets are prone to sniffing by any other node in the network. In a local area network running on the Ethernet, any computer can enter into a sniffing mode. In Ethernet protocol the packets are broadcast in the network with the MAC address of the recipient in the header. All the nodes receiving the packet read the header. If a node finds that the packet is meant for some other node, as the MAC address in the header does not match the MAC address of the node reading it, the packet is dropped. Only the node with the intended MAC address retains and processes the packet. However, a node may want to analyze all the packets flowing in the network. In that case it does not drop any packet, whether it is intended for it or not, and starts reading, storing, or analyzing all the packets. When a node enters such a state, it is known to be running in promiscuous mode. Once a promiscuous node has received a packet, it can read all the information contained in the packet, i.e. source, destination, port number, and actual data being transferred. Network sniffing can be used to extract login IDs and passwords from the data moving in the network.

Capturing all the data moving in the network and analyzing it may require tremendous processing and storage capabilities in the sniffer. Moreover, the sniffer may not be interested in all types of information and would be looking for specific data to support network attacks. In such scenarios, the sniffers are configured only to look for a specific category of data based on source IP, destination IP, or the port number being used to determine the specific network service such as FTP, Telnet, http, or email to which the data belongs. If the attacker is looking for an FTP password, it will only look inside the contents of an FTP packet and ignore the other data packets. The sniffer captures the network data in binary mode, but generally decodes it and provides the same in human readable form to the attacker using any related tool.

The network user is generally not aware of any ongoing sniffing in the network, as it is a passive attack. In a passive attack, the attacker does not cause any change in data, affect the network, or damage the networking devices. 'Snort' is one of the common tools used for sniffing. Snort works on IP networks performing real-time packet logging, traffic analysis, protocol analysis, and finally content analysis. Therefore, snort is suitable also for use in intrusion detection systems. When snort is not used for intrusion detection, which is a complex task, it may be configured to be used only in sniffing or packet logging mode. In the sniffing mode, the tool displays the information passing through the network as a continuous stream of data in text form. If the logging mode of snort is enabled, the sniffed data is logged and stored in the local storage of the computer running snort.

ARP spoofing. Sniffing of information in a network can be supported by ARP spoofing. In ARP spoofing, the attacker sends a forged IP packet to a node that is planning to be the source to send a packet to the destination. In the forged IP packet sent by the attacker to the sender, the IP address of the sender of the packet is changed to that of the intended destination and the MAC address is that of the attacker, as shown in Figure 13.1. When the sender receives this packet from the attacker, it thinks that the packet has been received from the intended destination and stores its IP address (the actual IP address of the receiver) and the MAC address (actually the MAC address of the attacker) in its ARP table. While sending the data to the intended destination, the source refers to this ARP table and sends the data in the network unknowingly with the MAC of the attacker. The attacker, on receiving this data, stores and analyzes it and forwards a copy of this data to the actual destination. As the actual destination gets the packet, it believes everything to be in order. A lot of preattack knowledge gathering about the network has to be conducted for successful ARP spoofing. The network is scanned using various tools and techniques, and the IP addresses of source, destination, and routers are obtained. The attacker also builds up a topological map of the network, giving the attacker an idea about the interconnections in the network.

DNS spoofing. A domain name server (DNS) is used to convert the name of the website or a server, which is known as the domain name, to the corresponding IP address of the server that hosts the website. The domain name service is neither centralized nor based on some predefined servers, but is a distributed system with a typical hierarchical structure and a high degree of scalability. Each domain name is

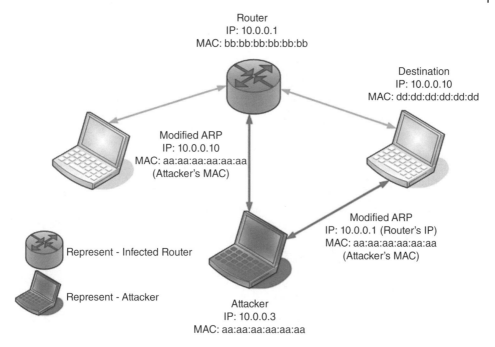

Figure 13.1 ARP spoofing attack.

resolved by at least two domain name servers – the primary DNS and the secondary DNS. Although there is only one primary DNS for a domain, there can be multiple secondary DNSs to provide redundancy and backup. A domain is further divided into subdomains, and a separate DNS server is responsible for each subdomain. The area of responsibility of a DNS server is known as its zone. When the domain name has to be translated into its equivalent IP address, the information about the same is searched in the primary DNS server. The secondary name servers are queried only if the primary name server fails.

As the DNS servers have a hierarchical setup, when the primary name server is unable to resolve the name query, the query is forwarded to the next DNS server in the hierarchy. This is done in two ways – iteratively or recursively. In the case of recursive name resolution, the name server itself queries the next name server up in the hierarchy for resolving the domain name until the DNS server is located, which finally resolves the name and the information is sent back to the client through the intermediate name servers that participated in the recursive name resolution. When the information is sent from the DNS server resolving the name back to the client, each intermediate DNS server sending it backwards keeps a copy of the DNS information in its cache for future name resolution. The process is shown in Figure 13.2. In the case of an iterative name resolution, as depicted in Figure 13.3, each server that is unable to resolve the domain name sends back information about non-availability of the name entry in it and also informs the client about the next DNS server in the hierarchy that may be contacted for name resolution, and the client contacts the next name server and the process continues until the name is resolved.

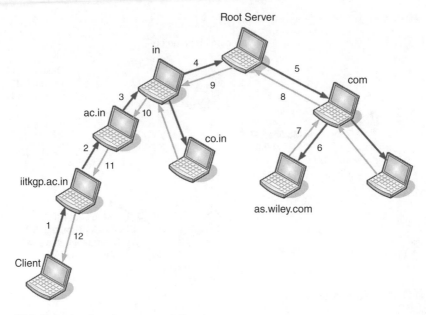

Figure 13.2 Recursive domain name resolution.

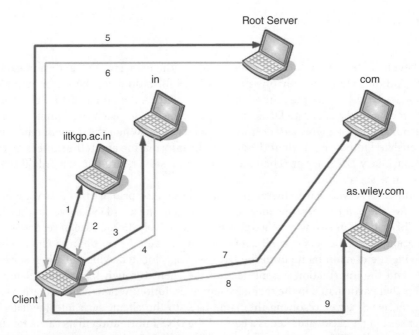

Figure 13.3 Iterative domain name resolution.

The attacker poisons the DNS cache of the intermediate DNS servers by using a fake DNS server in the DNS resolution path, whereby the name resolutions are manipulated to help the forwarding of packets to the attacker's computer instead of to the actual server hosting the domain. This is done by resolving the domain name to the IP address of the attacker's computer. The attacker activates the attack by sending a name resolution request to a target DNS server. The request is for a domain name unknown to the DNS server, say 'unknown.org'. However, the fake DNS has the entries for unknown. com, and so when the target DNS server cannot resolve unknown.com, it contacts the fake DNS for resolving the name. The fake DNS not only resolves unknown.com but also transfers information about other name resolution entries it has, most of which are manipulated. Now the target DNS server and all other DNS servers that participated in the recursive name resolution of unknown.com have manipulated DNS entries in their caches, which they use to resolve the domain names received by them after the DNS poisoning. In this way, the attacker not only can cause non-delivery of packets, it can also redirect all traffic through its server.

DNS spoofing is a common technique for man-in-the-middle attack, where the attacker logically positions its server between the sender and the receiver, connecting to both in an invisible manner and gathering all the traffic flowing from sender to receiver.

Trojan and backdoor. A Trojan in an unauthorized segment of code that generally resides within a legitimate program. Being a legitimate program, the user believes the program to be performing those operations it is intended to perform, but the legitimate program may also perform some other unknown or undesirable functions owing to the unauthorized code segment running in it. The Trojans generally connect to an attacker's remote server using TCP/IP or UDP to pass on the information gathered by the Trojan or to get instructions from the remote server.

There are wide varieties of Trojan, some of which have known attack patterns. Remote-access Trojans provide attacker access to the host computer to use it as a zombie. Key-logger Trojans pass on all keystrokes being typed in the attacked computer, while password-sending Trojans detect all the passwords stored and entered in the attacked machine and forward them to the remote server. FTP Trojans give the attacker access to the attacked machine through its FTP port. There are destructive Trojans and software detectors and killers, which can delete specific files, programs, antiviruses, and anomaly alarms. Destructive Trojans can be activated on detection of a specific event or by remote command. Alternatively, the activation can be time based or logic based. Attack Trojans enable the use of an infected machine as one of the launching platforms for various types of attack such as distributed denial of service attack (DDoS) attacks, and thus depict the infected machine as the attacker to the outside world and obfuscate the controller of the attack.

Certain free proxies are available on the Internet for use by anyone for any purpose. Such proxies are generally created by the Wingate Trojans, which convert the victim server into a proxy server for use by everyone or the attacker. These proxy servers can then further be used to launch various types of attacks and retain the anonymity of the attacker and the attack trail.

13.1.2 Denial of Service Attack

A denial of service attack can be launched from a single source or, to be more impactful, can be distributed. In DDoS, the attacker captures the control of hundreds of computers, which may be spread across a wide geographical region. These captured machines are then used to launch the DDoS attack on the target node. When such a huge number of compromised and controlled computers launch the attack simultaneously on the target victim, this leads to exhaustion of the link bandwidth and processing capability or the memory buffers of the intermediate network nodes, leading to non-availability of the resources to a legitimate user.

Some of the commonly known DoS attacks are as follows:

- *Ping of death.* The attacker sends to the victim an IP packet with a length larger than 65 536 bytes, which is beyond the permissible limit allowed by the IP protocol. This IP packet is fragmented by TCP/IP, and the smaller segments are forwarded to the destination where they are reassembled. The reassembled packet, which is too large to be handled, leads to buffer overflow.
- *Land exploit.* The attacker sends a number of spoofed SYN packets to one of the open ports of the victim. These SYN packets are modified by the attacker to have the same source IP as destination IP. Now, when the victim receives the SYN packets, it sends the SYN/ACK to itself for all the modified packets, thus exhausting the TCP queue for legitimate SYN packets from genuine users.
- *Smurf.* The smurf attack is set off by sending a series of ICMP echo packets to broadcast addresses. The source address of these ICMP echo packets is spoofed to that of the victim. When the nodes in the network receive these broadcast packets, they reply to it, and all these replies go to the victim node whose address has been spoofed as the source of the ICMP echo. Such a huge number of ICMP replies consume the network bandwidth or the computational resources of the victim.
- *Fraggle.* The signature of a fraggle attack is similar to that of a smurf attack, the only difference being in the use of UDP echo packets instead of ICMP echo packets.
- *SYN flooding.* A number of TCP SYN packets are sent to a victim computer after spoofing the source address of the SYN to a non-existing IP. This creates a huge number of half-open connections, each waiting for completion of the handshake between the source and the destination, and thereby the performance of the attacked computer slows down.
- *WinNuke.* The attacker sends an out-of-band network packet to the victim that the latter is unable to handle, and it may behave erratically, break the network connection, or crash.

The known denial of service attacks are of three genres: bandwidth attack, protocol attack, and software vulnerability attack. In the case of a DoS attack targeted towards bandwidth, the attacker tries to consume the entire bandwidth of the network or the throughput of the network equipment. DDoS attack generally causes the effect by operating in this category. The attacker consumes the entire link bandwidth, denying the legitimate traffic any passage across the network and leading to heavy packet drops. The protocol-based DoS attacks exploit the vulnerabilities of the network protocols and their inherent design. Examples of protocol attacks are SYN flood, smurf, and fraggle attack. Instead of attacking the protocol, if the DoS attack is based on the vulnerabilities of the network and application software, such as email and webserver, the class of attack is classified as software vulnerability attack. Examples of software vulnerability attacks are teardrop, land, naptha, and ping of death.

13.1.3 Social Engineering

Social engineering techniques are used to obtain information from the authorized holder of the information by using deceiving or persuasive action and thereafter using this information for network attack, information stealing, and cyber fraud. The network administrators, system administrators, or users are unaware that they have been scammed and parted with security information to an unauthorized person or system. The strongest security infrastructure in place can also be compromised by the use of such techniques, as it rarely uses technology and is heavily dependent on the human factor.

Information may not always be accessed from the administrators, but the security guard or helpdesk can be befriended or fooled into providing unauthorized access. Reconnaissance about the systems in use, the names of administrators, email addresses, and the stage of a project can be done from a telephone directory, published job openings, tender documents, or an organization's website. Social engineering can provide all the required information about a system without gaining any unauthorized access to the system. These attacks can either be completely human based or computer aided. Common examples of human social engineering attacks are through impersonation whereby the attacker tries to gain information by posing as an employer or any other known, authorized, or needy person for the information or by staging a third-party authorization scenario where the attacker acts as an authorized third party, such as an auditor, law agency, or maintenance agency to access information, or masquerades as a maintenance person.

There have been many instances where information has been successfully accessed by an unauthorized person in the guise of courier delivery, mail man, window cleaner, janitor, visitor, or pest controller. It is quite common to see people overhearing official conversations in public places, shoulder surfing, or even dumpster diving. Computer-aided social engineering attacks are conducted through popup screens asking for a password to restore connection, spurious or disguised email attachments from spoofed known senders, disguised websites similar to that of the victim organization asking the user for information, and phishing emails.

13.1.4 Packet Filtering

Packet filtering makes it possible to stop egress or ingress of a packet from or into a network. Some of the common packet filtering mechanisms are access control lists, egress filters, IP chains, and IP tables. Packet filtering should be properly implemented to avoid traffic from a spoofed trusted source and prevent fragment attacks. Firewalls are a commonly used device not only capable of packet filtering but also of inspecting the network flow and the packets. Proxy firewalls add a further layer of protection by preventing the establishment of a direct peer-to-peer connection between the attacker and the victim. The client intending to connect to the server has to establish connection with the proxy firewall, and the firewall in turn establishes connection with the server, fetches the data from the server, and provides the data to the client. In the process of data forwarding between client and servers, the proxy can also examine the type of data being exchanged and ensure compliance of security policy on the requests being made. Reverse proxy is a very reliable and commonly used server for security of websites in the website hosting design architecture.

Packet filtering is best done in a network by a router, as it is on the perimeter. A router also performs a number of other security roles, such as network address translation and virtual private network (VPN) connectivity, and it can also act as a firewall. However, in huge networks, it is preferred that the router should continue with its traditional role of routing and the security overlay should be provided by a separate infrastructure of firewalls, proxies, intrusion detection systems, intrusion prevention systems, and link encryptors.

13.2   Attack Surface

Attackers devise new techniques every day to compromise the security of computer systems and networks. One of the best ways to improve system security is to reduce the attack surface [2, 3]. The attack surface of anything is the area that is vulnerable, or is exposed to attack by an enemy. The concept of 'attack surface' can be easily understood by taking the example of a fighter jet. If an enemy wants to attack the jet from the front, the area visible to it for the attack will be less. But if it attacks by taking the bottom or top view of the aircraft, the area exposed for attack by the enemy will be more than in the previous case. This exposed area of the aircraft at any particular angle of visualization is its attack surface. It is quite obvious that, the greater the exposed part, the greater will be the attack surface and the easier it will be for the enemy to attack.

The attack surface of a system is the amount of application area that is exposed to adversaries. In the field of computing, the attack surface of a system may include protocols, services, interfaces, inputs entered by users, code processing those inputs, and even humans having access to vital information. The attack may be executed via the network also. Measuring the attack surface of a system is a good indicator of its security. The overall vulnerability can be lowered by reducing the attack surface of a system. Even after security compliance, a network system has an exposed vulnerable surface due to the mandatory access points left in the node to ensure accessibility over a network. For quality assurance, organizations use attack surface metrics to foretell vulnerabilities in the network nodes prior to deployment [4]. The users have an interest in considering the security of a network architecture and configuration when they have to choose between alternative designs. The user interface for network nodes through the web browser is one factor that is responsible for a node's attack surface. But this minimum attack surface has to be retained to ensure usability. Therefore, all the user interface components have to be considered for attack surface calculation.

The increasing use of network systems has made it important to analyze the system carefully for security and robustness flaws. The use of networks by organizations to run their applications is increasing day by day. For this reason, such systems have become the main target of attackers and consequently are the largest source of security vulnerabilities [5]. Identifying theft, phishing, malware, and other computer crimes is often costly to consumers and organizations and puts doubts in people's minds about trusting online applications [6]. Even though large security metrics [7] have been proposed, complete security of systems is not guaranteed.

The attack surface for the Windows operating system was introduced in the year 2005. This was liberal and informal. The attack surfaces of different versions of Windows [8] and Linux have been measured. Methods have been introduced to calculate the attack surface of various small applications as well as some large-enterprise systems implemented and coded in C and Java. Application vulnerability description language (AVDL) is a theoretical approach and is meant to realize and understand a unified data model. The AVDL is mainly based on technology that is independent and used for analysis of web applications [9]. The static analysis vulnerability indicator (SAVI) is a tool that links up several static analysis metrics and is used in ranking network application vulnerability.

Various methods have been proposed theoretically as well as on the basis of tools for attack surface calculation. The attack surface of a web application can be quantified by a method [10] that consists of a multidimensional metric for the attack surface of network applications. However, the rationale and principle behind the attack surface of network applications should be well understood. A scalar numeric indicator for easy comparison and a descriptive and detailed vector representation for deeper analysis can be used. The attack surface metric is the estimation of the amount of functionality and code in a network system exposed to outside attackers.

13.2.1 Types of Attack Surface

Attack surfaces can be divided into three categories:

- network attack surface,
- software attack surface,
- human attack surface.

Network attack surfaces are involved when the attacks are launched via networks. These kinds of attack surface involve sockets and open ports such as TCP ports and UDP ports. These attack surfaces mainly include tunnels running inside the networks. In tunneling, the packets that cannot be transferred through a network owing to security restrictions are encapsulated and put inside other transferable packets and are transported. On the destination side, the encapsulated packets are received and stripped to obtain the original data. These tunnels are very hard to identify, as they appear to be like normal traffic on the network.

The software attack surface comprises the code and functionalities that are available to unauthenticated users. It is an important problem, as the number of bugs and vulnerabilities found in software are on the rise owing to the proficiency and skills gained by attackers, and therefore the number of successful attacks is increasing at a high rate.

Human attack surfaces are quite different from the other two attack surfaces, as human attacks are led by unauthenticated users. A human attack may involve social engineering, the unavailability of an employee owing to a change in job, retirement, or death, or retention of a new employee when it is not clear how much trust can be placed in him. A human attack surface may involve dishonest or unscrupulous employees stealing or destroying data.

13.2.2 Attack Surface and System Resources

An attack surface is exposed to the attacker by the code, functionality, and interfaces of a system. An attack surface can be calculated in terms of system resources. The system resources are generally the data items, communication channels, and operating environment. The following describes the use [11] of attack surface metrics:

- The analysts use the attack surface metric to improve the design of the network.
- Testers use this metric to estimate the extent to which testing has to be done.
- Users use this metric to compare different architectures.
- Organizations use this metric to make proper investment in better systems.

When a network is reviewed, analyzed, and audited, a variety of problems can be unearthed that affect the security. By arranging these problems in some order, it will be easy to tackle them. This order of problems can be referred to as vulnerability categories [9]. Some of the vulnerability categories are session management, exception management, input validation, confidentiality authentication, authorization, and integrity.

13.2.3 Attack Surface Metric

An attack surface is measured in terms of the resources of a network that are exposed to adversaries. The more the resources are exposed to the user or attacker, the greater will be the attack surface of the network and hence the greater will be its insecurity. All the resources are not considered to have an equal effect on the security of a network. The entry points and exit points of a network are also considered to be part of the attack surface. An entry point is a point through which data can be entered into the system, and an exit point is a point through which data can be retrieved from the system [12]. The network links are also considered to be points of attack, as the attacker can connect to the application through a network link. Another basis for attack on a network is that attackers can use persistent data to attack an application. This persistent data can be referred to as untrusted data items [13]. The attack surface of an application can be reduced by reducing the amount of running code, reducing the network access by users/ attackers at entry points, and minimizing privileges to limit damage potential.

The attack surface vector (AS) represents the attack surface. According to the Eucledian norm, the attack surface indicator (ASI) is given by $ASI = |AS|$. Boolean values are raw measurements that show the presence or absence of a feature, given by a value of 1 or 0, and enumerations show multiple-choice measurements or non-negative integer values as a result of counting.

The concept of attack surfaces to measure the security level of a product lies in-between the code level and system level approaches. It is above the level of coding, thereby giving the bugs importance based on their weights on ease of exploitation, but at the same time it is below the level of system, and thus is able to identify specific system configurations. We can say that this approach is at the design level of the system. Thus, instead of counting bugs in the code and reading software vulnerability reports from the bulletins, the location of possibility of attack by an attacker should be identified, which collectively form the attack surface of a system.

An important point to make is that the attack surface approach measures the system security in relative terms, such as comparing two versions of the same product, and not as a metric that can give a particular value as a result of a system security measure.

13.2.4 Reduction in Attack Surface

The goal of reduction in the attack surface can be accomplished by fulfilling the following tasks:

- Reduction in the volume of code that runs by itself.
- Reduction in the exposed code that can be accessed by unauthenticated and untrustworthy users.
- Reduction in the attack surface of the node or the removal of unused protocols and functionalities.

- The security of a system can be increased by running processes having least privileges or access right, and hence the capability of the attacker to attack the node can be reduced.
- The developer should not trust the input of the user, as the user is considered to be the primary weapon of attack.
- Various researchers have proposed frameworks to avoid vulnerabilities. A framework based on access control policy description language and security policy description language [14] can be used to provide web security.
- Elimination of the entry points that can be easily attacked.
- Reduction in damage when the attack is done.

Not all areas of the attack surface are targeted equally. As a result, some attack vectors have more probability of being attacked than others. For example, it has been proved that files having full access rights, are more open to attack than files with restricted access rights. Services running with system privileges are more likely to be attacked than services running with other user privileges. Minimizing the use of various script engines is a good attack surface reduction measure, as these are quite frequently used as enablers for attack.

A good practice to reduce the attack surface is to reduce the code running by default. This reduction can be done by turning off features that are used very rarely while keeping the option of easily enabling them again if required. To reduce the code and functionalities for untrustworthy users, the access of the network endpoints of software applications can be restricted only to those users present in local groups or subnets. Simply asking for authentication from the users can also restrict the access possible from untrustworthy users, thereby reducing the attack surface of the system. Reducing the privileges for execution of the code is also a good step.

The value of the attack surface metric of a web application can be reduced, but not to zero. As a network cannot run in isolation without inputs and outputs from or to users, agents, and other applications or network connectivity, it is prone to attacks through these exposed surfaces. A network node generally has some degree of cohesion and coupling exposing the gaps. Besides ease-of-use calls for reduction in security, enhanced security reduces the ease of use. As the application has to be finally used by a person or by software, accessibilty to the network has to be provided, leading to a compromise between accessibility and security. Even a 'black body' application is exposed to attacks from the data that it receives. The greater the attack surface of the system, the greater is the effort designers have to put into testing, whereas if the attack surface of the system is smaller, the designer has to put less effort into testing.

13.3   Networked Battlefield

From the era of information warfare there has been a paradigm shift towards network-centric warfare, which has encapsulated information and network security within its fold. In a networked battlefield, the adversary always tries to detect, manipulate, or deny information flow, which is considered to be a prime resource and the deciding factor for supremacy of tactics and technology. Cyber terrorism is another area of concern.

Information is the backbone of the present-day warfare. In a battlefield scenario, the networks and information were traditionally related to communication capabilities. However, in the network-centric warfare scenario, command, control, tactics, camouflage, deception operations, weapons systems, platforms, fire control, and sensor-enabled soldiers are all dependent on the real-time availability of information. The use case diagram of a typical networked battlefield is depicted in Figure 13.4.

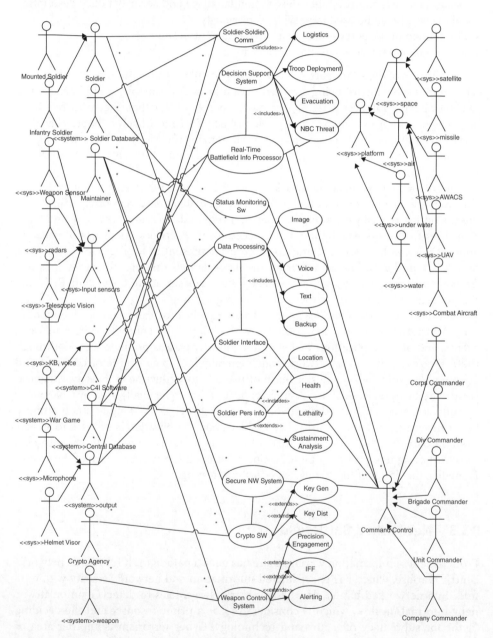

Figure 13.4 Use case diagram of a networked battlefield.

A combat group can successfully carry out an operation only if it is networked within itself as well as to its higher and lower formations and the information needs of the group are met. The combat group's capability to conduct network-centric joint operations with a distributed force structure can only be supported by the doctrine of a synchronized networked battle force covering all assets that are in the physical domain. This ensures speed of command with enhanced responsiveness, lethality, and survivability to accomplish the assigned mission with a leaner force strength that can travel lighter and faster. The fighting tactics of a networked force are remarkably different from those of a traditional military formation, as each soldier is aware of the situation in his surroundings and knows the deployment of other troops and the state of availability of the weapon systems. A networked battlefield helps in two-way information support. A smart soldier keeps sending his data to the command control center, and at the same time keeps regularly receiving data from various battlefield sensors as well as from the command centers as depicted in Figure 13.5. The sensor enables the smart soldier to perform onsite analysis of the data with cognitive capabilities and help in taking decisions. The sensor-to-shoot time is also reduced, with increase in lethality, owing to network integration as depicted in Figure 13.6.

13.4 Mobile Agents

The increased use of network resources and the need for its efficient distribution and access have emphasized the requirement for new technologies and approaches for information dissemination. Mobile agent technology is one such promising technology that uses mobile code to perform a designated task. The growth of networks and the Internet has led to an explosion in the field of information services. These technologies have revolutionized the growth and proliferation of network resources. An e-resource may be defined as a data source in electronic format, which may be accessed digitally. The data may be configuration files, access logs, or routing tables. An e-resource may be accessed by multiple users simultaneously. These inherent benefits have led to considerable growth in the available e-resources, which makes their distribution and proliferation an essential but challenging task. Several technologies have been proposed to achieve this task. Mobile agent technology is one such approach that is being used widely to implement e-resource distribution and access systems.

The mobile agent [15] is a program that performs a predefined computational task on behalf of a user by moving autonomously from one node to another in a network. It is a software agent that contains data and code. The mobile agent phenomenon creates a code and transports it to another machine to execute the code [16]. Mobile code technologies such as Java enable the creation of mobile agents that can move across heterogeneous entities and perform various functions. The increasing pervasiveness of the Internet and the development of scripting languages such as Perl and TCL have made mobile agents a promising technology. The technology is now a widely accepted design paradigm used in implementation of applications requiring distributed control.

As the computation code moves closer to nodes, use of mobile agents helps to reduce the network bandwidth requirements and latency. The approach is a step ahead of the

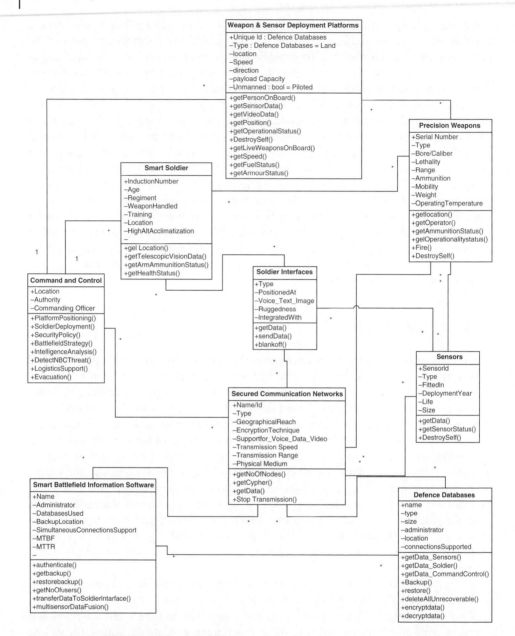

Figure 13.5 Class diagram of a smart soldier.

traditional client–server model. These features make mobile agent technology a useful alternative when dealing with various distributed systems. The applications of mobile agents are diverse, ranging from network management to Internet services. Mobile agents have been contributing to the proliferation of electronic resources by providing a platform to implement security services, information filtering, retrieval systems, and monitoring services.

Weapon Statechart

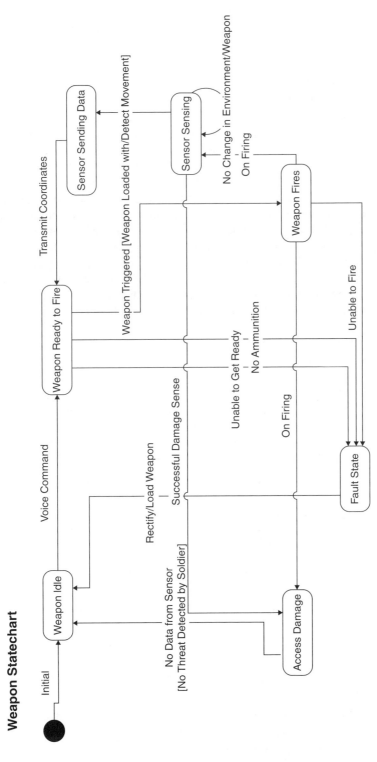

Figure 13.6 State chart diagram of a smart weapons system in a networked battlefield.

There are agents for both users and network nodes. The exact architecture of a mobile-agent-based network system may vary with the platform and specific requirements. Some of the agents, which may be part of a network, may include:

1) User presentation agent
2) Database query agent
3) Query management agent

Network systems function on structured blocks of data, filtering and utilising required data fields from distributed sources to complete a transaction. Mobile agents help in the execution of tasks autonomously without the need for client interference. They can interact with multiple nodes and make transaction execution flexible and simple [17]. Many networked systems use a distributed database management system (DDMS) to store and retrieve data. Traditional approaches exert a heavy load on DDMS systems owing to the centralized nature of their execution. Use of lightweight and portable mobile agents [18] for database connectivity makes the transactions faster and reduces the load on the servers. Applications require some data to be extracted from sources and thereafter processed. Mobile agent technology is now being widely used to implement information retrieval systems. Performance analysis of such applications has illustrated that there is a significant increase in performance and reduction in latency when these applications are incorporated using mobile agent technology.

13.4.1 Architecture and Framework

A mobile agent has a task encapsulated in it. The managing station dispatches the agent to a remote site. On reaching the site, the agent executes the code and performs the designated task. The agent can either return to the originating site or it may send the results in a message [16]. As each agent performs its function independently, a number of agents are deployed across different entities in a system to achieve a target. As the computation is now performed in parallel, the task can be executed in lesser time. A mobile agent may be characterized [19] by the following features:

1) the ability to execute the code automatically on reaching a remote site,
2) the ability to communicate with other agents,
3) the ability to perform functions delegated by the user,
4) the ability to function autonomously,
5) the capability of asynchronous functioning.

As depicted in Figure 13.7, a mobile agent comprises two parts and various stages [20] in the execution life cycle:

- *Code.* This comprises the instructions that define the agent, its functions, behavior, and intelligence.
- *Data.* Agent data comprises the global variables that characterize the behavior of the agent.
- *Execution state.* The agent may reside in one of several states. These states are stages in the life cycle of the agent.

The mobile agents require a network to communicate with each other. This network is often referred to as the agent communication network. An agent communication network [19] comprises the following components:

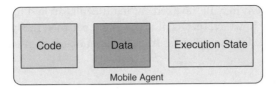

Figure 13.7 Representing a mobile agent.

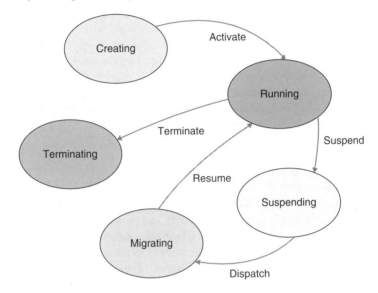

Figure 13.8 Markovian chain representing the mobile agent life cycle.

I) A host station – defined as a network node that houses an agent. A station may have multiple hosts at any instant of time.

II) A society – a group of agents that have a common goal.

III) A query station – an interface for a user to specify a task, which in turn is broken up into several smaller goals.

IV) A species – a society that has a common goal of identical priority.

13.4.2 Life Cycle

A mobile agent undergoes transition from its birth to termination. This evolution is represented by the mobile agent Markovian chain illustrated in Figure 13.8.

At any instant, a mobile agent can be in any of the states of the Markovian cycle, and their state transition diagram can be explained as follows:

I) In the 'creating state', the agent is created.

II) Once activated, transition occurs to the 'running' state. The agent executes the code and performs its functions in this state.

III) The agent may be deleted in the running state to enter into the 'terminating' state.

IV) A running agent may be suspended in the running state to enter into the 'suspending' state. The agent sits inactively on the agent server and waits to be activated again.

V) A suspended agent may be revoked. When it is dispatched to a host, it enters the 'migrating' state in which the agent is travelling between the hosts.

VI) On reaching the site, the agent resumes its functions and enters running mode.

13.4.3 Challenges

Mobile agent technology has been widely accepted as a design approach in implementing distributed systems. However, there are certain challenges [17] that the technology faces. Complexity in design and maintenance as compared with traditional approaches is one such road block. Security of mobile-agent-based systems is another major concern that hampers their widespread deployment. Mobile agents are susceptible to attacks from malicious users during transition. Host machines are also vulnerable to ghost agents, which may masquerade as genuine mobile agents. Integration and interoperability with existing applications are some other issues that should be addressed to make the technology widely acceptable.

13.5 Cognitive Security

The security systems are not capable of providing an end-to-end security solution and foolproof security against all threats to a network and the systems attached to it. Critical and strategic systems running on commercial off-the-shelf products and solutions need a layer of security that is superior to similar solutions being run for general day-to-day operations and transactions. There can be layers of security around a system. With increase in each layer of security around the system, the cost of implementing, managing, and maintaining the security layers increases. After a certain level of security enhancements, a tradeoff is achieved between the cost of the security and the associated tangible and intangible benefits in terms of protection of the networked system.

There are certain systems for which any level of security is not considered to be enough, and the security paraphernalia across these systems is increased with the passage of time to keep enhancing the robustness of the system against cyber attacks. Typical examples of these network systems are military networks, banking and stock exchange networks, fly-by-wire systems controlling aircraft, and networks controlling nuclear reactors or any other hazardous industry. No effort or cost is too great for the security of such critical systems. Security is also important for systems controlling natural resources, national assets, and government transactions. A cyber attack on these systems can be devastating.

Today, network threats have become smart and stealthy. Malware is intelligent and polymorphic and can deceive traditional security layouts. Advanced persistent threats can penetrate the network security or the nodes within it at various levels by circumventing detection by the security software. The commercial off-the-shelf tools and software, many of which are integrated parts of the network in the form of operating systems, user interfaces, and document readers, are used equally by trivial applications and by critical government, military, and commercial networks owing to wide acceptance, ease of availability, low cost, ready-to-use state, and ease of use.

The commercial software opens the way to undetectable zero-day attacks on national assets. However good the design of a security tool may be, it can be tested and benchmarked only for known threats. In fact, security tools are designed only for known threats. These tools are regularly patched for newly detected threats. The patching of deployed security software is a difficult task, as the deployed nodes may not all be reachable or available for updates and patching. Certain nodes that cannot be patched remain prone to attacks. This also leads to variation in the version of the security tool and the number of installed patches across the network. An intelligent attack can circumvent the static security architecture and remain undetected until it has already damaged the system. False alarms by the security system may also lead to lowering of the thresholds of a security system, making it easy for the attacker to evade it.

Network security systems can be classified into three categories:

  i) intrusion prevention systems,
 ii) intrusion detection systems,
iii) network behavior detection systems.

The most common examples of intrusion prevention systems are firewalls. The network behavior detection system may work in tandem with an intrusion detection or prevention system to trigger an action on detection of anomalous behavior in the network. Traditionally, security systems were rule based. This led to frequent security breaches based on flaws detected in the rules, leading to exploration of vulnerabilities in the rules. The only way to cope with these flaws was to keep analyzing the rules and patching up any uncovered cases. In the case of an attacker bringing down the system, the rules were analyzed and redone or improved. This was a regular activity without ensuring foolproof security.

The rule-based systems were followed by machine-learning-based systems and hybrid systems, which are a combination of rule-based and machine-learning-based systems. The present-day network behavior detection systems use artificial intelligence and cognition abilities to detect anomalies in the network. Cognitive network behavioral analysis (NBA) helps in detection and mitigation of unknown threats coming from new viruses, bots, worms, zero day attacks, modern malware, or internal attacks. The NBA system continuously monitors the traffic flowing through the system to detect and prevent on a real-time basis any advanced persistent threat that has just attempted an attack. The NBA systems are generally based on a multimobile agent framework running artificial intelligence and soft computing algorithms to have traffic visibility for security posture assessment.

13.5.1 Solution Concept

The solution to the intelligent patterns can be achieved by using multiple mobile agents that keep moving across the network, sampling data flows to detect any deviation in traffic pattern or anomalous node behaviors. The intelligent agents should also be capable of reconfiguring themselves to adapt to new security threats and detected attacks. The security system should be capable of storing and analyzing streams of data to ensure routine and normal network behavior and retrieve unknown and unforeseen attack patterns. As these types of attack are generally unknown to the network designers and the developers of the security solutions, the intelligent security solution should be capable of behavioral analysis of the network not only to detect but also to mitigate known as well as unknown attacks.

Network behavioral analysis uses artificial intelligence techniques such as pattern recognition, stylometrics, adaptive learning and cognition, soft computing techniques (such as fuzzy logic, genetic algorithms, and neural networks), and game theoretical techniques, in combination with traditional rule-based methods, to detect and thwart the exploration of vulnerabilities by attackers as well as organized hacking. A combination of these techniques is essential for developing intelligent network security systems because the present-day attackers also study network behavior using trend retrieval and analysis tools to design malware for successful penetration of the security of the network. The security software should also possess high storage and processing capabilities to enable storage of data traffic patterns for future case-based study, trend analysis, and machine learning.

The solution should work at different levels of abstraction and visibility. It should be configurable for protection and management of a node, LAN, or the complete enterprise connected across a WAN. The system has to be self adaptive to changing network architecture and variable link characteristics to enhance fault tolerance and reduce operational cost. The system should be trained enough to suppress false alarms.

However, as such systems have to deal with unknown and unbelievable threats and security attacks, they should possess inference and learning capabilities and should preferably be operational in a distributed environment for redundancy, reliability, coverage, and speed. The optimized distributed operation of such a detection system is best achieved by using multiple mobile artificial-intelligence-based agents patrolling the network. These multiple mobile agents are well integrated and self-configurable to ensure an overlapped and complete coverage of the network to avoid any shadow regions out of security analysis scans for future network defense.

13.5.2 Cognitive Capabilities

The integrated mobile agents in the system are capable of detecting conventional security attacks based on known patterns. These are also intelligent to detect advanced malware, morphed attacks, and masquerading behavior of bots by patrolling the network to detect anomalies in traffic patterns and statistically analyze them. As these next-generation security systems should be deployable in a variety of networks, they should be able to self-configure their parameters. One technique for doing so is for a security system to harvest data from the network and then stabilize itself in the new operational environment by fine-tuning the parameters. Very few rules are hardwired in these systems, to avoid bias and keep the systems open ended for deployment as well as learning. Game theoretical models are used to learn system behavior. The behavior of the agents is also dynamic and changes in response to change in network behavior to enable security agents to camouflage themselves in the network flow and avoid detection by malware and other threat agents. The system has a heuristic-based scanner to detect a potential malware footprint. It may also add a layer of automated fuzzy testing to monitor exceptions.

These systems have a trust engine and are robust to detect and mitigate a variety of attacks, the most dreadful of which are from the following categories:

- *Zero-day weakness and attacks.* These can be against in-house as well as commercial software.

- *Advanced persistent threats.* Buffer overflows, file format vulnerability, digital masquerade, attribution problems, identity mask, and remote administration tools (RATs).
- *Digital intruders.* Poison Ivy, spear-phish, DLL resident malware, and kernel mode rootkits.
- *Polymorphic malware threats.* Mutation-based, generation-based, polymorphic obfuscation, zombies, metamorphic code techniques, oligomorphic engines.

13.5.3 General Capabilities

A cognitive security system generally possesses the following capabilities:

- It complements an existing network security infrastructure and provides the necessary intelligence to address the growing complexity of future requirements.
- It provides self-configuration and self-management for easy administration by network administrators and security analysts.
- It has low integration and management costs.
- It has a rapid startup and is able to start detecting requirements in less than a predefined time after being installed.
- It has high sensitivity in detecting requirements at the granular level.
- It has low false alarms, by using artificial intelligence in the core processing engine.
- The cognitive security solution should have good trust models.
- The deployment structure should be hardened against attacks.
- Comprehensive reports in external industry-standard alert formats such as email, ticket reporting, file logs, or system logs (syslog) should be generated.
- It works in conjunction with other layers of the network.

References

1 S. Bosworth, M. E. Kabay, and E. Whyne (eds). *Computer Security Handbook, volume 1.* Hoboken, NJ, John Wiley & Sons, Inc., 4th edition, 2009.
2 S. Goswami, N. R. Krishnan, M. Verma, S. Swarnkar, and P. Mahajan. Reducing attack surface of a web application by Open Web Application Security Project compliance. *Defence Science Journal*, **62**(5):324–330, 2012.
3 G. Singh and S. Goswami. Reducing the attack surface of software. *PC Quest*, August 2012.
4 N. E. Fenton and M. Neil. A critique of software defect prediction models. *IEEE Transactions on Software Engineering*, **25**(5):675–689, 1999.
5 Y. Shin, A. Meneely, L. Williams, and J. A. Osborne. Evaluating complexity, code churn, and developer activity metrics as indicators of software vulnerabilities. *IEEE Transactions on Software Engineering Journal*, **37**(6):772–787, 2011.
6 J. Walden and M. Doyle. SAVI: static-analysis vulnerability indicator. *IEEE Security Privacy*, **10**(3):32–39, 2012.
7 P. K. Manadhata and J. M. Wing. An attack surface metric. *IEEE Transactions on Software Engineering*, **37**(3):371–386, 2011.
8 M. Howard, J. Pincus, and J. Wing. Measuring relative attack surfaces. *Workshop on Advanced Developments in Software and System Security*, Springer, 2005.

9 T. L. Ha and P. K. K. Loh. Evaluating AVDL descriptions for web application vulnerability analysis. *IEEE International Conference on Intelligence and Security Informatics ISI*, pp. 279–281, 2008.

10 T. Heumann, J. Keller, and S. Turpe. Quantifying the attack surface of a web application. *GI Sicherheit 2010: Sicherheit, Schutz und Zuverlässigkeit*, pp. 305–316, Bonner Kollen Verlag, 2011.

11 V. Lee and L. Shao. Estimating potential IT security losses: an alternative quantitative approach. *IEEE Security and Privacy*, **4**(6):44–52, 2006.

12 P. K. Manadhata, Y. Karabulut, and J. M. Wing. Report: Measuring the attack surfaces of enterprise software. *1st International Symposium on Engineering Secure Software and Systems – Lecture Notes in Computer Science, Volume 5429*, pp. 91–100, 2009.

13 P. K. Manadhata, J. M. Wing, M. A. Flynn, and M. A. McQueen. Measuring the attack surfaces of two FTP daemons. *2nd ACM Workshop on Quality of Protection, QoP '06*, pp. 3–10, New York, USA, October 2006.

14 T. Lv and P. Yan. A web security solution based on XML technology. *ICCT '06. International Conference on Communication Technology*, pp. 1–4, 2006.

15 A. R. Tripathi, T. Ahmed, and N. M. Karnik. Experiences and future challenges in mobile agent programming. *Microprocessors and Microsystems*, **25**(2):121–129, 2001.

16 O. K. Sahinkoz and N. Erdogan. A two-levelled mobile agent system for electronic commerce. *Journal of Aeronautics and Space Technologies*, **1**(2):21–32, 2003.

17 M Eid, H Artail, A Kayssi, and A Chehab. Trends in mobile agent applications. *Journal of Research and Practice in Information Technology*, **37**(4):323–351, 2005.

18 R. H. Glitho, E. Olougouna, and S. Pierre. Mobile agents and their use for information retrieval: a brief overview and an elaborate case study. *IEEE Network*, **16**(1):34–41, 2002.

19 W. Du, J. Deng, Y. S. Han, S. Chen, and P. K. Varshney. A key management scheme for wireless sensor networks using deployment knowledge. *IEEE Transactions on Dependable and Secure Computing*, **3**:62–77, 2006.

20 H. Al-Sakran. Developing e-learning system using mobile agent technology. *2nd Information and Communication Technologies Conference ICTTA '06, Volume 1*, pp. 647–652, 2006.

Abbreviations/Terminologies

ARP   Address Resolution Protocol
AS   Attack Surface
ASI   Attack Surface Indicator
AVDL   Application Vulnerability Description Language
CPU   Central Processing Unit
DDMS   Distributed Database Management System
DDoS   Distributed Denial of Service (Attack)
DLL   Dynamic Link Library
DNS   Domain Name Server
FTP   File Transfer Protocol
ICMP   Internet Control Message Protocol
IDS   Intrusion Detection System
IEEE   Institute of Electrical and Electronics Engineers

IP   Internet Protocol
LAN   Local Area Network
NBA   Network Behavioral Analysis
NBC   Nuclear, Biological, Chemical (Detection/Threat)
SAVI   Static Analysis Vulnerability Indicator
SCADA   Supervisory Control and Data Acquisition
SYN/ACK Synchronize/Acknowledge
TCL   Tool Command Language
TCP   Transmission Control Protocol
UDP   User Datagram Protocol
VPN   Virtual Private Network
WAN   Wide Area Network

Questions

1   How is a flaw in programming or misconfiguration of a network device considered as a security risk?

2   Explain the steps followed to poison a recursive DNS.

3   Explain five different types of denial of service attack.

4   How can social engineering techniques be used to gain access to a network?

5   Can the attack surface of web-based software with a user interface be reduced to zero? Give reasons for your answer.

6   Explain the various measures that can be taken to reduce the attack surface.

7   What is the essentiality of a cognitive security system?

8   Explain the following:
    A promiscuous mode,
    B ARP spoofing,
    C the life cycle of a mobile agent,
    D zero-day attack,
    E a network behavior detection system,
    F polymorphic malware,
    G advanced persistent threats.

9   Differentiate between the following
    A active attack vs passive attack,
    B iterative DNS vs recursive DNS,
    C key logger-Trojan vs password-sending Trojan,
    D network attack surface vs software attack surface,
    E intrusion detection vs intrusion prevention system,
    F smurf vs fraggle.

10 State whether the following statements are true or false and give reasons for the answer:
   A The physical security of the network and computational assets is important even when they are secured technically.
   B Cryptic/encoded data is resistant to any attack.
   C Snort is a network sniffing tool.
   D A secured system has a zero attack surface.
   E The attack surface does not have standard quantified metrics, but indicates system security in relative terms.
   F A basic firewall acts as an intrusion detection system (IDS).

Exercises

1 Install snort in your network and capture network data for 10 min. What are the network parameters that you can interpret from the captured traffic?

2 Use a suitable command to see the ARP table of your computer/server.

3 Try to write down DNS entries as written in a DNS server for a website/webserver, mail server, and other related entries in a DNS.

4 Write a program in a suitable language for a rule-based intrusion detection system for any specific protocol attack.

5 With respect to network-centric warfare with a smart soldier and smart weapons system, please draw the following (prior knowledge of unified modeling language is assumed):
   A activity diagram,
   B state transition diagram,
   C sequence diagram,
   D class diagram.

6 Capture the snort traffic for a day and then use a suitable machine-learning tool to classify the traffic into 'normal' and 'not normal' traffic profiles. There was no attack on the network during traffic capture, which indicates that there should be no 'not normal' traffic profiles, and hence the entire traffic should be 'normal'. Still, what is the percentage accuracy of predicting the 'normal' traffic?

14

Reliability and Fault-Tolerant and Delay-Tolerant Routing

14.1   Fundamentals of Network Reliability

A computer network finds widespread applications in multiple fields. One of the major criteria to determine whether a network system can be used for a certain purpose is the reliability of the system. It is essential to understand what reliability means for a network system, the idea behind various concepts related to it, and the different methods that can be adopted to calculate the reliability of various systems. Today, computer systems and networking have become the backbone of society. No matter if the field of application be big or small, networking and computer systems have found their way into each one of them. New technologies are being proposed and developed at a rapid pace, each with its own set of characteristics [1]. But for the new technologies to supersede the previous ones, it is important for them to be better off than previous technologies in terms of some measurable features. Reliability of a network is one such important feature. A new technology can only be useful if it is more reliable than the previous

Network Routing: Fundamentals, Applications, and Emerging Technologies, First Edition.
Sudip Misra and Sumit Goswami.
© 2017 John Wiley & Sons Ltd. Published 2017 by John Wiley & Sons Ltd.
Companion website: www.wiley.com/go/misra2204

technologies [2–5]. Many attempts are being made to devise new ways and means to develop a fairly consistent and accurate approach to calculating reliability.

Network is a very broad term. In the generic sense, it can be understood as a system in which all the components (usually called nodes) can interact with each other, either directly or via some other component(s) of the network. Formally, a network refers to interconnection of at least a pair of computers that are capable of communicating among themselves either for transfer of data or control signals [6]. If a computer network has been established, the processes on each of the interconnecting devices become capable of communicating, directly or indirectly, with the processes on one or more devices on the network [7].

Networks can also be categorized [6] on the basis of spread (LAN, MAN, WAN), topology (the geometric arrangement of a computer system), protocol (the common set of rules and signals that computers on the network use to communicate), or architecture (peer-to-peer or client/server architecture).

The reliability of a network $R(t)$ is the probability that the system is continuously up, i.e. all the nodes are continuously working and the communication links between them are active for the entire time period $[0, t]$. A reliable network gives a guarantee of successful communication between source and destination computer devices. Many methods have been presented to define reliability [8–13]. In one method, the reliability (R_{SN}) of a network node is predicted by a formula using Poisson's distribution:

$$R_{SN} = \exp(-\lambda t) \tag{14.1}$$

where λ is the failure rate of a node and t is the time period.

Equation (14.1) describes reliability and can also be used to measure the fault tolerance of sensor nodes. Reliability depends upon the exponential of two terms – failure rate and time.

The reliability of a network $R(t)$ is often confused with its availability $A(t)$ [8–12]. Reliability is the probability that the system is continuously up for the entire time period $[0, t]$, while availability is the average fraction of time over the interval $[0, t]$ for which the system is up.

For example, consider a system that fails every minute on average, but comes back after only a second. Such a system has a mean time between failures (MTBF) of 1 min, and therefore the reliability of the system $(= e^{-t/\text{MTBF}})$ for 60 s is $R(60) = e^{-60/60} = 0.368$ ($\approx 36.8\%$), which is low, whereas the availability = MTTF/MTBF = 59/60 = 0.983 ($\approx 98.3\%$) is comparatively high [11–13]. To clarify further the difference between reliability and availability, we can take the example of a cooling system. If we just want to reduce the temperature of a normal room, then the availability of the cooling system is more important, whereas the reliability of the cooling system is important if we want to keep the temperature of a drug constant.

14.1.1 Importance of Reliability Calculation

It will be of importance to us if the reliability of a network can be determined, as then we will have an approximate idea about the maintenance cost of the network, i.e. for a less reliable network the maintenance cost will be high compared with a reliable network. Further, we can plan accordingly whether to continue with the same network or incorporate some changes with which the reliability of the network could be increased [14]. Also, if we have to choose between some networks, we can choose the suitable network based on our needs (reliable or more readily available). As almost every system can be said to be a network, such as a railway network, a communication network, and other transport networks, knowing their reliability in terms of relevant parameters (such as links and nodes) can help to monitor and improve them effectively from time to time.

14.1.2 Methods to Calculate the Reliability of a Network

Various methods have been proposed for calculating the reliability of a network. Some of the major methods are as follows:

A. Using Probability

Suppose a network has nodes and for this network to be in working condition a minimum m nodes should be up. This kind of case can occur if the network has extra nodes as spares.

Then

$$R(t) = \sum_{i=m}^{n} Pi(t) \tag{14.2}$$

where $Pi(t)$ is the probability that exactly i nodes are up at time t [11].

Analysis of the method Calculation of the reliability of the network using probability is an easy and straightforward approach, but it cannot be widely used, as in most of the networks we cannot say that for a system to be up we need a fixed minimum number of nodes in working condition. Instead, the minimum number of nodes necessary to be up depends on the path taken by the data during communication between source and destination node. Hence, this method is generally not used for practical purposes.

Example Suppose a network has ten nodes and a minimum six nodes should be in working condition for the system to be up. Using equation (14.2),

$$R(t) = \sum_{i=m}^{n} Pi(t)$$
$$= \frac{{}^{10}C_6 + {}^{10}C_7 + {}^{10}C_8 + {}^{10}C_9 + {}^{10}C_{10}}{2^{10}}$$
$$= 0.377$$
$$\approx 37.7\%$$

B. Using a Constant Failure Rate

The failure rate (λ) [11–13] can be calculated through MTBF [15]; if it is nearly constant, then

$$\lambda = 1 / \text{MTBF}$$

To define the MTBF, we must know the mean time to failure (MTTF) and the mean time to repair (MTTR). The MTTF is the average time for which the system operates normally until a failure occurs, and MTTR is the average time taken to repair a failed system. Thus, the MTBF (the average time between two consecutive failures of the system) is

$$MTBF = MTTF + MTTR \qquad (14.3)$$

So, the reliability R is obtained as

$$R = e^{-t/MTBF} \qquad (14.4)$$

whereas the availability $A(t)$ is the average fraction of time over the interval $[0, t]$ for which the system is up. In other words, availability determines the fraction of time for which the system is up or available in a certain period of time. We obtain availability A as

$$A = MTTF / MTBF \qquad (14.5)$$

Analysis of the method   Calculation of the reliability of the network using a constant failure rate requires the data to calculate MTTF, MTBF, and MTTR, and can be used only if the failure rate is nearly constant, i.e. if the failure rate is not dependent on time and does not change with it. So, this method can be used only if the failure rate is constant or approximately constant.

Example 1   Consider a system that fails every minute on average, but recovers after a second. This implies that MTBF = 60 s, and for t = 1 h, using equation (14.4), we have

$$R(t) = e^{-t/MTBF}$$
$$\approx 8.76 \times 10^{-25}\%$$

Example 2   Table 14.1 presents data related to the uptime of an organizational network, collected over a period of 6 months. The exact date and time of failure of the network and its coming back into operation had been logged by the network management system.

Table 14.1 Network downtime statistics for the last 6 months of an organization.

Day	Down during a day (24 h)		Down during working hours (8 h)	
	No. of times	Total downtime (min)	No. of times	Total downtime (min)
Sunday	7	2130	0	0
Monday	9	1370	4	180
Tuesday	12	1220	2	50
Wednesday	10	1460	1	10
Thursday	4	1320	0	0
Friday	7	1140	1	40
Saturday	5	980	1	10

Table 14.2 Reliability and availability calculated for different time periods.

	8 h	1 day	1 week	1 month	6 months
Reliability	90.5%	74.1%	12.2%	0.012%	3.53×10^{-22}%
Reliability (working hours only)	95.1%	86.1%	35%	22.3%	1.88×10^{-10}%
Availability	96.3%	96.3%	96.3%	96.3%	96.3%
Availability (working hours only)	99.7%	99.7%	99.7%	99.7%	99.7%

Now, suppose we have to calculate the availability and reliability of this system for 8 h, 1 day, 1 week, 1 month, and 6 months. Using equations (14.4) and (14.5), we obtain the results given in Table 14.2.

C. Using the Circuit of the Network

If the reliability of all the links of the network is known, then we can calculate the reliability of that network for data to flow between source and destination node [11–13]. Suppose the reliability of the *i*th link at time *t* is $RI(t)$. Then, for series connection

$$Rs(t) = \prod_{i=1}^{N} RI(t) \tag{14.6}$$

where N represents the nodes connected in series, and for parallel connection

$$Rp(t) = 1 - \prod_{i=1}^{N} \left(1 - RI(t)\right) \tag{14.7}$$

where N represents the nodes connected in parallel.

Analysis of the method Calculation of the reliability of the network using the circuit of the network is an extremely simple, easy-to-use method and requires a similar effort to solving a current–resistance (IR) circuit. However, if the network is not completely reducible, then it becomes very difficult to solve the system and therefore it becomes complex to calculate the reliability of such a network. It would then require some other method to solve the problem completely. So, this method is not sufficient for large systems and would need some other method to assist in solving large networks.

Example Figure 14.1 shows five nodes, namely, a, b, c, d, and e, which are connected through different communication links and have different reliabilities, which are indicated against each link. This network is maintaining communication between points A and B.

Reducing Figure 14.1 with the series formula and with the parallel formula, equations (14.6) and (14.7) respectively, we obtain Figure 14.2.

Using equations (14.6) and (14.7),

$$
\begin{aligned}
R &= \left(1 - \left(1 - (0.8)(0.85)(0.6)\right)(0.2)\right)(0.94) \\
&= (1 - 0.118)(0.94) \\
&= 0.829 \\
&\approx 82.9\%
\end{aligned}
$$

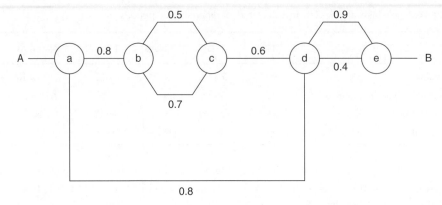

Figure 14.1 Sample network for reliability calculation using the circuit of the network.

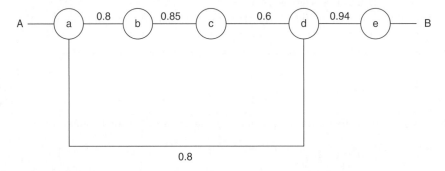

Figure 14.2 Reduced network for reliability calculation using the circuit of the network.

D. Using the Probability of the Path

This method [11, 16] requires a network to be visualized in the form of a circuit diagram that cannot be further reduced using series and parallel reduction formulas. The notation used in this method is as follows:

Let

N_s	source node
N_d	destination node
m	number of paths between N_s and N_d
P_i	i^{th} path between N_s and N_d
E_i	event in which P_i is working
R_{N_s,N_d}	probability that N_s and N_d can communicate (path reliability)
$\text{Prob}\{E_i\}$	probability of event E_i happening.

Then

$$R_{N_s,N_d} = \text{Prob}\{E1\} + \text{Prob}\{E2 \cap E1'\} + \ldots$$
$$+ \text{Prob}\{Em \cap E1' \cap E2' \cap \ldots \cap E(m-1)'\} \qquad (14.8)$$

Analysis of the method   Calculation of the reliability of the network using the probability of the path is very easy to understand and apply. However, this method is complex in the case of a huge network, as it will be difficult to reduce it and the calculation will be a bit tough.

Example   In Figure 14.3 the network shown has six nodes, namely a, b, c, d, S, and T, out of which a, b, c, and d are connecting nodes, while S and T are source and destination nodes respectively [18]. All the nodes are connected through different communication links and have different reliabilities, which are indicated against each link. The approach to calculating reliability is as follows:

- Reduce the figure until it cannot be further reduced from series parallel reduction.
- Find path reliabilities to obtain the reliability of successful communication between source and destination node.
- Reducing Figure 14.3 using equations (14.6) and (14.7) for series and parallel connection respectively [18], we obtain the network shown in Figure 14.4.

The edges are denoted by E1, E2, E3, ..., E7.

The paths possible (assuming cycles are not allowed) from source (S) to destination (T) are:

P1 = {E1, E3, E7}
P2 = {E2, E6}
P3 = {E1, E4, E6}
P4 = {E1, E3, E5, E6}
P5 = {E1, E4, E5, E7}
P6 = {E2, E4, E3, E7}
P7 = {E2, E5, E7}

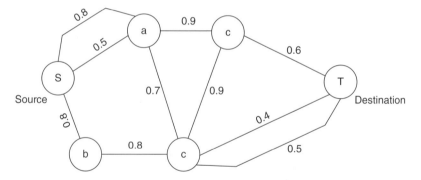

Figure 14.3 Sample network for reliability calculation using the probability of the path.

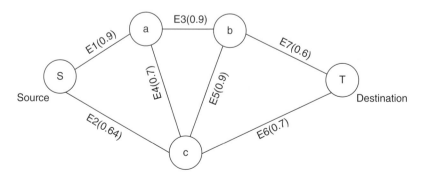

Figure 14.4 Reduced network for reliability calculation using the probability of the path.

$$R_{Ns,Nt} = Prob\{P1\} + Prob\{P2 \cap P1'\} + Prob\{P3 \cap P1' \cap P2'\}$$
$$+ Prob\{P4 \cap P1' \cap P2' \cap P3'\} + Prob\{P5 \cap P1' \cap P2' \cap P3' \cap P4'\}$$
$$+ Prob\{P6 \cap P1' \cap P2' \cap P3' \cap P4' \cap P5'\}$$
$$+ Prob\{P7 \cap P1' \cap P2' \cap P3' \cap P4' \cap P5' \cap P6'\}$$

$$= \{E1.E3.E7\} + (E2.E6.E1'.E3'.E7') + \{E1.E4.E6.E3'.E7'.E2'\}$$
$$+ \{E1.E3.E5.E6.E7'.E2'.E4'\} + \{E1.E4.E5.E7.E3'.E2'.E6'\}$$
$$+ \{E2.E4.E3.E7.E1'.E6'.E5'\} + \{E2.E5.E7.E1'.E3'.E6'.E4'\}$$

$$= (0.9)(0.9)(0.6) + (0.64)(0.7)(0.1)(0.1)(0.4)$$
$$+ (0.9)(0.7)(0.7)(0.1)(0.4)(0.36)$$
$$+ (0.9)(0.9)(0.9)(0.7)(0.4)(0.36)(0.3)$$
$$+ (0.9)(0.7)(0.9)(0.6)(0.1)(0.36)(0.3)$$
$$+ (0.64)(0.7)(0.9)(0.6)(0.1)(0.3)(0.1)$$
$$+ (0.64)(0.9)(0.6)(0.1)(0.1)(0.3)(0.3)$$

$$= 0.486 + 0.001792 + 0.0063504 + 0.02204496 + 0.00367416$$
$$+ 0.00072576 + 0.00031104$$

$$= 0.52089832$$
$$= 0.521$$
$$\approx 52.1\%$$

This implies that this system of communication between source (S) and destination (T) is 52.1% reliable.

E. Poisson Processes

The probability of exactly k nodes failing within time interval t is given by the equation [11, 13, 17]

$$P\{k(t)\} = \frac{(\lambda t)^k e^{(-\lambda t)}}{k!} \quad \left(\text{for } k = 0, 1, 2, \ldots\right) \tag{14.9}$$

where λ is the constant failure rate and k is the number of events occurring within time interval t.

The convergence factor (c) is the probability of successful detection and repair of the defective node, and c^k is the probability that the system will survive the failure of k nodes. Using equation (14.9), we have

$$\text{Reliability in time interval} \left[0, t\right] = R(t) = \sum_{k=0}^{\infty} Pk(t) c^k \tag{14.10}$$

Analysis of the method   Discussion of the calculation of network reliability using the Poisson process requires knowledge of redundancy. Redundancy [8, 11] occurs when there are a greater number of components in the system than needed for the system to work successfully. In this method of reliability calculation, as we are using infinite spares, the network requires an infinite degree of redundancy, which is not necessarily available in a real-life network. Further, some may opt not to increase the level of redundancy of the network as this would increase the cost of the network, while others may increase the level of redundancy of the network, as increasing the redundancy level will surely make the system more reliable. So, this type of method will be useful when the organisation has data regarding the convergence factor and redundancy.

Example   Let us consider that there are ten identical nodes in the system, with infinite spares. The spares are assumed to be resistant to failure (until active), while the rate of failure for active nodes is 1 min per processor.

The convergence factor $(c) = 0.7$. Here, as we have ten nodes, $\lambda = 10 \times 1 = 10$ min per processor. From equation (14.10)

$$
\begin{aligned}
R(t) &= \sum_{k=0}^{\infty} Pk(t) c^k \\
&= \sum_{k=0}^{\infty} \frac{(\lambda t)^k e^{-\lambda t} c^k}{k!} \\
&= e^{-\lambda t} \sum_{k=0}^{\infty} \frac{(\lambda t c)^k}{k!} \\
&= e^{-\lambda t} e^{\lambda t c} \\
&= e^{-\lambda t (1-c)}
\end{aligned}
$$

The reliability of the system over 10 min is

$$
\begin{aligned}
R(60) &= e^{-10 \times 10 \times 0.3} \\
&= e^{-30} \\
&= 9.358 \times 10^{-14} \\
&\approx 9.36 \times 10^{-12}\%
\end{aligned}
$$

F.   Markov Model

Markov models [11–13] are used to calculate system reliability where combinatorial arguments are insufficient to discuss reliability issues. This method is different from the Poisson process as in this method we can include coverage factors and repair the process without assuming an infinite amount of redundancy.

The Markov chain is said to be $X(t)$ if

$$
\text{Prob}\{X(tn) = j \mid X(t0) = i0, \ldots, X(tn-1) = in-1\} = \text{Prob}\{X(tn) = j \mid X(tn-1) = in-1\}:
$$

$$(14.11)$$

where $X(tn) = j$ implies that at time tn the process is in state j.

In this method we assume that we are dealing with discreet Markov chain states and continuous time.

The variables used are as follows:

λi the rate of leaving state i,
$\lambda i,j$ the rate of transition from state i to state j,
$Pi(t)$ the probability that the process is in state i at time t.

Using the above variables, and calculating accordingly, we obtain

$$\frac{d}{dt}Pi(t) = -\lambda i Pi(t) + \sum_{j!=1} \lambda j,i Pj(t) \qquad (14.12)$$

Initial conditions:

For an n-processor system, $Pn(0)=1, Pn-1(0)=0,\ldots, P0(0)=0$
Reliability $= R(t)=1-P0(t)$

Analysis of the method   Calculation of the reliability of the network using the Markov model requires information on the failure and repair rate. The concept that is used in this method is quite simple to understand, but the calculation involved in this method is difficult to solve, and hence the method is not applicable if a large number of processors are being considered.

Example   In a triplex system we have three active network devices, each having a failure rate of λ and a repair rate of μ. The Markov model for this system is shown in Figure 14.5.

Using equation (14.12) for this system, we obtain the following:

$$\frac{d}{dt}P3(t) = -3\lambda P3(t) + \mu P2(t)$$

$$\frac{d}{dt}P2(t) = -(2\lambda + \mu)P2(t) + 2\mu P1(t) + 3\lambda P3(t)$$

$$\frac{d}{dt}P1(t) = -(\lambda + 2\mu)P1(t) + 2\lambda P2(t) + 3\mu P0(t)$$

$$\frac{d}{dt}P0(t) = -3\mu P0(t) + \lambda P1(t)$$

Initial conditions:

$$P3(0)=1, P2(0)=0, P1(0)=0, P0(0)=0$$

Solving the above equations with these initial conditions, we can obtain the reliability of the system.

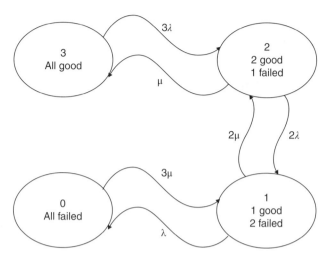

Figure 14.5 Sample network for reliability calculation using the Markov model.

G. Using Minterms

In this algorithm [18], a probabilistic graph is used to represent the network as a collective unit, and the branches of the graph represent the specific links. The algorithm makes use of three assumptions:

- The only existing states of the branches are UP (operating) and DOWN (failed).
- The nodes are 100% reliable.
- All the branches under consideration are undirected.

The basis of this algorithm is to determine the non-reliability of the network and then to use this to calculate the reliability. The general representation of the branch is through a binary variable xi:

> If $xi = 1$, branch i is UP;
> If $xi = 0$, branch i is DOWN.

The minterms that are applicable are identified and enumerated using the minimal cut method. The minimal cut of the graph is the set of minimum branches that on removal disrupt all paths between the two specified nodes. The first step of the algorithm is the series-parallel reduction procedure. Here, the series, series-parallel, and parallel sets of branches are reduced to a single branch to simplify the computations.

The notation used in this method is as follows:

> Qs,t the probability that all paths between the source node and the terminal node are disrupted, where s is the source node and t is the terminal node;
> xi the binary variable to indicate the state of branch i;
> b the number of branches present in the graph;
> Xx $x1x2 \dots xb$, the b-tuple minterm indicating the state of a graph with b binary-state channels;
> Xax the ba-tuple minterm indicating the state of subgraph α, $\alpha = $ I, II;
> ba the number of branches present in subgraph α, $\alpha = $ I, II.
> ri the reliability of branch i;

Nc   the set of nodes on the boundary between subgraphs I and II;

k the number of nodes on the boundary between subgraphs I and II;

Ai' the complement of Ai, a subset of NC:$Ai \cup Ai'$ = Nc.

Φ an empty set;

Sx the set of all b-tuple minterms XX;

$S\alpha x$ the set of all $b\alpha$-tuple minterms $X\alpha x$, α = I, II.

The flowchart depicting the algorithm is as given below.

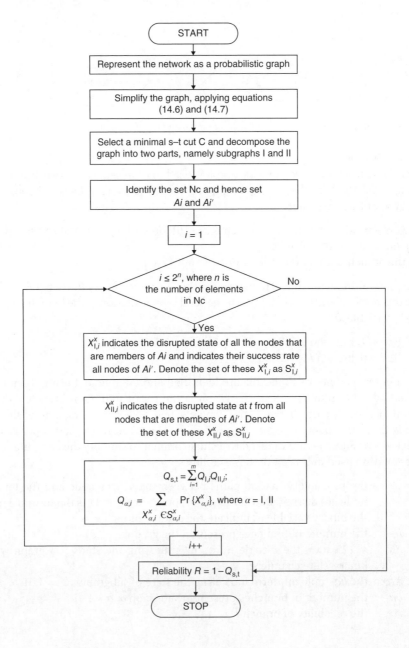

Analysis of the method   Calculation of the reliability of the network using minterms will be tough to use in the case of large graphs, as the number of states increases exponentially with the number of branches.

Example   Figure 14.6 shows a network having six nodes, namely a, b, c, d, S, and T, out of which a, b, c, and d are connecting nodes, while S and T are source and destination nodes respectively. All the nodes are connected through different communication links and have different efficiencies, which are indicated against each link.

Step 1

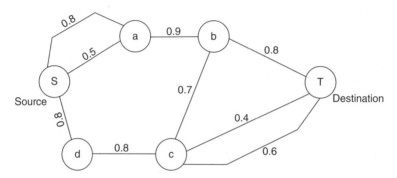

Figure 14.6 Sample network for reliability calculation using minterms.

Step 2
Reducing Figure 14.6 using equations (14.6) and (14.7) for series and parallel connection respectively, we obtain the network shown in Figure 14.7.

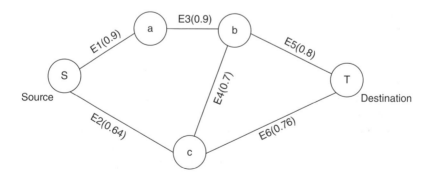

Figure 14.7 Reduced network for reliability calculation using minterms.

Step 3
Let us select a minimal cut as {E2, E3}. Based on the minimal cut, we obtain subgraph I and subgraph II, as shown in Figure 14.8.

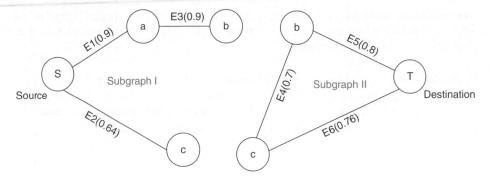

Figure 14.8 Subgraph I and subgraph II for minimal cut b–c.

Step 4

Nc = {b, c}

$A1 = \Phi, A2 = \{b\}, A3 = \{c\}, A4 = \{b, c\};$

$A1' = \{b, c\}, A2' = \{c\}, A3' = \{b\}, A4' = \Phi$

Steps 6 and **7** are shown in Table 14.3.

$$Q_{s,t} = 0.0248832 + 0.01284096 + 0.02729376 + 0.0684$$
$$= 0.13341792$$

$$\text{Reliability} = 1 - Q_{s,t}$$
$$= 0.86658208$$
$$\approx 86.6\%$$

The analysis of various reliability methods is summarized in Table 14.4.

14.2 Fault Tolerance

Computer networks are used in many important applications in industries, hospitals, financial institutes, power generation, e-governance, telecom, telematics, agriculture, the transportation sector, and many more. As of today we can claim these networks to be critical, the malfunction of which can lead to breaches of national security, loss of human life, or financial loss. Failure of networks controlling a nuclear reactor can lead to a catastrophe leading to loss of human life. Failure of a network in a stock exchange or banks can lead to financial losses. A failed supervisory control and data acquisition (SCADA) network can lead to financial loss as well as to loss of human comfort or life. The failure cannot always be attributed to software faults, configuration errors, or hardware failure, as the network might be subjected to operations in a rough terrain with adverse environmental conditions in terms of operating temperature, electromagnetic radiation, dust, obstacles, and pressure. The failure may not necessarily lead to the termination of network operation, but it may lead to unpredictable behaviors, such as delayed operation or unexpected outputs.

Table 14.3 Implementation of steps 5, 6, and 7 of the algorithm used in reliability calculation using minterms.

i	A_i	A_i'	$SI_{j,ix}$ branch 1 2 3	$Q_{i,j}$	Subgraph II for calculating $Q_{i,j}$	$Q_{II,j}$	$Q_{I,j} Q_{II,j}$
1	Φ	{b, c}	1 1 1	(0.9) (0.64) (0.9) = 0.5184	A$_1'$ —(0.8)— t, —(0.76)—	0.2×0.24 = 0.048	0.5184×0.048 = 0.0248832
2	{b}	{c}	1 1 0 0 1 1 0 1 0	(0.9) (0.64) (0.1) = 0.0576 (0.1) (0.64) (0.9) = 0.0576 (0.1) (0.64) (0.1) = 0.0064 = 0.1216	(0.7) b (0.8); A$_2'$ (0.76) t	$(1 - (0.8)(0.7))(0.24)$ = 0.1056	$0.1216 \times$ 0.1056 = 0.01284096
3	{c}	{b}	1 0 1	(0.9) (0.36) (0.9) = 0.2916	A$_2'$ (0.8) t; (0.7) c (0.76)	$(1 - (0.7)(0.76))(0.2)$ = 0.0936	$0.2916 \times$ 0.0936 = 0.02729376
4	{b, c}	Φ	0 0 0 0 0 1 1 0 0	(0.1) (0.36) (0.1) = 0.0036 (0.1) (0.36) (0.9) = 0.0324 (0.9) (0.36) (0.1) = 0.0324 = 0.0684	(0.7) b (0.8) t; c (0.76)	1.000	0.0684×1.000 = 0.0684

Table 14.4 Pros and cons of reliability methods.

Method	Pros	Cons
Probability	An easy and straightforward approach.	As this method is based on the minimum number of nodes required to be up, it is not useful for most of the systems.
Constant failure rate	Useful for systems in which data for MTTF, MTBF, and MTTR can be easily calculated.	Works only if the failure rate is constant or approximately constant.
Circuit of the network	Extremely simple method.	Does not work for large systems.
Probability of a path	Easy to understand.	Calculation is a bit tough.
Poisson process	Useful if the convergence factor and redundancy level of the system are known.	Does not work for non-redundant systems.
Markov model	Very simple concept to understand.	Requires an n-degree polynomial equation to be solved, so the calculation is tough; n is the number of states a system can have.
Minterms	Can be used to solve large systems.	Difficulty level of the calculations increases rapidly as the size of the system increases.

Network systems are complex and comprise active components such as network devices (NIC, multiplexers, modems, repeaters, switches, and routers) and hosts, as well as passive components such as cables for interconnection and interfaces. Further, the network devices have millions of transistors and interconnections inside them, each with a probability of failure. The larger the size of a network, the greater is the number of nodes and links, leading to an increase in the points of failure. Failure is a deviation from the expected functioning of the system on account of hardware, software, configuration, user, or network errors leading to the inability to deliver the desired optimum results [11, 19]. Faults cause errors, and errors lead to failures.

Based on the time faults remain in the system and the reoccurrence of faults, there are three types of fault:

1) *Transient faults.* A transient fault occurs all of a sudden and then disappears, not to reoccur again. An example of a transient fault is an error in message delivery owing to rebooting of a router in the network as a result of a power surge, which becomes operational again once the router boots back into operation.
2) *Intermittent faults.* An intermittent fault keeps reoccurring. The frequency of fault reoccurrence as well as the time duration of the fault may be regular or irregular. An example of intermittent failure is regular connection and disconnection of the network owing to a loose connection of the cable in the switch.
3) *Permanent faults.* A permanent fault occurs once and is not rectified of its own, and the system continues to be in the faulty state until it is repaired by external intervention. An example of a permanent fault is a cut in the network cable or switch failure due to burnout.

Traditionally, a network fault indicated a change from the connected state to the disconnected state. However, when a network moves from the connected state to the disconnected state, there are intermediate network degradation stages. The final stage, where the system ends up, and the way it handles the processes running in it after a fault occurrence may also be different. A fault may lead to complete cessation of network operation, or the output may clearly indicate that the network has failed. This category of network fault is known as a *fail silent fault*, indicating that the network becomes silent (stops responding) in the case of a fault. Alternatively, the network may keep running even after a fault, but behave unexpectedly or produce incorrect outputs. This category of fault is known as a *byzantine fault*. In the case of a *fail safe fault*, the network moves to a safe state. Still, there are systems designed to continue operation and provide correct output even after the occurrence of a fault, and this category of system is known as a *fail operational system*. However, a fail operational system may not continue to operate at its optimum performance level and may slow down in its operations, leading to *graceful degradation*.

The fault, which may either be in the software or in the hardware, may also be injected into a system at its design stage, owing either to malicious intent or to incompetence of the designer, followed by lack of verification and quality checking. After design, the system enters into a manufacturing stage where again it is prone to fault injection. As the system is deployed, the probability of configuration faults exists. Once the system is operational, it faces communication faults, maintenance faults, and system attacks and intrusions causing faults. To cope with these varieties of faults, a system is designed with physical (hardware) redundancy, software redundancy, information redundancy, and time redundancy. Software faults are generally induced at the time of design and development in the form of bugs, while hardware faults may be induced at the design stage, at the manufacturing stage, or even at the operational stage owing to wear and tear.

Information redundancy is achieved by adding extra bits in the data for error detection and correction, information replication, or coding. Time redundancy provides sufficient slack time and time to repeat a process to regenerate output and retransmit data not only to overcome transient or intermittent faults but also to keep the system operational during graceful degradation. Fault prevention and recovery [20] are generally addressed by incorporating either redundancies or replication. Redundancy refers to the availability of a standby system, which takes over the work if the system under operation fails. Replication refers to the same work being performed by two separate systems in parallel, operating on the same input. The output from one of the systems is selected on the basis of the majority or polling.

To enhance the trustworthiness of a system, fault tolerance is not the sole approach. The system should be designed for fault prediction and fault avoidance. Still, in the case of a fault there should be enough redundancies in place for fault removal. Fault taxonomy can be based on various reasons for the occurrence of the fault, the background of the fault, and the severity of the fault. The criteria of fault classification [21] for the creation of a fault tree, as can be visualized for a network system, are based on the following: the stage of inception (development or operational), the place of occurrence (internal or external), the cause (man made or natural), the system affected (hardware or software), the harmful effect (malicious or non-malicious), the skill of the designer and operator (accidental fault or incompetence fault), the purpose (deliberate or non-deliberate), and the life of the fault (permanent or transient).

14.2.1 Fault-Tolerant Network

The scale of the network for designing a fault-tolerant system ranges from the interconnection of processors and memory in a distributed system to a wide area network of independent systems. There may be one or more paths between the source and destination and the interconnection links may be unidirectional or bidirectional. Network resilience is the capability of the network to ascertain connectivity between nodes and avoid partition even after faults in certain nodes and links. The simplest approach to measuring network resilience is based on the graph theoretical approach based on connectivity and diameter stability.

Node connectivity refers to the minimum number of nodes that should become faulty to disrupt the network in terms of failed connectivity to any node. Link connectivity is defined on similar lines and refers to the minimum number of links that should become faulty to disconnect the network. The second measure of network resilience is diameter stability, which is the rate of increase in the diameter of the network in the case of any node failure. The distance between two nodes is defined as the minimum number of links between those two nodes, and the diameter of a network is the largest distance between nodes present in the network. As a few links start to fail in the network, the distance between nodes starts to increase for those source–destination pairs on the path between which is located any one of the failed nodes, and hence an alternative longer path is now necessary. With increase in the distance between nodes, there is a probability that the diameter of the network will also increase if the failed links are in that path. Diameter stability indicates the level of interconnection in the network such that it should increase at the minimum possible rate.

Fault tolerance can be incorporated into a network by having either multiple paths between all source–destination pairs or spare nodes and redundant links that can replace the failed nodes or links. There may be more than one alternative for designing a fault-tolerant topology, and it can be based on redundant links or spare nodes. Some of the common network topologies and their associated fault-tolerant topologies [11] are indicated in Table 14.5.

14.2.2 Autonomic Network

Owing to escalation of market requirements and advancement in technologies, there has been a reduction in the time available for development and testing. With every new system being introduced, the complexity of the system in terms of hardware components, interconnections, and lines of code increases, and the size of the smallest building block of the system decreases, going to nanoscale in the present day. This leads to a high probability of inception of faults at some stage or other of network design, configuration, or operation. As the probability of faults increased, the design of network systems went through a paradigm change from being simply fault tolerant to being self-healing. Systems are being designed to be aware of themselves as well as of their operating environment. An autonomic network system has the capability to continue operation even after fault detection, and also rectifies the fault itself without human

Table 14.5 Network topologies with non-fault-tolerant and fault-tolerant architecture.

Type of Network	Non-Fault-Tolerant Architecture	Fault-Tolerant Architecture
Butterfly network		
Crossbar network		

(Continued)

Table 14.5 (Continued)

Type of Network	Non-Fault-Tolerant Architecture	Fault-Tolerant Architecture

Mesh network

Hypercube

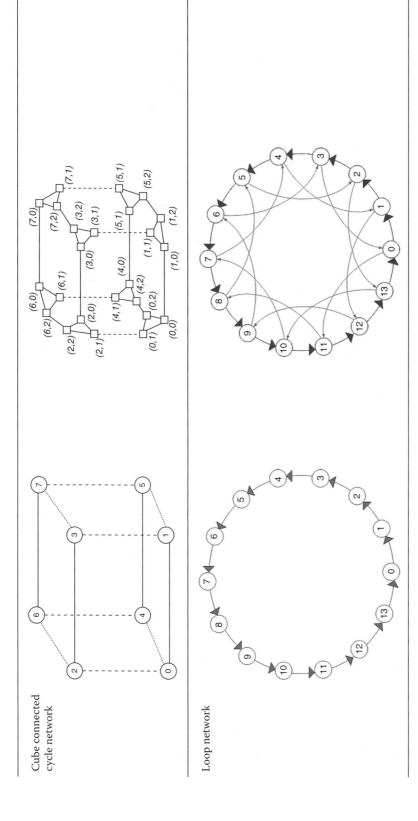

Cube connected cycle network

Loop network

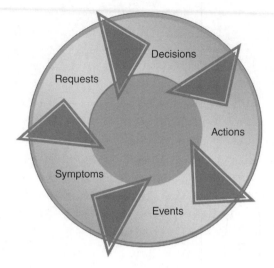

Figure 14.9 Adaptation processes in a self-adaptive system.

intervention. Although most of these features are difficult to incorporate in a wired network, these are common features of present-day wireless networks in general and of ad hoc and sensor networks in particular.

An autonomic network [22] has self-configuring, self-optimizing, self-managing, self-protecting, and self-healing capabilities. A network system, as is common in the case of sensor networks, is self-configuring if after deployment it discovers its neighbors, detects its environment, becomes aware of the communication requirement, and configures itself not only to be a part of the network but also to continue optimum operation by updating its configuration to support mobility and security. The network systems, being resource constrained with high operational requirement, are also *self-optimizing*, reducing power, computation, and bandwidth requirement and increasing the life of the network and its utilization. The *self-protecting* feature of the network relates to prevention of network intrusion and security attacks on the network of links in order to ensure availability, integrity, and confidentiality of the network. Further, if the network is compromised or disrupted, the network system possesses the self-healing capabilities to detect the intrusion and the points of failure and to continue operation from an alternative segment or recover the compromised or damaged links and nodes. An autonomic network system may run through one or more of these capabilities to ensure adaptability and continued operation, as indicated in Figure 14.9.

14.3 Network Management for Fault Detection

Communication networks have acquired increased significance today in the wake of network-centric operations. Decision-making in the field by the network devices as well as by the administrator is based on information about resources, node location, and other vital information obtained using the latest means of communication. The need for real-time and accurate information makes pressing demands on today's networks.

It is important to understand the challenges involved in the management of communication networks and some of the approaches involved in the implementation of management systems for such networks. The use of mobile agent technology as a framework for implementing network management systems is a recent trend. The use of policy-based network management to delegate management functions and automate decision-making is also an upcoming area.

Future network application, which may be in disaster management, health monitoring, industrial application, or warfare, envisages rapid engagement in the operational scenario and requires a resilient, high-capacity backbone extending its reach up to the tactical deployment area. Changes in planning, monitoring, and management in operationally focused tactical communications networks are warranted to match the rapid engagement as well as the advancement in communication technology, which has taken a great leap forward. Advanced network management techniques are now being adopted to manage tactical communication networks.

In this digital-application field scenario, information holds the key to success. In a battlefield scenario managed either locally or remotely across continents, the commanders and the troops are increasingly becoming dependent on information to execute operations successfully. The dynamics of the tactical requirements and the commander's intent distinguishes modern tactical networks from commercial networks. These modern tactical networks have constraints in terms of bandwidth and intermittent connections and are subject to changes driven by the commander's intent, which affects the network behaviors and characteristics and the use of network resources.

14.3.1 Traditional Network Management

Traditional network management systems are based on the manager agent (MA) paradigm. The agent is a software code that resides on the network device and collects management information and sends it to the manager. The manager collects the information from agents, processes it, and generates reports. The interactions between the manager and agent are governed by management protocol. Simple Network Management Protocol (SNMP), standardized by the Internet Engineering Task Force (IETF), is the most commonly used management protocol. It has three versions – SNMP v1, v2, and v3. Some of the other protocols used for network management are netconf, syslog, and net flow. The architecture of a traditional network management system is illustrated in Figure 14.10.

Management functions are often characterized using reference models. The most common reference model is the FCAPS model proposed by IETF. It covers the following management functions:

1) fault management,
2) configuration management,
3) accounting management,
4) performance management,
5) security management.

Traditional network management systems, which are often characterized by a centralized architecture, prove incapable when used in the tactical environment. Management

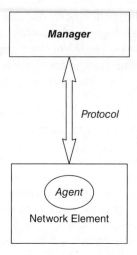

Figure 14.10 Architecture of a traditional network management system.

of tactical networks poses several challenges owing to the inherent nature of the operations involved. Some of the challenges are listed below [23]:

- critical response time,
- rapidly changing topology,
- limited resources,
- flexibility.

The periodic polling of the agents by the manager in traditional systems and regular exchange of management information put a strain on the limited bandwidth and add to the delay in transmission and congestion [24]. Owing to these challenges, management of tactical networks calls for a paradigm shift in design and implementation.

14.3.2   Mobile Agent

The MA approach uses a mobile code as an agent instead of a static agent. A single agent now traverses the network and collects management information. This approach significantly reduces the bandwidth consumed for management purposes. It also increases the performance, as only the final result is sent to the manager and most of the processing is done by the MA at the local level. The MA technology has several features that make it promising for use in tactical environments. The decentralization and delegation of management functions ensure that, if a part of the network is compromised, the management of the entire network is not affected.

The architecture of a mobile-agent-based tactical network management system is illustrated in Figure 14.11.

The system consists of a mobile agent generator (MAG), which generates MA with the functionality to migrate across the network. The agent traverses the network, collects the management information, processes it, and moves to the next element. The manager controls the movement of the agent, manages its security, and stores the results.

MA technology is now increasingly being used in new implementation strategies to monitor networks. However, there are certain challenges that should be addressed

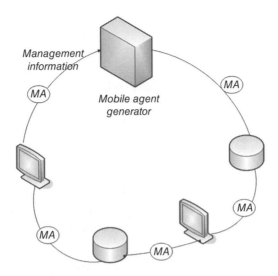

Figure 14.11 Architecture of an MA-based tactical network management system.

to utilize its full potential. The primary concern in all instances of MA deployment has been security. Given that MAs may traverse hostile tactical networks with classified management information on them, it is imperative to ensure their security. Robust security frameworks have been suggested with advanced authentication and cryptographic techniques [25]. Coordination of agents and prevention of conflicts and deadlocks are among the other challenges that a MA-based management system has to address.

14.3.3 Policy-Based Network Management

Policy-based management is another approach that is being adopted over static management systems in tactical network management. Tactical networks are often complex and heterogeneous, comprising a myriad of components ranging from laptops to management consoles to wireless radios. The topology of such networks is often varying, necessitating the need for adaptive configuration management. Fault management in field networks also requires automation owing to the critical nature of the operations involved. Policy-based network management systems provide automated response to situations based on predefined policies. The policies are stored in a repository in a policy server. The architecture of a policy-based network management (PBNM) system is illustrated in Figure 14.12.

The system has the following components [26]:

1) The policy repository is a database of predefined directives that enable the system to take decisions depending on the situation.
2) The policy decision point (PDP) performs the task of choosing the policy from the database depending on the inputs from the clients.
3) The policy enforcement point (PEP) acts as an interface between the PDP and the clients or the network elements.

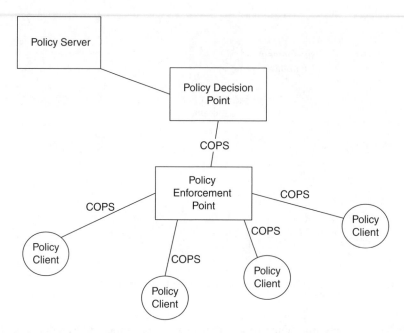

Figure 14.12 Architecture of a policy-based network management (PBNM) system.

The communication between the PDP and the PEP is governed by the Common Open Policy Service Protocol (COPS). Policy-based systems minimize manual intervention and automate most of the functions of network management, thereby facilitating their use in the mission and time-critical tactical networks.

14.4 Wireless Tactical Networks

Wireless networking has changed the rules of communication, security, and interception in a tactical network. Resourceless communication on the move without any physical footprint is a success factor in today's scenario. Wireless communication protocols, models, and equipment are available for mobile as well as static users and systems on any platform – aerial, land, water, and even underwater. The range of present-day wireless networks used in defense applications as well as in civilian applications varies from a few nanometers to submarine–satellite communication. However, the technologies and protocols change with distance. Theoretically, an ad hoc network is generally of higher communication range with better power availability than a wireless sensor network. The rate of data transmission within a similar type of network makes it difficult to standardize any crypto or cryptanalysis engines. For example, a personal area network can be categorized into WPAN, high-rate WPAN, low-rate WPAN, mesh networking, and body area networks. Personal area networking, which supports the smart soldier, encompasses the wireless technologies of Bluetooth, Z-Wave, ZigBee, 6LoWPAN, ISA-100, RFID, and Wireless HART. The upcoming standards in the field of personal area networks, which are yet to be explored for critical applications, are Internalnet, Skimplex, and Dash7. The key distribution in a wireless network is a challenge in itself, which is yet to be resolved.

The applications of wireless tactical networks are also increasing day by day. Wireless sensor networks and ad hoc networks are technologies of the past. They have graduated to vehicular ad hoc networks (VANETs), underwater sensor networks, and networks within the human body – from sensor-enabled patients to robotic soldiers and their weapon systems. These scenarios have made the management of wireless networks even tougher because there are deviations in the wireless networks in addition to mobility and localization problems. The nodes can connect to any network at any time and again leave the network. Identifying friend, foe, and captured sensors is a challenge assigned to the network management system.

Transformation of defense forces to meet the threats of fourth-generation warfare will be heavily dependent on technology as a key enabler and force multiplier. The tactical commander's aims and objectives of battle and consequent design of battle will be greatly dependent on the flow of information from the digital battlefield. Despite all the technological advancements, the real-time flow of communication from the tactical battle area is a major concern.

Net-centric warfare command, communication, computation, information, surveillance, and reconnaissance capability should be improved in order to enhance situational awareness and the capability to identify, monitor, and destroy targets in real time. These activities have to be coordinated, which will further ensure enhanced battlefield transparency at each level of command, leading to responsive decision-making in near real time. A robust backbone network will therefore act as a force multiplier that will facilitate cumulative employment of destructive power at the most vulnerable point of the enemy.

14.5 Routing in Delay-Tolerant Networks

The present-day Internet connects computers across continents over TCP/IP through different types of link providing point-to-point connectivity. The Internet was designed to work on symmetric bidirectional links, comparatively low error rates, enough bandwidth for effective data transmission, and short delay between data transmission and data reception.

The evolution of wireless networks added a new variety of network environments over which data communication was to occur. The wireless network has asymmetric links, high error rates due to loss and collisions, intermittent connectivity due to link disruptions and reconnections, higher error rates, and longer and variable delays in data transmission. Examples of such wireless networks are the networks used in network-centric operations connecting the soldier, combat vehicles, airborne systems, satellites, and surface as well as submerged vessels in the sea. The operating scenario of mobile land networks for civilian usage is no different.

There may be a number of reasons for intermittent connectivity in a network, some well known, including moving out of reception range on account of mobility, node shutdown to preserve power, maintain electromagnetic silence, or preserve secrecy, and configuration for only opportunistic communication or the effect of jamming on the network. Besides the unavailability of the next hop, there may also be long queuing times for messages before delivery. A delay-tolerant network enables interoperability among heterogeneous networks or Internet buffering of the long delays in communication among them, which otherwise would be a mismatch and hence unacceptable.

14.5.1 Applications

Delay-tolerant networks are a requirement in non-legacy-type networks with extreme constraint on bandwidth or huge transmission delays arising because of processing or distance factors. Mobility factors further complicate the networking issues, as the nodes can change address and neighbors. A battlefield network is an example of a highly mobile network without any predefined orientation. Satellite links, underwater networks, and networks for wildlife monitoring are examples of low-bandwidth and high-delay networks. Some typical examples of existing networks or future networks prone to delays are as follows:

- Interplanetary networks: network connectivity from earth to space missions and missions on other planets, and Internet connectivity in space.
- Wildlife/habitat monitoring: tracking sensors in the body of animals or tracking collars.
- Vehicular networks: drive-through Internet and vehicle-to-vehicle networks.
- Connectivity to inhospitable terrains and inaccessible geographic regions.
- Internet connectivity to mobile communities such as theater groups, circuses, jamborees, and cruise liners.
- Underwater networks: communication between autonomous underwater vehicles, submarines, buoys, and surface vessels.
- Battlefield networks.

14.5.2 Routing Protocols

The routing protocols for a delay-tolerant network require an agent to interconnect incompatible networks, including the rate of communication between the systems.

Bundle protocol. An agent that provides a protocol overlay to heterogeneous networks is known as a bundle layer because it bundles messages and transmits them across the networks, and these bundles may or may not be acknowledged [27, 28]. A bundle consists of the user data, control information regarding handling of data, and the bundle header. The bundle protocol works on the principle of store-carry-and-forward. The custodian of the bundle holds the bundle for a very long period until the connectivity is available to forward the bundle to the next custodian.

In a delay-tolerant network with a bundle layer, the nodes are classified as hosts, routers, and gateways. A host is capable of receiving or sending bundles. Routers can forward bundles among hosts within the same delay-tolerant network (DTN) region or forward bundles to gateways if they have to be sent to another DTN region. Gateways support the forwarding of bundles between DTN regions. Each DTN region has a unique identifier and has homogeneous communication patterns. A host is identified by a combination of its region identifier and entity identifier. Asymmetric key cryptography is used for authentication, integrity, and confidentiality of the bundles while they are being forwarded among the nodes.

The bundle layer uses the concept of custody transfer progressively to move the bundle from source to destination. The gateways should essentially support custody transfer, while this is optional for routers and hosts. Initially, the sender holds the custody of the bundle. It transmits the bundle to the next recipient and waits for acknowledgement from the recipient. If an acknowledgement is received from the recipient, the custody is transferred to the recipient. If an acknowledgement is not received by the sender within the expiry of 'time-to-acknowledge', the sender retransmits the bundle to the recipient. If an acknowledgement is not received in the bundle's 'time-to-live', the custody of the bundle can be dropped.

The two prime strategies for routing in DTN are forwarding and flooding. In flooding, the message is forwarded to more than one node. Hence, the intermediate nodes receive more than one copy of the message. The probability of message delivery is high, as a number of relay nodes have a copy of the message and each of them attempts to forward it to the destination on availability of a link to the next hop. Flooding strategies not only relieve the node from having a global view of the network, it also frees the nodes from having a local view too. A variation of flooding is tree-based flooding, where the message is forwarded to a number of relay nodes that forward it to the previously discovered nodes in their neighborhood.

Store-and-forward is one of the most commonly used strategies for message relay in delay-tolerant networks. The major requirement of the store-and-forward technique is a huge memory in the interconnecting devices because the communication link may not be available for long and hence all the data available for transmission at the source should be buffered in the intermediate routing terminal at the earliest. Furthermore, the data transmitted to the destination may have to be retransmitted after a long time delay owing to error in the data received by the destination or the data not reaching the destination at all. As the status of data receipt is expected to be available very late, the data should be retained in the buffer of the transmitting intermediate routing terminal for long durations.

The major routing protocols suggested for DTNs can be classified into three categories – flooding-based protocols, knowledge-based routing, and probabilistic routing.

A few common DTN routing protocols are as follows:

- Epidemic,
- Erasure Coding,
- Oracle,
- Message Ferrying,
- Practical Routing,
- Probabilistic Routing Protocol using History of Encounters and Transitivity (PRoPHET)
- RPLM,
- MaxProp,
- MobySpace,
- Resource Allocation Protocol for Intentional DTN (RAPID),
- SMART,
- Spray and Wait.

References

1 A. Conti, M. Guerra, D. Dardari, N. Decarli, and M. Z. Win. Network experimentation for cooperative localization. *IEEE Journal on Selected Areas in Communications*, **30**(2):467–475, 2012.

2 J. L. Guardado, J. L. Naredo, P. Moreno, and C. R. Fuerte. A comparative study of neural network efficiency in power transformers diagnosis using dissolved gas analysis. *IEEE Transactions on Power Delivery*, **16**:4, 2001.

3 N. O. Sokal and A. D. Sokal. Class E – a new class of high-efficiency tuned single-ended switching power amplifiers. *IEEE Journal of Solid-State Circuits*, **10**(3): 168–176, 1975.

4 Y. Han and D. J. Perreault. Analysis and design of high efficiency matching networks. *IEEE Transactions on Power Electronics*, **21**(5):1484–1491, 2006.

5 M. P. van den Heuvel, J. S. Cornelis, R. S. Kahn, and H. E. Hulshoff Pol. Efficiency of functional brain networks and intellectual performance. *The Journal of Neuroscience*, **29**(23):7619–7624, 2009.

6 Network, http://www.webopedia.com/TERM/N/network.html.

7 Computer Network, http://en.wikipedia.org/wiki/Computer_network.

8 A. Avizienis and J. C. Laprie. Dependable computing: from concepts to design diversity. *Proceedings of the IEEE*, **74**, May 1986.

9 P. Jalote. *Fault Tolerance in Distributed Systems*. PTR Prentice Hall, 1994.

10 M. L. Shooman. *Reliability of Computer Systems and Networks: Fault Tolerance, Analysis, and Design*. Wiley-Interscience, 2001.

11 I. Koren and C. M. Krishna. *Fault-Tolerant Systems*. Morgan Kaufmann, 2010.

12 C. E. Ebeling. *An Introduction to Reliability and Maintainability Engineering*. Tata McGraw-Hill Education, 1997.

13 K. S. Trivedi. *Probability and Statistics with Reliability, Queuing, and Computer Science Applications*. John Wiley, 2002.

14 S. McLaughlin, P. M. Grant, J. S. Thompson, H. Haas, D. I. Laurenson, C. Khirallah, Y. Hou, and R. Wang. Techniques for improving cellular radio base station energy efficiency. *IEEE Wireless Communications*, **18**(5):10–17, 2011.

15 E. O. Schweitzer III, B. Fleming, T. J. Lee, and P. M. Anderson. Reliability analysis of transmission protection using fault tree methods. *24th Annual Western Protective Relay Conference*, pp. 1–17, 1997.

16 T. Korkmaz and K. Sarac. Characterizing link and path reliability in large-scale wireless sensor networks. *IEEE 6th International Conference on Wireless and Mobile Computing, Networking and Communications (WiMob)*, October 2010.

17 H. Chin-Yu, M. R. Lyu, and K. Sy-Yen. A unified scheme of some nonhomogenous Poisson process models for software reliability estimation. *IEEE Transactions on Software Engineering*, **29**(3):261–269, 2003.

18 J. DeMercado, N. Spyratos, and B. A. Bowen. A method for calculation of network reliability. *IEEE Transactions on Reliability*, **R-25**(2):71–76, 1976.

19 Fault Taxonomy. iitkgp.intinno.com/courses/738/wfiles/92209.

20 Electronic Architecture and System Engineering for Integrated Safety Systems (EASIS). http://www.transport-research.info/Upload/Documents/201007/20100726_144640_68442_EASIS-.pdf.

21 A. Avizienis, J.-C. Laprie, B. Randell, and C. Landwehr. Basic concepts and taxonomy of dependable and secure computing. *IEEE Transactions on Dependable and Secure Computing*, **1**(1):11–33, 2004.

22 J. O. Kephart and D. M. Chess. The vision of autonomic computing. *Computer*, **36**(1):41–50, 2003.

23 D. Vergados, J. Soldatos, and E. Vayias. Management architecture in tactical communication networks. *21st Century Military Communications Conference Proceedings MILCOM, Volume 1*, pp. 100–104, 2001.

24 F. Guo, B. Zeng, and L. Z. Cui. A distributed network management framework based on mobile agents. *3rd International Conference on Multimedia and Ubiquitous Engineering MUE '09*, June 2009.

25 M. A. M. Ibrahim. Distributed network management with secured mobile agent support. *International Conference on Electronics and Information Engineering (ICEIE), Volume 1*, pp. V1-247–V1-253, 2010.

26 K.-S. Ok, D. W. Hong, and B. S. Chang. The design of service management system based on policy-based network management. *International Conference on Networking and Services ICNS '06*, pp. 59–64, July 2006.

27 Delay tolerant networking architecture, IETE Trust, *RFC 4838*. http://tools.ietf.org/html/rfc4838.

28 Bundle Protocol Specification, IETE Trust, *RFC 5050*. http://tools.ietf.org/html/rfc5050.

Abbreviations/Terminologies

COPS	Common Open Policy Service Protocol
DTN	Delay-Tolerant Network
HART	Highway Addressable Remote Transducer (Protocol)
IETF	Internet Engineering Task Force
IP	Internet Protocol
LAN	Local Area Network
MA	Mobile Agent
MAG	Mobile Agent Generator
MAN	Metropolitan Area Network
MTBF	Mean Time Between Failures
MTTF	Mean Time to Failure
MTTR	Mean Time to Repair
NIC	Network Interface Card
PBNM	Policy-Based Network Management
PDP	Policy Decision Point
PEP	Policy Enforcement Point
RFID	Radio Frequency Identification
SCADA	Supervisory Control and Data Acquisition
SNMP	Simple Network Management Protocol
TCP	Transmission Control Protocol
UPS	Uninterrupted Power Supply
VANET	Vehicular ad hoc Network
WAN	Wide Area Network
WPAN	Wireless Personal Area Network
6LoWPAN	IPv6 over Low-Power Wireless Personal Area Networks

Questions

1 What is software reliability and how do we determine the reliability of software?

2 In the earlier days, systolic arrays were used for fault tolerance. Study about systolic computing and describe the role played by the systolic arrays in fault tolerance.

3 What is autonomic computing and what are the features of an autonomic system?

4 Differentiate between availability and reliability. A network goes down for 1 s every hour. What is the reliability, availability, MTBF, and MTTF for this network over a period of 1 day?

5 Explain the process of reliability calculation using minterms.

6 Describe the three types of fault based on the frequency of reoccurrence of the fault.

7 Explain the difference between a fail silent system and a gracefully degraded system. Mention a scenario of application for each where a fail silent system will be required and a gracefully degraded system cannot work, and vice versa.

8 Draw various fault-tolerant network architectures and their equivalent non-fault-tolerant designs.

9 Describe the working of a mobile agent.

10 Explain the policy-based network management approach.

11 Why are delay-tolerant networks required? Justify your answer giving examples.

12 Explain the working of a bundle protocol.

13 There can be a number of reasons, other than security attacks or battery failure, for link disruptions and intermittent connectivity between wireless nodes. Mention a few such reasons.

14 State whether the following statements are true or false and give reasons for the answer
A Reliability is the average time over a given period when the system is operational.
B For a network that goes down very frequently, reliability and availability are the same.
C MTTF = MTBF – MTTR.

D If $F(t)$ is the probability that a component will fail at or before time t, and $\lambda(t)$ is the failure rate of the component at time t, then $F(t) = 1 - e^{-\lambda t}$.

E If a transient fault occurs in the network, the network goes down for a long period until it is repaired and made operational manually.

F A fail silent system does not stop working immediately on detection of a fault, but keeps working with lower performance for a very long period and then silently goes down.

G In a policy-based network management, PDP acts as an interface with the clients or the network elements.

H Bluetooth, Z-Wave, ZigBee, 6LoWPAN, ISA-100, and RFID are wireless technologies.

I Internalnet, Skimplex, and Dash7 are security standards.

J Bundle protocol is a routing protocol used in delay-tolerant networks.

Exercises

1 Calculate the reliability of communication between points A and B of the network given below. The reliability of each communication link is indicated with each of the links.

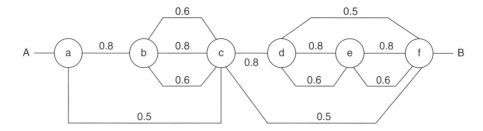

2 A certain cryptographic system uses a shared key. The key is spread across four different smart cards, and all the smart cards are required to decrypt the information. All the smart cards are similar and the reliability of a smart card is 0.7 What is the reliability of the information being decrypted? What will the new reliability be if USB crypto tokens with a reliability of 0.9 are used in place of smart cards?

3 A data center is said to have an availability of four 9s if it has an availability of 99.99%, five 9s if it has an availability of 99.999%, and so on. What is the permissible downtime per year permitted in data centers with two 9s (99%), three 9s, four 9s, five 9s, and six 9s?

4 A data center provides power to the servers through its uninterrupted power supplies (UPSs), and each UPS has a reliability of 99%. The designers of the data center want to provide power to the servers with a reliability of 99.671%. A minimum of how many UPSs have to be placed in parallel to ensure this reliability. By how many does

the number of UPSs in parallel have to be increased at each stage if the designers plan to increase the reliability in stages to 99.741% and then to 99.982% and then finally to 99.995%?

5 A network is depicted below as a probabilistic graph. Calculate the reliability of the network using minterms for communication between X and Y.

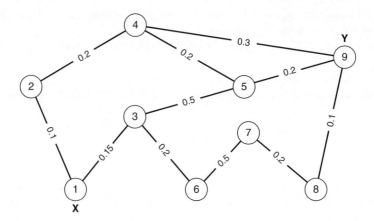

Index

Network Routing: Fundamentals, Applications, and Emerging Technologies, First Edition.
Sudip Misra and Sumit Goswami.
© 2017 John Wiley & Sons Ltd. Published 2017 by John Wiley & Sons Ltd.
Companion website: www.wiley.com/go/misra2204